Achim Struve

Analysis of a Rotatable Wind Turbine Tower by means of Aero-Servo-Elastic Load Simulations

BAND 11

Versuchsanstalt für Stahl, Holz und Steine
Berichte zum Stahl- und Leichtbau

Analysis of a Rotatable Wind Turbine Tower by means of Aero-Servo-Elastic Load Simulations

by
Achim Struve

KIT Scientific Publishing

Karlsruher Institut für Technologie
Versuchsanstalt für Stahl, Holz und Steine

Analysis of a Rotatable Wind Turbine Tower
by means of Aero-Servo-Elastic Load Simulations

Zur Erlangung des akademischen Grades eines Doktor-Ingenieurs
von der KIT-Fakultät für Bauingenieur-, Geo- und Umweltwissenschaften
des Karlsruher Instituts für Technologie (KIT) genehmigte Dissertation

von Achim Struve, M.Eng.

Tag der mündlichen Prüfung: 14. Juli 2020
Hauptreferent: Prof. Dr.-Ing. Thomas Ummenhofer
Korreferent: Prof. Dr.-Ing. Torsten Faber
Korreferent: Prof. Dr.-Ing. Peter Knödel

Impressum

KIT Scientific Publishing

Karlsruher Institut für Technologie (KIT)
KIT Scientific Publishing
Straße am Forum 2
D-76131 Karlsruhe

KIT Scientific Publishing is a registered trademark
of Karlsruhe Institute of Technology.
Reprint using the book cover is not allowed.

www.ksp.kit.edu

Print on Demand 2021 – Gedruckt auf FSC-zertifiziertem Papier

ISSN 2198-7912
ISBN 978-3-7315-1045-1
DOI 10.5445/KSP/1000123255

Acknowledgement

This work was written throughout my time as scholarship holder of the Energie und Klimaschutzgesellschaft Schleswig-Holstein (EKSH) and scientific assistant of the Wind Energy Technology Institute (WETI) of the University of Applied Sciences Flensburg. Besides, I was external postgraduate at the Karlsruhe Institute of Technology (KIT).

Special thanks go to Prof. Dr.-Ing. Torsten Faber, who gave me the theme for this work and provided me an employment as scientific assistant at the WETI. His trust in me, the access to his network of professionals, and his support have been crucial for the success of this work.

Another special thanks go to Prof. Dr.-Ing. Thomas Ummenhofer, who accepted me as an external postgraduate at the KIT. His trust in me and the many advices regarding light weight steel structures have been important for the success of this work.

Thanks go to Jason Jonkman, Ph.D. and Rick Damiani, Ph.D from the National Renewable Energy Laboratory (NREL) U.S. for provision of an internship in their institution from February to April 2017. Furthermore, they helped me in the understanding of the aero-servo-elastic load simulation environment FAST and its submodule SubDyn.

Thanks for technical and encouraging conversations throughout the writing of the dissertation go to Dr. Peter Maulshagen, Prof. Dr. Clemens Jauch, Prof. Dr. Alois Schaffarczyk, and Prof. Dr.-Ing. Peter Knödel.

Thanks for the provision of a three year postgraduate scholarship goes to the EKSH and especially to Dr. Klaus Wortmann and Stefan Sievers. The postgraduate scholarship enabled me to work much more on the dissertation in a shorter period of time, than it would have been without it.

Moreover, I thank my wife Nadine and my daughter Elisa that they gave me time to work on the dissertation and that they supported me with encouragement and love. Thanks go to my parents Elke and Jens for their encouragement and especially to my joy and saviour, Jesus Christ.

Abstract

This work contributes to the improvement of renewable energy competitiveness over conventional fossil and nuclear energy sources. It highlights how the costs and CO_2-emissions of land-based wind turbines can be reduced as a result of an innovative and material efficient support structure concept. The fundamental approach for the concept is to place the yaw system at the tower base, allowing the whole wind turbine tower to be rotated so that it always has the same alignment as the rotor. This achieves a load-direction-derived design for the tower through which material is saved. This work indicates that the material saving potential of a rotatable elliptical tower compared to a conventional circular one is limited. The reason for that is found in the transport constraints through bridges and tunnels, which apply likewise to elliptical tubular towers. A rotatable lattice tower is more convenient because its size is not limited through transport, and lattice towers are material saving concepts for the trend to larger hub heights. This work shows that the ability to rotate the tower enables wider leg distances along the blades passing lattice tower section, whereby local member loads are reduced, and material can be saved. Furthermore, a tower inclination against the main wind direction was introduced to induce counter acting moments, which reduce the extreme loads and achieve more material savings. Moreover, a vertical rotor alignment was realized through the increased blades tip distance to the inclined tower and the wind turbine became more efficient. The potential of the rotatable inclined lattice tower concept was analysed by means of aero-servo-elastic load simulations in the FAST environment. Since the convenient linear beam finite element submodule, SubDyn, had no capability to represent other member cross sections than circular ones, it was adjusted. In the present case, the beam element stiffness and mass matrix formulations had to be changed to account for arbitrary cross sections. Furthermore, the direction cosine matrices had to be expanded to provide the control over lattice member alignments around their length axes. A validation of the general aero-servo-elastic analysis procedure and the changes in SubDyn ensured reliable results.

The aero-servo-elastic load simulations and an appropriate load post processing are the main tools for a rotatable inclined lattice tower design according to IEC 61400-1 and to DNV GL standards. This includes the consideration of relevant design load cases and the ultimate, the fatigue, and the serviceability limit states. A balance between saved material costs, tower top yaw system costs, increased wind turbine efficiency, and increased tower base yaw system and foundation costs revealed significant cost savings. Under the assumptions made, the savings of the rotatable inclined lattice tower amounted to 45 k€ at 80 m to 768 k€ at 175 m hub height compared to the most cost-effective conventional alternatives for a 5 MW reference wind turbine. This implies 8 % and 49 % in cost savings and indicates the big economical potential of land-based innovative rotatable inclined lattice towers for larger hub heights under transport constraints.

Kurzfassung

Diese Arbeit trägt zur Verbesserung der Wettbewerbsfähigkeit erneuerbarer Energien gegenüber konventionellen fossilen und nuklearen Energiequellen bei. Sie zeigt auf, wie die Kosten und CO2-Emissionen von Onshore-Windenergieanlagen durch ein materialeffizientes und innovatives Tragstrukturkonzept reduziert werden können. Der grundlegende Ansatz für das Konzept besteht darin, das Giersystem am Turmfuß zu platzieren, sodass der gesamte Turm der Windenergieanlage gedreht werden kann und er immer dieselbe Ausrichtung wie der Rotor hat. Dadurch wird für den Turm ein von der Lastrichtung abgeleitetes Design möglich, welches Materialeinsparungen zur Folge hat. Diese Arbeit offenbart, dass das Materialeinsparpotenzial eines drehbaren elliptischen Turms im Vergleich zu einem konventionellen runden Turm begrenzt ist. Der Grund dafür liegt in den Transportbeschränkungen durch Brücken und Tunnel, die ebenfalls für elliptische Rohrtürme gelten. Ein drehbarer Gitterturm ist somit eine gute alternative, da seine Größe nicht durch den Transport begrenzt ist und Gittertürme ohnehin materialsparende Konzepte für den Trend zu größeren Nabenhöhen sind. Die Arbeit zeigt, dass solch ein Turm größere Stielabstände entlang des oberen Turmabschnitts, an dem die Rotorblätter passieren, ermöglicht. Hierdurch wird die Belastung der Stäbe reduziert und Material eingespart. Darüber hinaus wurde eine Turmneigung gegen die Hauptwindrichtung eingeführt, um gegenwirkende Momente zu induzieren, die die Extremlasten reduzieren und noch mehr Materialeinsparungen erzielen. Außerdem wurde eine vertikale Ausrichtung des Rotors durch den erhöhten Blattspitzenabstand zum geneigten Turm erreicht, wodurch die Effizienz der Windenergieanlage zunahm. Das Potenzial des drehbaren geneigten Gitterturmkonzepts wurde mittels aero-servo-elastischer Lastsimulationen in der dafür gut geeigneten FAST-Umgebung analysiert. Da das Subdyn-Modul mit seinem Linearen-Balken-Finite-Elemente-Ansatz keine anderen Querschnitte als kreisförmige repräsentieren konnte, wurde es angepasst. Im vorliegenden Fall mussten die Balken Steifigkeits- und Massenmatrixformulierungen geändert werden, um beliebige Querschnitte zu berücksichtigen. Außerdem mussten die Richtungskosinusmatrizen erweitert werden, um die Ausrichtung der Balkenelemente um ihre Längsachsen zu steuern. Eine Validierung der aero-servo-elastischen Lastberechnung und der nötigen Änderungen in SubDyn stellten die Zuverlässigkeit der Ergebnisse sicher.

Die aero-servo-elastischen Lastsimulationen und eine geeignete Transformation der Lasten in Ausnutzungsgrade sind die Hauptwerkzeuge für die Auslegung eines drehbaren geneigten Gitterturms gemäß IEC 61400-1 und den DNV GL-Standards. Dies beinhaltet die Berücksichtigung relevanter Lastfälle und der Grenzzustände der Tragfähigkeit, der Ermüdung und der Gebrauchstauglichkeit. Eine Bilanzierung zwischen eingesparten Turmmaterialkosten, eingesparten Kosten für das Giersystem am Turmkopf, Mehreinnahmen durch die erhöhte Effizienz der Windenergieanlage sowie erhöhten Kosten für das Giersystem am Turmfuß und den erhöhten Fundamentkosten ergab signifikante Kosteneinsparungen. Unter den gemachten

Annahmen beliefen sich die Einsparungen des drehbaren geneigten Gitterturms auf 45.000 Euro bei 80 m bis 768.000 Euro bei 175 m Nabenhöhe, im Vergleich zu den kostengünstigsten konventionellen Alternativen, für eine 5 MW Referenzanlage. Dies bedeutet entsprechend 8 % und 49 % Kosteneinsparungen und zeigt das enorme wirtschaftliche Potenzial von innovativen drehbaren geneigten Onshore-Gittertürmen mit großen Nabenhöhen unter den üblichen Transportbeschränkungen.

Contents

Nomenclature

The following list shows the abbreviations, greek symbols, latin symbols, mathematical symbols, and relevant FAST environmental Parameters, which were used and likewise described in this work. Not contained abbreviations and symbols are explained throughout the text.

Abbreviations

1p, 3p	one-per-revolution and three per revolution
ASE	aero-servo-elastic
BEM	blade element momentum
BU	buckling
C1-C6	design cases as given in Table 9.3
DNV GL	Det Norske Veritas Germanischer Lloyd
DOF	degree of freedom
ECD	extreme coherent gust with wind direction change
EDC	extreme direction change
EOG	extreme operating gust
ETM	extreme turbulence model
EWM	extreme wind model
EWS	extreme wind shear
FA	fore-aft
FAST	fatigue, aerodynamics, structures, and turbulence
FMD	flexible multibody dynamics
GEBT	geometrically exact beam theory
HiFIT	high frequency impact treatment
LDD	load-direction-derived
NREL	National Renewable Energy Laboratory
NTM	normal turbulence model
NWP	normal wind profile
RILT	rotatable inclined lattice tower
RNA	rotor nacelle assembly

SS	side-side
TP	transition piece
UIT	ultrasonic impact treatment
WndDir	horizontal wind direction with 0 deg along the global x-axis and mathematical positive anti-clockwise rotation around the upwards pointing z-axis
YS	yield strength

Greek Symbols

α	power law exponent / buckling curve dependent imperfection coefficient
α, β, γ	beam element node rotations
α_1	curve linear coordinate
α_{AoA}	angle of attack on an aerodynamic profile
α_{B}	bracing angle
α_{cr}	relation of the design load to the ideal buckling load
α_{PC}	blade precone angle
α_{ST}	rotor shaft tilt
α_{TI}	tower inclination angle
α_{TT}	tower taper angle
α_{ω}	steepness of prescribed yaw acceleration curve
β_{FTB}	flexural-torsional buckling coefficient
χ	column buckling reduction factor regarding the considered buckling curve
$\Delta\sigma$	stress range
$\Delta C_{\mathrm{found.}}$	foundation cost increase
ΔI_{RILT}	additional money income through vertical rotor alignment
$\eta_{\mathrm{mech.}}, \eta_{\mathrm{elec.}}$	mechanical and electrical efficiencies
γ_{f}	patial safety factor for the type of loads
$\gamma_{\mathrm{M3(,ser)}}$	partial safety factor to account for the kind of a joint connection
γ_{m}	patial safety factor for the material
γ_{n}	patial safety factor for the consequences of failure
$\gamma_{\mathrm{zx}}, \gamma_{\mathrm{zy}}$	cross sectional shear strains in along its x and y directions
$\kappa_{\mathrm{x}}, \kappa_{\mathrm{y}}$	beam curvature around x and y axes
κ_{z}	beam torsional twist rate
λ	lattice structure slenderness
Λ_1	turbulence length parameter
λ_1	slenderness to evaluate the slenderness ratio
μ	friction coefficient / expected statistical value / viscosity coefficient
ν	kinematic viscosity

ω	angular frequency / sectorial area
$\overline{\lambda}$	slenderness ratio of a column buckling member
Φ	function for evaluation of χ / Euler angle
$\phi(x, y, z, t)$	velocity potential function
Ψ	Euler angle
ψ_λ	reduction factor to account for finite tower slenderness
ρ	density
σ^2	statistic variance
σ_1, σ_2, σ_3	standard deviation of turbulence in longitudinal, lateral, and upward direction
$\sigma_{\theta,\mathrm{Rd}}$	circumferential buckling stress
$\sigma_{\mathrm{m,T}}$	mean normal stress on the towers tensile side
$\sigma_{\mathrm{m,W}}$	mean normal stress due to structural weight
σ_{m}	mean normal stress
σ_{v}	standard deviation of the turbulence
$\sigma_{\mathrm{x,d}}, \sigma_{\mathrm{y,d}}$	design normal stresses in x- and y-direction
$\sigma_{\mathrm{x,Rd}}$	meridian buckling stress
τ_{d}	design shear stress
$\tau_{\mathrm{x}\theta,\mathrm{Ed}}$	shear buckling stress
Θ	Euler angle
θ_{cg}	wind speed increase for an ECD
θ_{e}	extreme direction change angle
$\underline{\underline{\Omega}}_m$	$m \times m$ diagonal matrix, which represents the eigenfrequencies of the retained Craig-Bampton eigenmodes
$\underline{\underline{\Phi}}_{\mathrm{R}}$	transformation matrix for the interior node DOFs with respect to the rigid body interface motions
$\underline{\underline{\Phi}}_m$	transformation matrix between the Craig-Bampton reduced generalized DOFs and the original interoir nodes
$\underline{\varphi}$	eigenvector of a system
ε	factor in dependence of f_{y}
ε_{z}	cross sectional axial strain
φ	lattice structure member density / leg cross sectional angle / angle of attack on an aerodynamic profile
$\Theta_{\mathrm{b},1}$	demanded blade one pitch angle positive clockwise around the blades pitch axis looking from its tip to its root
ζ	critical damping ratio

$\zeta^{(\mathrm{HD})}$	damping ratio for the platform-heave DOF in HydroDyn

Latin Symbols

$\widetilde{M}^{(\mathrm{ED})}$	RNA and tower mass, modelled in ElastoDyn
$\mathcal{E}_\mathrm{e}$	beam element orthonormal basis
$\mathcal{I}$	global coordinate system orthonormal basis
$\mathcal{L}_1, \mathcal{L}_2$	line 1 and 2
$\mathcal{P}_\mathrm{proj.}$	beam element cross sectional plane at its start position
$\mathcal{P}_\mathrm{XY}$	global coordinate system XY-plane
$\mathrm{d}m/\mathrm{d}t$	mass flow
$\overline{A}_\mathrm{Life}$	average lifetime availability of a WT
$\overline{b}$	mean lattice tower member width
$\overline{E}_\mathrm{Life}$	average lifetime energy yield of a WT
$\overline{I}, \overline{J}, \overline{K}$	global coordinate system orthonormal basis unit vectors
$\overline{i}_\mathrm{e}, \overline{j}_\mathrm{e}, \overline{k}_\mathrm{e}$	beam element orthonormal basis unit vectors
$\overline{n}_1, \overline{n}_2$	direction unit vector of line 1 and 2
$\overline{P}$	averaged WT rotor power
$\overline{S_i}$	mean of load cycle i
$\overline{S}_\mathrm{F}$	prescribed fixed load mean
$\overline{U}_\mathrm{L}$	SubDyn internal DOFs
$\overline{U}_\mathrm{R}$	SubDyn interface DOFs
$\overline{x}$	arithmetic mean
$\mathbf{A}$	beam element alignment point in space
$\mathbf{A}'$	projected beam element alignment point in space
$\mathbf{E}$	end point position vector of a beam element in space
$\mathbf{O}$	global coordinate system origin
$\mathbf{S}$	start point position vector of a beam element in space
$\underline{\dot{r}}$	velocity vector of a material point in a beam
$\underline{\underline{\widetilde{K}}}_\mathrm{BB}$	substructures equivalent stiffness matrix, referred to the transition piece reference point
$\underline{\underline{\widetilde{M}}}_\mathrm{BB}$	substructures equivalent mass matrix, referred to the transition piece reference point
$\underline{\underline{\widetilde{M}}}_{\mathrm{B}m}$	matrix partition after Craig-Bampton reduction, referred to the transition piece reference point
$\underline{\underline{B}}$	strain-displacement matrix
$\underline{\underline{C}}$	global damping matrix of a finite element beam system

$\underline{\underline{C}}^{(\mathrm{HD})}$	HydroDyn 6×6 linear damping matrix for the platform DOFs
$\underline{\underline{C}}_1, \underline{\underline{D}}_1, \underline{\underline{F}}_{\mathrm{Y1}}$	SubDyn model modal parameter containing matrices for transition piece load state-space formulation
$\underline{\underline{C}}_2, \underline{\underline{D}}_2, \underline{\underline{F}}_{\mathrm{Y2}}$	SubDyn model modal parameter containing matrices for substructure movement state-space formulation
$\underline{\underline{D}}_{\Phi\Theta\Psi}$	direction cosine matrix for rotation about all three Euler angles
$\underline{\underline{D}}_{\Phi}, \underline{\underline{D}}_{\Theta}, \underline{\underline{D}}_{\Psi}$	direction cosine matrices for rotation about different Euler angles
$\underline{\underline{D}}_{\mathrm{sys}}$	damping matrix of a system
$\underline{\underline{I}}$	identity matrix
$\underline{\underline{K}}$	global stiffness matrix of a finite element beam system
$\underline{\underline{K}}_{\mathrm{c}}$	beam cross sectional stiffness matrix
$\underline{\underline{K}}_{\mathrm{el}}$	beam element stiffness matrix
$\underline{\underline{K}}_{\mathrm{sys}}$	system stiffness matrix
$\underline{\underline{M}}$	global mass matrix of a finite element beam system
$\underline{\underline{M}}_{\mathrm{c}}$	beam cross sectional mass matrix
$\underline{\underline{M}}_{\mathrm{el}}$	beam element mass matrix
$\underline{\underline{M}}_{\mathrm{sys}}$	system mass matrix
$\underline{\underline{N}}(z)$	polynomial matrix
$\underline{\underline{N}}_{\alpha}$	transformation matrix for the generalized beam degrees of freedom
$\underline{\widetilde{F}}_{\mathrm{TP}}$	transition piece forces and moments after Craig-Bampton reduction
$\underline{\widetilde{F}}_{m}$	interior forces and moments, which account only for the Craig-Bampton retained modes
$\underline{F}$	nodal external load vector of a finite element beam system / external force vector of a node
$\underline{f}$	internal moment vector of a system / beam element load vector
$\underline{F}_{\mathrm{HDR}}$	hydrodynamic forces on boundary nodes in SubDyn
$\underline{F}_{\mathrm{L}}$	hydrodynamic forces on every interior node in SubDyn
$\underline{F}_{\mathrm{TP}}$	transition piece loads
$\underline{g}$	angular moment of inertia vector
$\underline{h}$	linear moment of inertia vector
$\underline{M}$	applied external moments
$\underline{m}$	internal moment vector of a system
$\underline{P}_{\mathrm{x}}$	x-bracing intersection point position vector
$\underline{q}_{m}$	modal coefficients for the retained modes

$\underline{U}$	nodal displacement vector of a finite element beam system
$\underline{u}$	SubDyn input vector / beam element node displacement and rotation vector
$\underline{U}_{\mathrm{R}}$	boundary DOFs
$\underline{x}$	system displacement vector
$\underline{x}_0$	position vector of a point on a beam curve
$\underline{x}_{\mathrm{leg,left}}$	position vector of a leg joint on the left side of the considered x-bracing
$\underline{x}_{\mathrm{leg,right}}$	position vector of a leg joint on the right side of the considered x-bracing
$\underline{x}_{\mathrm{node,incl.},i}$	position vector of inclined tower node i with respect to the global coordinate system
$\underline{x}_{\mathrm{node},i}$	position vector of tower node i with respect to the global coordinate system
$\underline{Y}_2$	SubDyn output vector
A	aerodynamically related profile shape area / cross sectional area
a	axial induction factor for an actuator disc / amount of support reactions
a_{B}	lattice tower bottom length
A_{disc}	rotor disc area
a_{M}	intermediate tower length
A_{NREL}	cross sectional area of the bottom reference NREL WT tower
A_{ref}	lattice tower reference area: projected wind attack area
A_{R}	rotor area
A_{T}	projected tower area
a_{T}	lattice tower top length
A_{x}, A_{y}, A_{xy}	corrected shear areas with respect to cross sectional reference axes x and y
$A_{\mathrm{el},m,i}$	area of cross sectional element i of member m
$A_{\mathrm{g},k}$	projected area of joint steel sheet k
B	length of an elliptical tower profile
b	outer dimension of a bracing member / side length of a lattice tower cross section
b_{B}	lattice tower bottom width
b_{M}	intermediate tower width
b_{T}	lattice tower top width
b_i	width of member i / width of the lattice tower at joint i
$b_{\mathrm{L,p},i}$, $b_{\mathrm{B,p},i}$	projected leg and bracing widths in air flow direction of segment i
C	Weibull scale parameter / constant
c/t	fineness or chord to thickness ratio of an aerodynamic profile
c_1	length of one-side supported plate section
C_1, C_2	integration constants

c_2	length of two-side supported plate section
c_D	drag coefficient
$c_{f,0}$	basic force coefficient
c_f	force coefficient
c_L	lift coefficient
c_M	moment coefficient
c_P	power coefficient
$c_s c_d$	structural coefficient
c_T	thrust coefficient
$C_{\text{yaw bear.}}$	yaw bearing costs
$C_{\text{yaw sys. conv.}}$	costs of a conventional yaw system
$C_{\text{yaw sys. RILT}}$	costs of a yaw system at the tower base
D	outer tower diameter
d	lattice structure width / dimensionless radial distance from the tower center
D_c	characteristic cumulative damage
D_{extr}	maximum extrusion moulding profile diameter
D_{rot}	rotor diameter
$D_{T,NREL}$	baseline NREL reference WT tower top diameter
d_T	horizontal distance from tower center line to outer tower surface
D_T, D_B	top and bottom diameter for a tower with circular cross section
$D_{\text{yaw bear.}}$	yaw bearing diameter
DFF	design fatigue factor
dnL	amount of elements along a straight part of a cross section
dnt	amount of elements along the thickness of a straight cross sectional part
E	Young's modulus
e	distance from an aerodynamic profile tip to its pressure point
E_{kin}	kinetic energy
E_{pot}	potential energy
f	frequency
$f_{b,Rd}$	buckling resistance load of the most unfavourable load combination and imperfections
f_b	buckling load of the most unfavourable load combination and imperfections
F_B, F_C	tower base yaw bearing reaction forces
F_{cr}	ideal buckling load
F_d	design load
F_{FTB}	maximum shear force on a beam until flexural-torsional buckling starts

f_{L}	dimensionless frequency
f_{m}	reduction factor for hot spot stress ranges
$F_{\mathrm{p,C}}$	tension force in bolt
$F_{\mathrm{s,Rd(,ser)}}$	slip resistance of bolt connection per bolt
$F_{\mathrm{t,Ed(,ser)}}$	tension force in bolt connection per bolt
$F_{\mathrm{v,Ed(,ser)}}$	shear force in bolt connection per bolt
f_{v}	vortex creation frequency
F_{w}	wind force on a structure
f_{y}	material yield strength
$F_{\mathrm{z,ref}}$	Extreme axial tower force from the reference ASE simulation
F_{z}	force in z-direction - this is often the axial force
$f_{0,j}$	structural eigenfrequency j
$F_{\mathrm{k},i}$	characteristic load i
$f_{\mathrm{R},i}$	excitation frequency i
$F_{\mathrm{u}i}$, $F_{\mathrm{v}i}$, $F_{\mathrm{w}i}$	beam element shear forces and axial force of node i
G	shear modulus
g	amount of equilibrium conditions per node
g_x, g_y, g_z	gravity components in $x-$, $y-$, and $z-$directions
H	width of an elliptical tower profile
h	outer dimension of a bracing member
H_{F}	hat profile one side supported flap length
H_{L}	hat profile two side supported length
$H_{\mathrm{T,aero}}$	artificially increased tower height for aerodynamic tower loads
$H_{\mathrm{T,NREL}}$	tower height of the reference NREL WT
H_{T}	tower height
h_i	vertical joint distance between joint $i-1$ and i
$h_{i,\mathrm{a}}$	vertical joint distance between joint $i-1$ and i at x-axis parallel side
$h_{i,\mathrm{b}}$	vertical joint distance between joint $i-1$ and i at y-axis parallel side
i	radius of gyration
i^2	area stiffness or radius of gyration
I_{min}	minimum second area moment of inertia of a cross section
i_{min}	minimum radius of gyration of an assembled profile with small spread
I_{p}	polar moment of inertia
I_{ref}	reference turbulence intensity
I_{t}	torsional constant

I_v	turbulence intensity
I_x, I_y, I_{xy}	second area moments of inertia with respect to cross sectional reference axes x and y
J_ω	warping constant
k	Weibull shape parameter / amount of all nodes inclusive support nodes / wall roughness / slope of a tapered tower
$k_{\sigma,min}$	minimum buckling value
k_s	geometry coefficient to account for the clearance between a hole and a bolt
k_x, k_θ, k_τ	buckling interaction parameters
L	integral length dimension / beam length
l	lattice structure height / related body dimension for aerodynamic profiles
L_1	tower base yaw bearing axial support distance
L_{cr}	buckling length of a column buckling member
L_{eXY}	beam element length, projected to the global XY plane
L_e	beam element length
L_F	flap length of a leg profile
L_L	side length of a leg profile
L_L*	scaled side length of a leg profile
l_{OH}	overhang length from the tower top center at height $H_T + l_{TS}$ to the blade root center in the hub
l_{shell}	unsupported shell length
l_{TS}	vertical distance from tower top to the rotor shaft
L_t	reference length
l_i	length of member i
m	Wöhler exponent
m_1, m_2	inverse Wöhler curve slopes for the finite life fatigue strength and fatigue limit cycle ranges
$M_{Manwell}$	gyroscopic moment by Manwell et al.
M_{max}	maximum overturning moment at tower base
M_T	torsional moment
$M_{W,y}$	bending moment due to tower wind loads in x direction
$M_{x,ref}$, $M_{y,ref}$	Extreme side-side and fore-aft tower bending moments from the reference ASE simulation
M_x, M_y	bending moments around x- and y-axes
$m_{yaw\ bear.}$	yaw bearing mass
$M_{z,q}$	Vlasov torsional moment

$M_{z,SV}$	Saint-Venant torsional moment
$M_{\alpha i}$, $M_{\beta i}$, $M_{\gamma i}$	beam element bending moments and torsional moment of node i
N	number of system degrees of freedom
$N(S)$	amount of load cycles until failure under load range S
$N_{b,Rd}$	member column buckling resistance against axial force
N_B	number of bins
n_{el}	number of cross sectional finite elements
$N_{FL,B1}$	fatigue limit lifetime cycles for detail class B1
N_{Leg}	number of lattice tower legs
n_M	number of lattice tower members
N_{nodes}	number of tower nodes
N_{seg}	number of bracing segments of a lattice tower
n_{ST}	amount of short time load cycles
n_S	number of cross sectional spots, where stresses were calculated
P	rotor power
p	amount of all lattice members
$p(V_H)$	probability density function for wind speed occurrances
$p-1$	highest power of shape function polynomials
P_r	rated WT power
P_w	wind power
$P_W(V_H)$	probability that a wind speed is below V_H
q	pressure
q_p	gust velocity pressure
$Q_{W,x}$	accumulated shear force distribution in x direction
$q_{W,x}$	shear force distribution in x direction
r	sum of all constraints between the members
R_d	design resistance
R_k	characteristic resistance
R_{rot}	rotor radius
Re	Reynolds number
S	load range / Strouhal number
s	amount of independent load components per member
S_0	theoretical laod range at $N(S) = 0$ to model the S-N curve
S_d	design load effect
$S_{FL,B1}$	fatigue limit stress range for detail class B1

S_k	characteristic load effect
S_L	gust energy as dimensionless spectral density
$S_{RILT,mat.}$	material related cost savings of the RILT
s_x, s_y	coordinates on a cross sections where stresses were calculated
S_i	load range i
$S_{G,i}$	Goodman corrected load range i
SF_{tm}	tower mass scaling factor
T	thrust force / time period
t	time / wall thickness
T_{DL}	wind turbine design lifetime
T_{ECD}	time period of extreme coherent gust with direction change
T_{EDC}	time period of extreme direction change
T_{EOG}	time period of the extreme operating gust
T_{EWS}	time period of extreme wind shear
T_L	calculated lifetime of a structural component
t_L	wall thickness of a leg profile
$t_{min,Brace}$	minimum wall thickness of the bracing profile to be not classified as class 4
$t_{min,Leg}$	minimum wall thickness of the leg profile to be not classified as class 4
T_{PII}	periodical inspection interval
t_p	thickness of an junction plate
u, v, w	beam element node displacements / flow velocities
u_{wake}	dimensionless wake velocity deficit
V	wind speed / beam volume
$v(z_e)$	current wind speed at the reference height z_e
V_∞	free stream velocity
V_1	free stream wind speed before wind turbine
V_2	wind speed at wind turbine rotor disc
V_4	wind speed downstream of a wind turbine rotor disc
V_{ave}	annual mean wind speed at hub height
V_{cg}	wind speed increase for an ECD
V_{div}	critical divergence wind speed
V_{e1}, V_{e50}	extreme wind speed with a recurring period of 1 and 50 years
V_{eq}	equivalent wind speed at the towers weighted wind attack point
V_{gust}	extreme operating gust wind speed
V_H	wind speed at hub height

V_{in}, V_{out}	cut-in and cut-out wind speed
V_m	mean wind speed, dependent of the basis wind speed with a mean recurring period of 50 years
V_{ref}	10 min mean of reference wind speed
V_r	rated wind speed
V_w	current wind speed
V_x, V_y	current stream velocity in x and y direction
w_p	filler plates representative continous width
W_x, W_y	section modules around x and y axes
x, y, z	Cartesian coordinates
x_s, y_s	cross sectional shear center coordinates with respect to the reference point
$x_{c,m,i}, y_{c,m,i}$	geometrical center point coordinates of cross sectional element i of member m
z_0	roughness height
z_H	hub height
z_{min}	minimum height
z_{Tip}	height coordinate of the rotor blade being at the lowest postion
z_t	reference height

Mathematical Symbols

$\ddot{\bullet}$	second derivative with respect to time
$\dot{\bullet}$	first derivative with respect to time
∇	nabla operator
$\bar{\bullet}$	unit vector
$\underline{\bullet}$	vector
$\underline{\underline{\bullet}}$	matrix
$\tilde{\bullet}$	skew symmetric matrix of a vector
^{T}	transpose
$E\langle\bullet\rangle$	expected value
S_i, C_i	sine and cosie function of argument i

Relevant FAST Environmental Parameters

BldPitch*n*	blade *n* pitch angle
GenPwr	generator power
IntfFXss	transition piece interface reaction force in global X-direction
IntfFYss	transition piece interface reaction force in global Y-direction
IntfFZss	transition piece interface reaction force in global Z-direction
IntfMXss	transition piece interface reaction moment around global X-axis

IntfMYss	transition piece interface reaction moment around global Y-axis
IntfMZss	transition piece interface reaction moment around global Z-axis
NacYIner	nacelle yaw inertia
PtfmCMxt	initial platform displacement in local tower x-direction
PtfmCMyt	initial platform displacement in local tower y-direction
PtfmCMzt	initial platform displacement in local tower z-direction
PtfmHeave	initial platform heave displacement
PtfmPIner	platform pitch inertia
PtfmPitch	initial platform pitch rotation
PtfmRIner	platform roll inertia
PtfmRoll	initial platform roll rotation
PtfmSurge	initial platform surge displacement
PtfmSway	initial platform sway displacement
PtfmTDxi	global tower top displacement in X-direction
PtfmTDyi	global tower top displacement in Y-direction
PtfmYaw	initial platform yaw rotation
RootMxc1	in-plane blade root bending moment
RootMyc1	out-of-plane blade root bending moment
RotSpeed	rotor speed
RotThrust	rotor thrust
RotTorq	rotor torque
TowerBsHt	height of the ElastoDyn tower base
TowerHt	height of the ElastoDyn tower
TwrBrMxt	tower top bending moment about its x axis
TwrBrMyt	tower top bending moment about its y axis
TwrBsMxt	tower base bending moment about its x axis
TwrBsMyt	tower base bending moment about its y axis
Wind1VelX	wind speed in global X-direction
Wind1VelY	wind speed in global Y-direction

1 Introduction

1.1 Historical Background

For thousands of years, humans have been using the wind to sail, pump water, or mill grain with windmills. Post windmills, such as in Figure 1.1 were exclusively built to mill grain. A yaw bearing was placed at the top of the first third of the whole support structure. It was required for the correct rotor alignment with respect to the wind direction. The upper rotatable support structure part was used to provide enough space for the millwork(8), the meal floor(14), and other mill components(7, 9-13). Thereby, the rotatable support structure was simultaneously working as the nacelle to contain the cog wheel (1), the wind shaft (3), and other drive train components (2, 3-6).

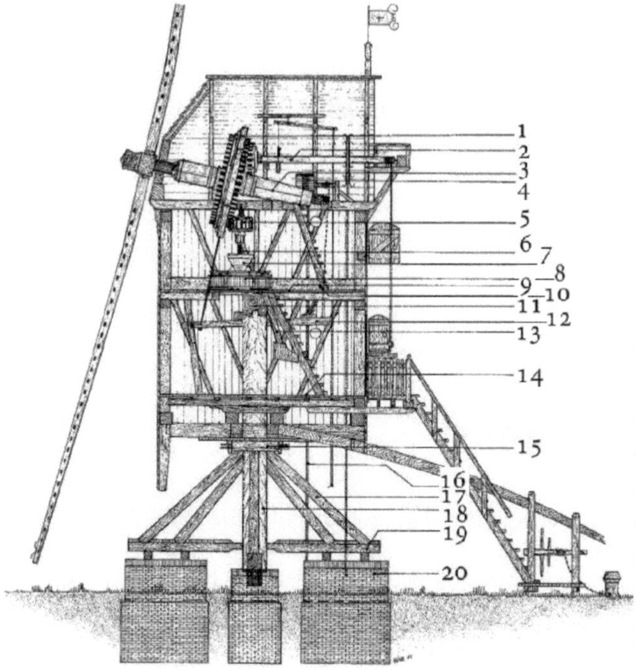

Figure 1.1: Construction of a post windmill [75]

The first systematically developed electricity-producing wind turbine (WT) was built in year 1891 by Poul La Cour in Askov, Denmark [92]. Its support structure consisted of a wooden lattice tower on top of a house. Because the meal room was not more necessary, the yaw bearing could be put on top of the tower. Therefore, the ratio of tower height to nacelle height increased drastically. This work focusses on WTs with rotatable support structures in which historical grain milling windmills may be seen as precedent case for this approach.

1.2 Motivation

The overarching motivation is to increase WTs competitiveness to conventional fossil and nuclear using energy sources. A major aspect to make renewable energies more attractive is their economical feasibility. Therefore, their efficiency has to be improved and their costs have to be reduced. System costs of multimegawatt WTs are assembled out of costs for components, their maintenance, erecting, and transport. Considering the favouring of larger hub heights in tendering procedures for modern land-based WTs in Germany [191], rising system costs are the consequence. Especially the support structure contribution can amount to more than 30 % of the WT system costs [77]. Problems arise with transport of conventional steel towers where the diameter restriction of 4.3 m caused by bridges and tunnel clearance constraints forces uneconomical great wall thicknesses to meet the design requirements. Besides economic aspects, the environmental impact of WT support structures has likewise to be considered. This becomes clear by assuming for example 2.8 t of CO_2-emissions per produced ton of steel [148]. Answers of the industry to these challenges may include alternative tower concepts, such as hybrid steel-concrete, pre-tensioned concrete, guyed, wooden, and lattice towers. Each of them has its own advantages and disadvantages, such as discussed in the books from Faber et al. [71], Hau [92], Gasch [77], and Manwell, McGowan, and Rogers [137]. The lack of sufficient conceptual answers to economic and environmental requirements motivates the call for new material saving WT support structures. Therefore, the next section introduces to the concept of a rotatable tower for land-based WTs and shows, how it should meet these cost and environmental requirements.

1.3 Problem Description

The concept of a rotatable WT tower requires the yaw system to be at the bottom of the tower instead on the top of it. This implies that the rotor nacelle assembly (RNA) aligns to the wind direction together with the whole tower. For convenience one can identify two directions: the fore-aft (FA) and the side-side (SS) direction where FA is orthogonal to the rotor plane and SS is parallel to it, such as shown in Figure 1.2.

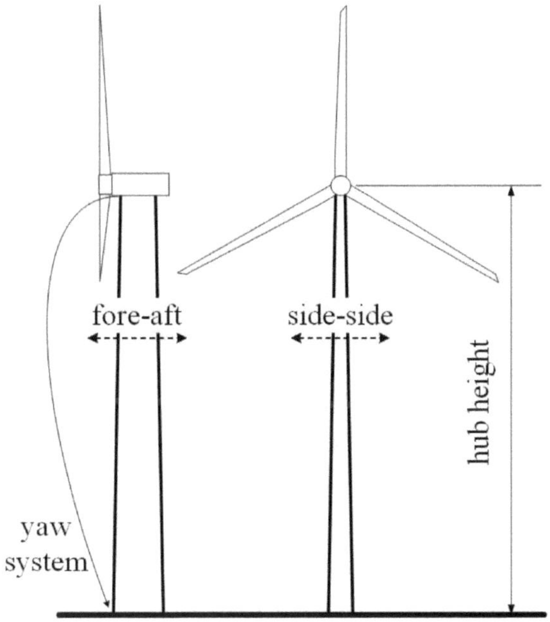

Figure 1.2: Moving the yaw bearing of a WT form the tower head to the tower base.

Such a fixed tower-to-RNA configuration achieves several benefits, but has likewise its challenges. One of the big issues with this approach lies in the concept and design of the yaw system, especially by considering the huge thrust forces at the top of the tower and the long lever arm, which result in big bending moments at the tower base. The yaw bearing must not only withstand these enormous bending moments and the self-weight of the structure, but must also provide sustainable rotation, combined with minimum tower tilting for all design load cases (DLCs). On the other hand, material savings of the rotatable tower, which achieve cost and CO_2-emission savings compared to a fixed tower may overcome the increased costs for such a yaw system. Material savings can be achieved in the case of different shear forces and bending moments with respect to the local rotatable tower coordinate system. Aero-servo-elastic (ASE) load simulations by Jonkman [110] for the land-based 5 MW National Renewable Energy Laboratory (NREL) definition WT show such different ultimate load magnitudes for the FA direction compared to the SS direction. This NREL WT is a conventional three-bladed upwind variable-speed variable blade-pitch-to-feather-controlled turbine. ASE investigations of the 10 MW Denmark Technical University (DTU) reference WT indicate the same, even for damage equivalent fatigue loads of the tubular tower [5]. In these studies, the shear forces in FA direction and the bending moments around the SS axis, defined as FA loads, were larger as the shear forces along the SS direction and the bending moments around the FA axis, defined as SS loads. Both studies consider a set of DLCs, which are identified to be tower design driving for the most horizontal WT configurations. But other results are possible, especially if vortex-induced vibrations of tubular towers lead to great SS loadings. However, the assumption of lower SS loads than FA loads indicates that smaller second area moments of inertia around the FA axis are sufficient, thus material can be saved. Such direction dependent stiffness and changed mass distributions along the tower lead to two different bending modes with lower natural frequency around the FA axes. This property has to be mentioned for the proposed operational range and within corresponding control strategies. Other advantages of rotatable towers are better transportability through bridges and tunnels due to their tendency for elongated more flat tower profiles and improved yaw system maintenance accessibility. Dependent on the yaw system design it is imaginable to exchange the whole bearing without dismounting the tower and RNA. This can be done by heaving the whole turbine with hydraulic presses, while the tower is guyed and would save crane costs. Lower aerodynamic tower drag may also be possible along the FA direction by means of aerodynamic profiles, but can lead to additional lift forces in case of skewed inflow. DLCs which account for a yaw system failure and the wind has a skewed or perpendicular inflow angle to the WT may be dangerous and have to be analysed within ASE load simulations. Nevertheless aerodynamic cross sections have the potential to reduce the tower dam effect and improve the tower shadow behaviour in case of downwind WTs. This work discusses these aspects in more detail and shows the development and analysis of an even more effective rotatable tower concept, than the typical suggestions of an elliptical or aerodynamically shaped tower cross section.

1.4 Limitations

The goal of the present work is to discuss the technical and economical feasibility of WTs with rotatable support structure. Since not many scientific works mentioned this concept before, this work has first to deal with a conceptual comparison of potential rotatable support structure solutions. Afterwards a deeper analysis is presented for the rotatable inclined lattice tower as one of several possible approaches. The preliminary nature of this work and its limited scope hindered more detailed investigations with respect to the yaw system, the foundation, the machinery bed at the tower top and the lattice tower joints, even though each of these aspects is mentioned throughout the work. Since the applied complex ASE simulations and their results depended not only on the support structure, but also on other parts, the environment, and the control system, many more short cuts had to be introduced to keep the size of the work within certain borders. For example, not all design load cases were considered, but the ones, which pointed out by the literature and own experience as being relevant. Furthermore, the tower assessment was primary done according to the IEC 61400-1 and only partially according to the more sophisticated DNV GL-standards. Control system adjustments for transient WT events were carried out on a simple level to decrease the loads for a rotatable WT support structure. Therefore, many improvements and deeper investigations in following scientific works are conceivable.

1.5 Document Structure

This section provides a short overview of the content in each chapter.

Chapter 1 - Introduction
Chapter 1 contains the historical background of grain milling windmills as precedents of the proposed rotatable WT tower concepts. A following motivational part explains why new material and cost saving tower concepts are required. A subsequent problem description introduces to some fundamental aspects of rotatable WT tower concepts. In the end, some limitations and the document structure of this work are presented.

Chapter 2 - State of the Art
Chapter 2 presents all fundamental basics and knowledge requirements for this work. It contains a short introduction to wind energy conversion and repeats the necessary tower and foundation assessment aspects. Afterwards, some WT tower related subjects are introduced, such as its manufacturing technology, transport and erection. The yaw system is likewise explained on the surface, because it is not the focus in this work. A more detailed introduction is given to lattice towers, because the more investigated concept in this work is a structure out of lattice members. For the structural analysis, the load evaluation is presented as one important aspect and followed by a literature research about rotatable support structures for WTs and related technology trees of other industry sectors.

Chapter 3 - Load Analysis of Conventional Designs
Chapter 3 shows the methodological approach for the chapter and partially for this work. It focusses on conventional WTs and introduces the reference WT and their analytically evaluated loads. After a complementary literature research about tower loads, the own ASE load analysis of the reference WT is carried out and validated.

Chapter 4 - Load-Direction-Derived Support Structures
Chapter 4 introduces to fundamental aspects of load-direction-derived support structures with the focus on rotatable WT towers. After a definition of the term load-direction-derived, tower base yaw systems and corresponding foundations are presented and discussed. Afterwards, the structural and aerodynamic properties of rotatable towers are explained. The end of the chapter shows the concept evaluation method for a rotatable tower.

Chapter 5 - Rotatable Tubular Concepts
Chapter 5 analyses tubular rotatable tower concepts and starts with a comparison of possible shapes. Afterwards, a more detailed analysis of an elliptical tower shape reveals its potential in terms of material savings. In the end, some additional tubular concepts are introduced.

Chapter 6 - Rotatable Inclined Lattice Tower Concept
Chapter 6 explains the favoured rotatable inclined lattice tower concept in this work. It starts with the cross sectional concept of the lattice tower, followed by the new aspect of tower inclination, and ends by other possible concept proposals.

Chapter 7 - Rotatable Inclined Lattice Tower Design
Chapter 7 introduces to the more detailed rotatable inclined lattice tower design and explains its parametric specifications. The end contains a short subsection about the general design as an iterative process.

Chapter 8 - Modelling the Rotatable Inclined Lattice Tower in FAST
Chapter 8 treats the modelling of the tower in the ASE load simulation environment. Therefore, the necessary adjustments in the source code are explained and the afterwards modelled tower in SubDyn is validated through a parallel modelling approach. The end of the chapter contains the implementation of the SubDyn model in the ASE load simulation framework of FAST, the corresponding adjustments in FAST, and the SubDyn simulation settings.

Chapter 9 - Structural Analysis of a Rotatable Inclined Lattice Tower
Chapter 9 treats the whole structural analysis of the rotatable inclined lattice tower. It starts with the definition of design parameter values. A following modal analysis of the structure is carried out before the actual ASE simulation of IEC 61400-1 load cases. After the post processing of the loads, the ultimate, fatigue, accidental, and serviceability limit state are checked to ensure the integrity of the structure over its lifetime.

Chapter 10 - Potential of Rotatable Support Structures
Chapter 10 provides the potential analysis of the rotatable inclined lattice tower with respect to its costs. Therefore, material savings are evaluated first to calculated the corresponding CO_2 and cost savings. The next two subsections determine the increased WT efficiency through tower inclination and provide an estimation of tower base yaw system costs. In the end, the economical potential is analysed by balancing the cost savings, additional money income, and additional costs of the rotatable inclined lattice tower.

2 State of the Art

2.1 Fundamentals of Wind Energy Conversion

This section gives a brief overview about WT technology, whereby subsection 2.1.1 gives a short introduction to wind as energy resource, subsection 2.1.2 explains the fundamental physics of wind energy conversion, and subsection 2.1.3 presents some main WT components and its functions. More detailed introductions into wind energy technology may be read in [77], [92], [71], and [137].

2.1.1 Wind

According to Manwell, Mcgowan, and Rogers [137, p. 24] the sun is one origin for wind, which heats earthly surfaces differently due to its solar radiation. A corresponding pressure gradient in the air with different temperatures forces the air to move from high pressure zones to low pressure zones. Another reason for air to move is the earth's rotation. Its circumferential speed amounts to 1670 km/h at the equator and decreases to zero at the poles [ibid.]. In the smaller scale the earth's surface roughness influences the vertical gradient of the horizontal wind speed, such as being relevant for WTs assessment, discussed in section 2.2. An appropriate measure for wind resource potential gives the wind power density $P_\mathrm{w}/A_\mathrm{disc}$, which results out of the wind mass flow $\mathrm{d}m/\mathrm{d}t = \dot{m} = \rho_\mathrm{air}\, A_\mathrm{disc}\, V_\mathrm{w}$ through a disc with an assumed area of $A_\mathrm{disc} = 1\ \mathrm{m}^2$. P_w is the rotor power, m is the mass of an air particle, t is time, ρ_air is the air density, and V_w is the wind speed.

$$P_\mathrm{w} = \frac{1}{2}\frac{\mathrm{d}m}{\mathrm{d}t}\, V_\mathrm{w}^2 \rightarrow \frac{P_\mathrm{w}}{A_\mathrm{disc}} = \frac{1}{2}\, \rho_\mathrm{air}\, V_\mathrm{w}^3 \tag{2.1}$$

Since Equation 2.1 shows that the wind speed contributes to the power of 3 to the wind power density, it is one of the most important factors to chose an appropriate WT site. Figure 2.1 shows therefore the mean wind speed of some European countries in altitudes of 50 m above the ground. From that it is clear that sea sites provide the highest wind power potential, but they require cost intensive WT support structure concepts, expensive mounting on site, and they introduce higher maintenance costs to the WT. However, land-based WTs have still their justification, especially if they use innovative more economic tower concepts for larger hub heights, which overcome transportability problems, such as discussed throughout this work.

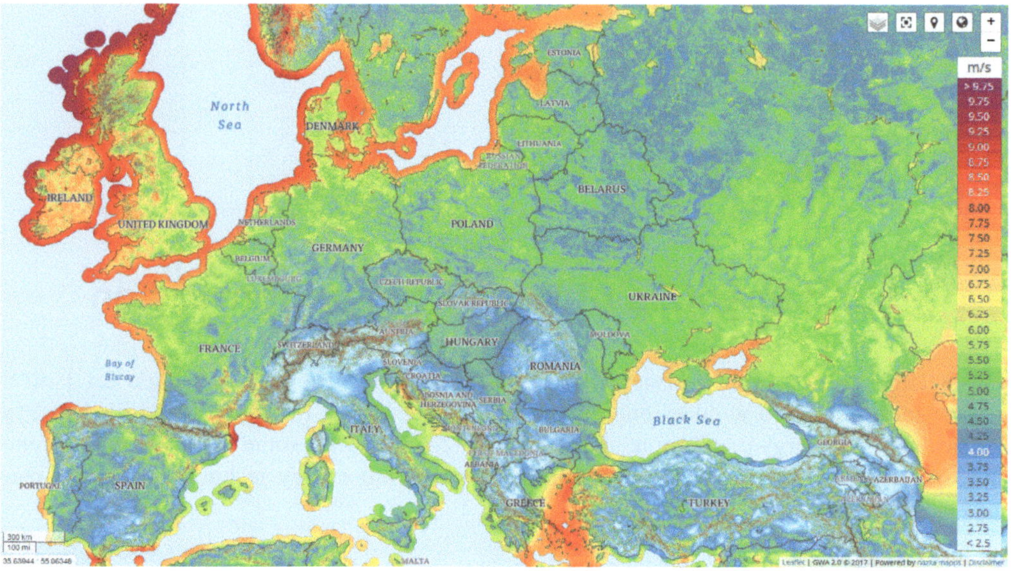

Figure 2.1: Mean wind speed over 10 years in 50 m above ground of European countries [66]

2.1.2 Wind Energy Conversion

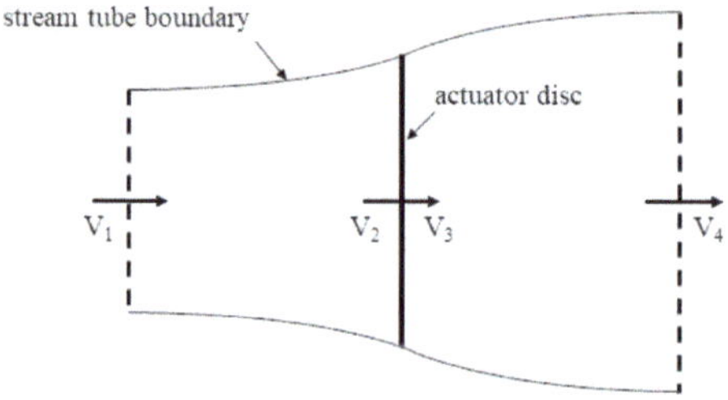

Figure 2.2: Actuator disc model of a WT in stream tube according to Betz [12]

The Betz [12] model, indicated in Figure 2.2, may be used to describe rotor thrust, rotor power, and the effect of the local wind field before and behind an ideal rotor disc. A corresponding derivation of the one-dimensional Momentum Theory by Betz can be reviewed in Manwell, McGowan, and Rogers [137, p. 92-96]. It is limited to homogeneous, incompressible, steady state fluid flow where its static pressure far before and behind the WT is equal to the ambient pressure. Furthermore, no frictional drag, an infinite number of blades, uniform thrust over the rotor area, and a non-rotating wake is assumed. The thrust force T is obtained by conservation of the linear momentum in the mass flow $\dot{m} = \rho_{\text{air},1}\ A_1\ V_1 = \rho_{\text{air},4}\ A_4\ V_4$ where $\rho_{\text{air},i}$ is the air density, A_i is the cross sectional area, and i denotes the corresponding section in the stream tube so that

$$T = \dot{m}\,(V_1 - V_4) \tag{2.2}$$

By application of the Bernoulli function, one can write Equation 2.2 with respect to the actuator disc area A_2 as

$$T = \frac{1}{2}\rho_2\, A_2 \left(V_1^2 - V_4^2\right) \tag{2.3}$$

In this simple model the wind speed at the actuator disc is the average between the upstream and downstream wind speeds according to

$$V_2 = \frac{V_1 + V_4}{2} \tag{2.4}$$

An axial induction factor a is introduced to describe the wind speed as fractional decrease of the wind speed V_1 in free stream condition with respect to the wind speed V_2 at the rotor disc according to Equation 2.5.

$$a = \frac{V_1 - V_2}{V_1} \tag{2.5}$$

Using Equations 2.4 and 2.5 in 2.3 results to

$$T = \frac{1}{2}\rho_2\, A_2\, V_2^2\, [4\, a\,(1 - a)] \tag{2.6}$$

and the thrust coefficient c_{T} is defined as

$$c_{\mathrm{T}} = \frac{T}{\frac{1}{2}\rho_2\, V_2^2\, A_2} = \frac{\text{Thrust Force}}{\text{Dynamic Force}} = 4\, a\,(1 - a) \tag{2.7}$$

The rotor power P is calculated by multiplication of the thrust force with the wind speed at the rotor disc and gives

$$P = \frac{1}{2}\rho_2\, A_2\, V_2^3\; 4\, a\,(1 - a)^2 \tag{2.8}$$

where the power coefficient c_{P} is defined as

$$c_{\mathrm{P}} = \frac{P}{\frac{1}{2}\rho_2\, V_2^3\, A_2} = \frac{\text{Rotor Power}}{\text{Wind Power}} = 4\, a\,(1 - a)^2 \tag{2.9}$$

The zero equalized derivative of the power coefficient in Equation 2.9 with respect to a reveals the maximum[1] theoretical power coefficient of $c_{\mathrm{P,max}} = 16/27 \approx 0.59$ at an axial induction factor of $a = 1/3$. In reality properties, such as the rotation of the wake behind the WT, tip losses, associated with a finite blade number, and non-zero aerodynamic drag cause a reduced power coefficient. Using an induction factor of $a = 1/3$ in Equation 2.7 results to a corresponding thrust coefficient of $c_{\mathrm{T,max}} = 8/9$. Note that the overall WT efficiency can be calculated by

[1] This theoretical maximum power coefficient is likewise known as Betz limit.

multiplication of the power coefficient with the mechanical efficiency $\eta_{\text{mech.}}$, which accounts for gear box and bearing friction heat dissipation and with the electrical efficiency $\eta_{\text{elec.}}$, which accounts for losses in the generator, power electronics, and cables.

A rotating wind wake behind the WT is not assumed in the previous consideration. More realistic approaches include the kinetic energy, which is lost by the rotor and given to the now partial rotating wind field behind the WT. This wind field may be seen as annual control volume around the rotor where the angular momentum loss is calculated for. Detailed explanations may be read in [137, p. 96-101] and are not given in this work.

Current rotor designs are established by means of the more exact Blade Element Momentum (BEM) Theory. In this approach the conservation of linear and angular momentum is combined with an analysis of forces and moments at rotor blade sections [137, p. 117]. Since it reveals more information about blade loads and provides flexibility to include more sophisticated aerodynamic models, such as tip loss as result of a finite blade number, wake rotation, and radial airflow along the blade, it is used in many aero-servo-elastic (ASE) simulation tools, such as explained in subsection 2.8.2. Detailed explanations of the previously mentioned aspects may be reviewed in [137, p. 115-145].

2.1.3 Wind Turbine Components

A conventional modern WT consists out of the RNA and its support structure, shown in Figure 2.3. From a constructive point of view the support structure is counted to the discipline of civil engineering. Possible and already built tower concepts are friction joint tubular steel, slipformed concrete, hybrid concrete, lattice steel, covered lattice wood, guyed tubular steel, and the most common tubular steel towers. A review of the literature, proposed in section 2.1 and the CO_2/weight/cost comparison in subsection 4.5 for a comparison between different non-rotatable tower concepts is encouraged. The foundation of the support structure must resist the overturning moment caused by the aerodynamic loads on the WT and must have a sufficient strength and stiffness, which depends amongst other things on the soil properties. Depending on the soil properties different foundation types are conceivable. Flat slab foundations are the standard with round or polygonal concrete footings. For weak soils pile foundations gain stability by reaching load-bearing ground layers [92].

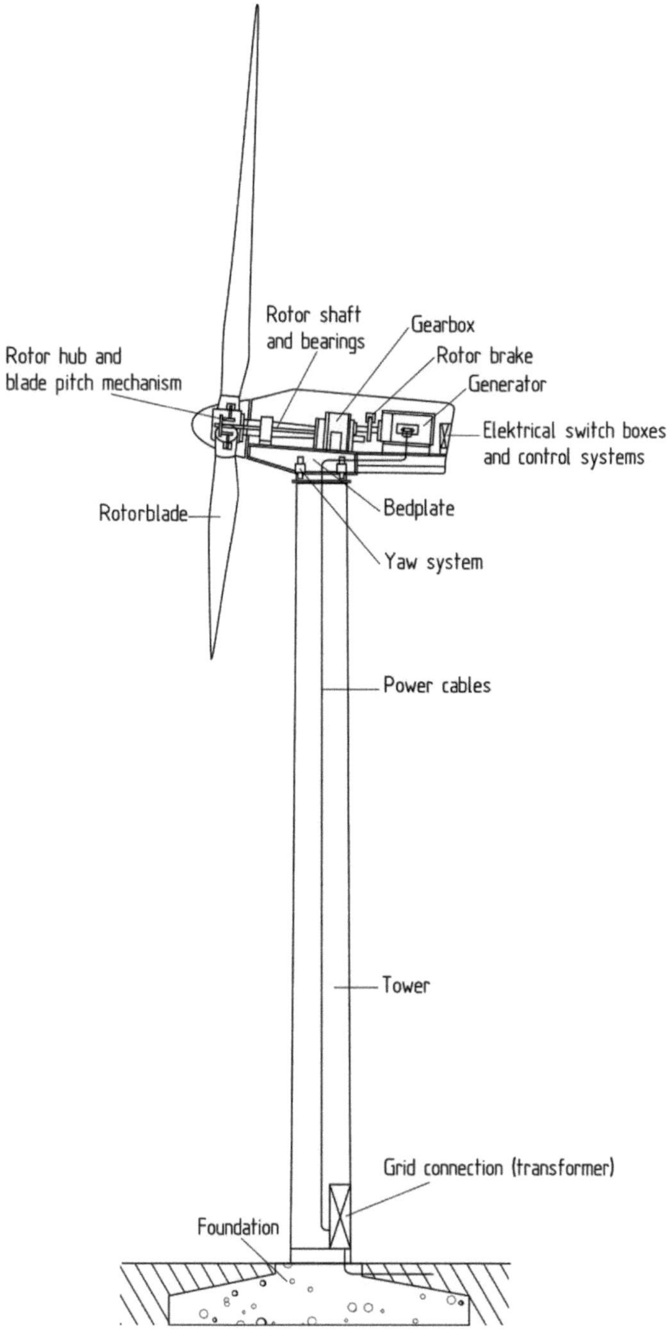

Figure 2.3: Complete exemplary wind turbine with conventional tubular steel tower and gravity slab foundation from Hau [92, p. 73]

A more detailed picture of the RNA is given in Figure 2.4. The machine foundation (17) is mounted on the yaw bearing (16), which is bolted to the top of the tower. As reaction to changing wind directions the necessary RNA's alignment to the wind will be realized by the yaw drives with its gears (18). To control the power output and the loads modern WTs have the additional ability to pitch their blades (12) with help of their pitch mechanism (1, 2, and 13). The blade carrying hub (3) rotates relatively slow with high aerodynamic torque. From the rotor the power is led into the gearbox (6) over the main shaft (4) where it is transformed to a lower torque with higher rotational speed on the high speed shaft (19) for the generator (20 and 21). The transformer brings the electric voltage on a reasonable level for the whole wind farm to feed it into the grid. A breaking (7, 14, and 15) and cooling (5) system ensures operational safety and ultra-sonic sensors (10) help to measure the wind properties. Modern WTs are connected to a computational network (9) to collect data and to provide an interface to external WT control. For maintenance services a crane (8) helps to manage heavy weight components.

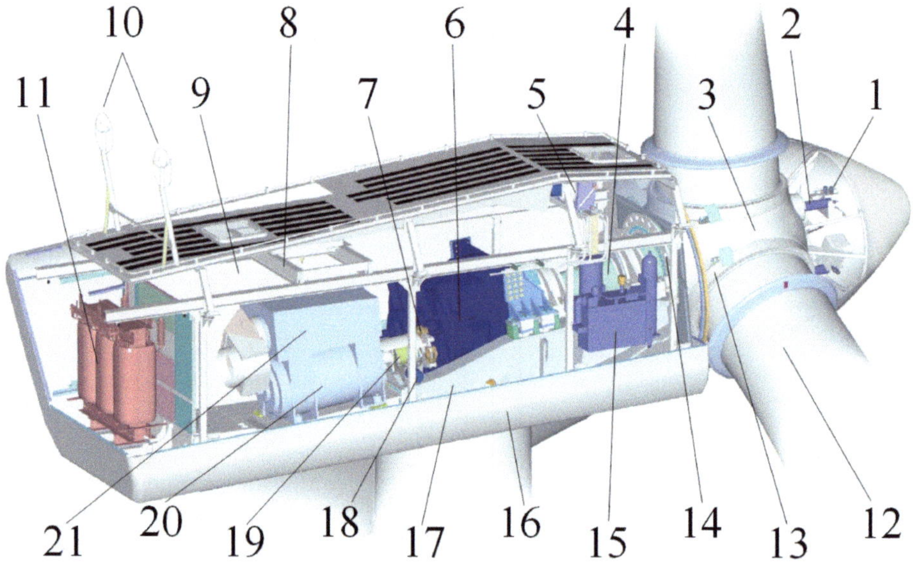

Figure 2.4: Vestas V80-2 MW wind turbine nacelle components. 1. hub controller, 2. pitch cylinder, 3. blade hub, 4. main shaft, 5. oil cooler, 6. gearbox, 7. parking brake, 8. service crane, 9. VMP-top controller with converter, 10. ultra-sonic sensors, 11. transformer, 12. blade, 13. blade bearing, 14. rotor lock system, 15. hydraulic unit, 16. yaw ring, 17. machine foundation, 18. yaw gears, 19. composite disk coupling, 20. OptiSpeed^{TM}-generator, 21. generator cooler. Picture from [28] and description of the components from [190]

2.2 Tower Assessment

Standards build the base for WT tower assessment, whereby some of them are introduced in subsection 2.2.1. On the loads and site conditions side, subsection 2.2.2 presents the DLCs, demanded by the IEC 61400-1 [50]. The following subsection 2.2.3 summarizes additional DLCs from other standards and their difference with respect to the IEC 61400-1. Furthermore, WT tower assessment ensures the observance of the following limit states throughout all considered DLCs:

- Ultimate Limit State (ULS), see subsection 2.2.5
- Fatigue Limit State (FLS), see subsection 2.2.6
- Accidental Limit State (ALS), see subsection 2.2.7
- Serviceability Limit State (SLS), see subsection 2.2.8

Each of these limit states can be evaluated under the partial safety factor method, implied by IEC 61400-1 [50] and explained in [63]. It demands that a design load effect S_d must not exceed the design resistance R_d

$$S_\mathrm{d} \leq R_\mathrm{d} \tag{2.10}$$

while both sides are scaled by means of certain partial safety factors. S_d of the design criterion in Equation 2.10 may represent internal or external forces and moments, a deformation, or stresses in a cross section. On the other side R_d may be used as capacity, a critical deformation, or a yield stress. A discussion about corresponding partial safety factors (PSFs) to be applied in different situations is given in subsection 2.2.4.

2.2.1 Prevailing Standards

Faber [71] gives an overview about different design requirements for support structures and lists some corresponding standards. The IEC 61400-1 standard [50] contains a minimum of load cases and requirements, especially for land-based WTs.[2] Owing to its fundamental character it is often used in scientific investigations and serves as base for the more practical relevant DIBt [152] and DNV GL standard. The DNV GL standard is distributed over several specialized documents, whereby the loads and site conditions standard [64], support structure standard [63], and the recommended practice report for fatigue design [62] are the most important guidelines for the design of WT support structures. Other standards, such as the Eurocode 3 for general requirements - DIN EN 1993-1-1 [42], for plated structural elements - DIN EN 1993-1-5 [43], for strength and stability of shell structures - DIN EN 1993-1-6 [44], and for fatigue assessment - DIN EN 1993-1-9 [47] must often be consulted if it comes to the direct verification of structural components. The tower and masts part of the Eurocode 3, DIN EN 1993-3-1 [48], may also be relevant together with the wind actions part of Eurocode 1, DIN EN 1991-1-4 [67]. As the most

[2] In this work edition 3 of IEC 61400-1 was used, because it was valid for the time when this work was written. The next edition 4 [51] was in its draft stage at this time

support structures have bolted or welded joints, DIN EN 1993-1-8 [45] is likewise important. A comprehensive overview about all relevant standards and guidelines for WT design is given in Table 1-1 of [63].

2.2.2 Design Load Cases of the IEC 61400-1

For the design of rotatable support structures the knowledge about acting loads on a WT is essential. Origins of loads can be distinguished with respect to Table 2.5, but its magnitudes should be investigated throughout ASE load simulations, such as introduced in subsection 2.8.2. The IEC 61400-1 [50] requirements contain the minimum load cases, which have to be considered in these simulations. They are divided into several operating conditions, such as shown in Tables 2.1 and 2.2.

Design situation	DLC	Wind condition	Other conditions	Type of analysis	Partial safety factor
1) Power production	1.1	NTM $V_{in} < V_H < V_{out}$	For extrapolation of extreme events	U	N
	1.2	NTM $V_{in} < V_H < V_{out}$		F	*
	1.3	ETM $V_{in} < V_H < V_{out}$		U	N
	1.4	ECD $V_H = V_r - 2$ m/s, V_r, $V_r + 2$ m/s		U	N
	1.5	EWS $V_{in} < V_H < V_{out}$		U	N
2) Power production plus occurrence of fault	2.1	NTM $V_{in} < V_H < V_{out}$	Control system fault or loss of electrical network	U	N
	2.2	NTM $V_{in} < V_H < V_{out}$	Protection system or preceding internal electrical fault	U	A
	2.3	EOG $V_H = V_r \pm 2$ m/s and V_{out}	External or internal electrical fault including loss of electrical network	U	A
	2.4	NTM $V_{in} < V_H < V_{out}$	Control, protection, or electrical system faults including loss of electrical network	F	*

Table 2.1: Design load cases according to IEC 61400-1 [50] - first part

Design situation	DLC	Wind condition	Other conditions	Type of analysis	Partial safety factor
3) Start-up	3.1	NWP $V_{in} < V_H < V_{out}$		F	*
	3.2	EOG $V_H = V_{in}, V_r$ ± 2 m/s and V_{out}		U	N
	3.3	EDC $V_H = V_{in}, V_r$ ± 2 m/s and V_{out}		U	N
4) Normal shut down	4.1	NWP $V_{in} < V_H < V_{out}$		F	*
	4.2	EOG $V_H = V_r \pm 2$ m/s and V_{out}		U	N
5) Emergency shut down	5.1	NTM $V_H = V_r \pm 2$ m/s and V_{out}		U	N
6) Parked (standing still or idling)	6.1	EWM 50-year recurrence period		U	N
	6.2	EWM 50-year recurrence period	Loss of electrical network connection	U	A
	6.3	EWM 1-year recurrence period	Extreme yaw misalignment	U	N
	6.4	NTM $V_H < 0.7\ V_{ref}$		F	*
7) Parked and fault conditions	7.1	EWM 1-year recurrence period		U	A
8) Transport, assembly, maintenance and, repair	8.1	NTM V_{maint} to be stated by the manufacturer		U	T

Table 2.2: Design load cases according to IEC 61400-1 [50] - second part

Column five of Table 2.1 and 2.2 refers to the type of analysis to be performed for a certain DLC. An U declaration requires a material strength, tip to tower clearance, and stability analysis, while a F declaration requires a fatigue limit state analysis. The sixth column contains information about the PSFs to be applied, whereby N refers to a normal design state, A to an abnormal design state, * to a fatigue design state and T to a transport and erection design state. A more detailed discussion about the PSFs is given in subsection 2.2.4.

Many of these DLCs result in stochastic sensitive tower loads caused by wind turbulence and coincidentally unfavourable coupling effects of different components. Therefore, it is required to calculate multiple timeseries with different random seeds[3] of DLCs with stochastic wind fields. It follows a summarized explanation of the different wind models, given by abbreviations in the wind condition column of Tables 2.1 and 2.2. The detailed explanations can be found in the IEC 61400-I [50], but a summarized repetition is useful for later discussion about certain load cases in the following ASE load simulations.

NWP (Normal Wind Profile)
The NWP is the basic underlying wind speed profile. It describes the wind speed along the height coordinate z with respect to the hub height z_H. It is used for some wind field models and leads to different wind speeds across the swept rotor area.

$$V_w(z) = V_H \left(\frac{z}{z_H}\right)^{\alpha}, \tag{2.11}$$

The NWP exponential Equation 2.11 contains the power law exponent α, which is prescribed with 0.2 for land-based turbines [50].

NTM (Normal Turbulence Model)
The NTM uses the NWP and introduces an additional stochastic variation of the wind speed. This variation is prescribed by the representative standard deviation of turbulence σ_1 with respect to the hub height of the WT. It depends on the WT class according to Table 2.3.

WT Class	I	II	III
V_{ref}/m/s	50	42.5	37.5

A $I_{ref}/-$	0.16
B $I_{ref}/-$	0.14
C $I_{ref}/-$	0.12

Table 2.3: Wind turbine classes according to IEC 61400-1 [50]

The Table 2.3 gives expected values of turbulence intensity I_{ref}, which can be used to calculate σ_1 in Equation 2.12.

$$\sigma_1 = I_{ref}\,(0.75\,V_H + b)\,;\; b = 5.6\frac{m}{s} \tag{2.12}$$

σ_1 is valid for the longitudinal wind direction, while the transversal and upwardly directed standard deviations of turbulence σ_2 and σ_3 are scaled values of it.

$$\sigma_2 \geq 0.7\,\sigma_1; \quad \sigma_3 \geq 0.5\,\sigma_1 \tag{2.13}$$

ETM (Extreme Turbulence Model)
An ETM differs from the NTM only regarding σ_1 by

$$\sigma_1 = c\,I_{ref}\left(0.072\left(\frac{V_{ave}}{c}+3\right)\left(\frac{V_H}{c}-4\right)+10\right);\; c = 2\frac{m}{s} \tag{2.14}$$

[3] Random variates require the definition of random seeds for the pseudo-random number generator in the FAST environment [110, p. 21]

ECD (Extreme Coherent gust with Direction change)
In this special wind situation a time dependent wind speed increase up to $V_{cg} = 15\frac{\text{m}}{\text{s}}$ is combined with a wind direction change of θ_{cg}. The gust appears over $T_{ECD} = 10$ s and is superimposed with the NWP wind field. Modelling equations are

$$V(z,t) = \begin{cases} V_w(z) & \text{for} \quad t < 0 \\ V_w(z) + 0.5\, V_{cg} \left(1 - \cos\left(\frac{\pi\, t}{T_{ECD}}\right)\right) & \text{for} \quad 0 \leq t \leq T_{ECD} \\ V_w(z) + V_{cg} & \text{for} \quad t > T_{ECD} \end{cases} \tag{2.15}$$

$$\theta_{cg}(V_H) = \begin{cases} 180° & \text{for} \quad V_H < 4\,\frac{\text{m}}{\text{s}} \\ \dfrac{720°\,\frac{\text{m}}{\text{s}}}{V_H} & \text{for} \quad 4\,\frac{\text{m}}{\text{s}} \leq V_H \leq V_{ref} \end{cases} \tag{2.16}$$

$$\theta(t) = \begin{cases} 0° & \text{for} \quad t < 0 \\ \pm 0.5\,\theta_{cg} \left(1 - \cos\left(\frac{\pi\, t}{T_{ECD}}\right)\right) & \text{for} \quad 0 \leq t \leq T_{ECD} \\ \pm\theta_{cg} & \text{for} \quad t > T_{ECD} \end{cases} \tag{2.17}$$

EWS (Extreme Wind Shear)
The EWS model refers to extreme wind speed difference with respect to different positions across the rotor plane. These wind shear situations occur over a time period of $T_{EWS} = 12$ s and are distinguished between a vertical wind shear

$$V(z,t) = \begin{cases} V_w(z) \pm \left(\frac{z - z_H}{D_{rot}}\right)\left(2.5\,\frac{\text{m}}{\text{s}} + 0.2\,\beta\,\sigma_1 \left(\frac{D_{rot}}{\Lambda_1}\right)^{\frac{1}{4}}\right)\left(1 - \cos\left(\frac{2\,\pi\,t}{T_{EWS}}\right)\right) & \text{for} \quad 0 \leq t \leq T_{EWS} \\ V_w(z) & \text{otherwise} \end{cases} \tag{2.18}$$

and a horizontal wind shear

$$V(y,z,t) = \begin{cases} V_w(z) \pm \left(\frac{y}{D_{rot}}\right)\left(2.5\,\frac{\text{m}}{\text{s}} + 0.2\,\beta\,\sigma_1 \left(\frac{D_{rot}}{\Lambda_1}\right)^{\frac{1}{4}}\right)\left(1 - \cos\left(\frac{2\,\pi\,t}{T_{EWS}}\right)\right) & \text{for} \quad 0 \leq t \leq T_{EWS} \\ V_w(z) & \text{otherwise} \end{cases} \tag{2.19}$$

with $\beta = 6.4$ and D_{rot} as rotor diameter. Both wind shears have to be investigated separately and with different signs. The turbulence length parameter Λ_1 is given with

$$\Lambda_1 = \begin{cases} 0.7\,z & z \leq 60\text{ m} \\ 42\text{ m} & z \geq 60\text{ m} \end{cases} \tag{2.20}$$

EWM (Extreme Wind Model)
The EWM can be modeled either as a stationary wind field or as a turbulent one. In case of a stationary wind field, additional yaw misalignments of $\pm 15°$ have to be accounted for. In this model two wind speeds have to be investigated. One with an extreme wind speed V_{e50} with a recurring period of 50 years and the other with a recurring period of 1 year V_{e1}. The equations are

$$V_{e50}(z) = 1.4\ V_{ref} \left(\frac{z}{z_H}\right)^{0.11} \tag{2.21}$$

and

$$V_{e1}(z) = 0.8\ V_{e50}(z). \tag{2.22}$$

For the turbulent case a standard deviation of the longitudinal turbulence of

$$\sigma_1 = 0.11\ V_H \tag{2.23}$$

should be assumed. Interestingly, the international standard EN 61400-1:2005 [103] prescribes Equation 2.21 as

$$V_{e50}(z) = V_{ref} \left(\frac{z}{z_H}\right)^{0.11} \tag{2.24}$$

in case of turbulent wind. This means, that the extreme wind speeds are smaller than in the German standard. In this work, the international assumption for the extreme wind speeds are applied to achieve comparability.

EOG (Extreme Operating Gust)
The EOG is an extreme wind speed change, whose function over time looks like a Mexican hat[4]. The maximum gust wind speed V_{gust} of it is

$$V_{gust} = \min \left\{ 1.35\ (V_{e1} - V_H);\ \ 3.3 \left(\frac{\sigma_1}{1 + 0.1 \left(\frac{D_{rot}}{\Lambda_1}\right)} \right) \right\} \tag{2.25}$$

and the time and height coordinate dependent function is

$$V(z,t) = \begin{cases} V_w(z) - 0.37\ V_{gust}\ \sin\left(\frac{3\,\pi\,t}{T_{EOG}}\right)\left(1 - \cos\left(\frac{2\,\pi\,t}{T_{EOG}}\right)\right) & \text{for}\quad 0 \le t \le T_{EOG} \\ V_w(z) & \text{otherwise} \end{cases} \tag{2.26}$$

with $T_{EOG} = 10.5$ s.

[4] This name is widely used in the wind energy sector to describe the EOG wind speed change. See for example the article of Gillebaart, Bernhammer, Zuijlen, and van Kuik [82]

EDC (Extreme Direction Change)
The EDC θ_e is given by

$$\theta_e = 180^\circ \geq \pm 4 \arctan\left(\frac{\sigma_1}{V_H\left(1 + 0.1\left(\frac{D_{rot}}{\Lambda_1}\right)\right)}\right) \geq -180^\circ \tag{2.27}$$

and modeled for a time period of $T_{EDC} = 6$ s by

$$\theta(t) = \begin{cases} 0^\circ & \text{for} \quad t < 0 \\ \pm 0.5\, \theta_e \left(1 - \cos\left(\frac{\pi\, t}{T_{EDC}}\right)\right) & \text{for} \quad 0 \leq t \leq T_{EDC} \\ \theta_e & \text{for} \quad t > T_{EDC} \end{cases} \tag{2.28}$$

This model is superimposed with the NWP.

2.2.3 Other Design Load Cases

The DIBt standard [152] demands for additional statistical extrapolation of loads (DLC D.2 and D.3), accounting for ice loads (DLC D.4), earthquakes (DLC D.5 and D.6), and another PSF in extreme wind parking situations (DLC D.7). Furthermore, extreme wind speeds should be assumed with respect to DIN EN 1991-1-4 [67] and its National Annex [41] instead of IEC 61400-1 [50]. In terms of fatigue DIBt defines numbers of occurrences for start-ups, normal shut-downs, losses of electrical network, and additional events, such as production in overspeed and extreme yaw misalignments. DNV GL [64] provides likewise some additional DLCs to consider temperature, ice formation, earthquakes, and wind farm effects. Such site specific load cases are not part of this work, because they have to be accounted for in a future realization phase. Other additional DLCs of DNV GL [64], which refer not to these site specific conditions, such as the investigation of natural frequencies during idling (DLC 6.4) and vortex-induced vibrations caused by the wind, waves or currents in installation phase (DLC 8.3) may be relevant for a rotatable WT support structure and should be considered.[5]

2.2.4 Partial Safety Factors

The IEC 61400-1 standard introduces three types of PSFs:

- γ_f: PSF for the type of loads
- γ_n: PSF for the consequences of failure
- γ_m: PSF for the material

[5] The later is only relevant for offshore WTs.

Each PSF may change for certain design situations. γ_f depends on the design situation in combination of Tables 2.1, 2.2, and Table 2.4

Normal (N)	Abnormal (A)	Transport and Erection (T)	Favourable Loads
1.35*	1.1	1.5	0.9

Table 2.4: Partial safety factor for loads γ_f according to table 3 in IEC 61400-1 [50]

Table 2.4 should be used for yield strength, buckling, and critical deflection ULS validations. The *-marked value should be reduced to $\gamma_f = 1.25$ for statistical extrapolated loads of DLC 1.1. Note that table 3 in IEC 61400-1 [50] provides optional reductions of γ_f for gravity caused loads in some cases. For the FLS, SLS, and ALS, the PSF for loads becomes $\gamma_f = 1.0$. The PSF for the consequences of failure is $\gamma_n = 1.0$ in the ULS/SLS/ALS and $\gamma_n = 1.15$ in the FLS for support structural components. Material uncertainties of steel are covered by PSF $\gamma_m \geq 1.1$ for the ULS/SLS/ALS and if no other regulation than the IEC 61400-1 is available. γ_m should be used for characteristic material properties with validated probabilities to survive of 95 % [50]. DNV GL [63] prescribes the material safety factor for steel to be $\gamma_m = 1.1$ for each cross sectional type and buckling analyses. Moreover, welded details have to be validated with $\gamma_m = 1.25$ in the ULS. In case of the FLS the PSF for material uncertainties becomes $0.9 \geq \gamma_m \leq 1.1$ in dependence of inspection interval frequency and with SN curves, which are related to a 97.7 % survival probability. The FLS concept of DNV GL [63] may use a material PSF $\gamma_m \in \{1.0, 1.15, 1.25\}$ depending on a design fatigue factor DFF. But this approach is only applicable to stress cycles in the $m = 5$ regime of a S-N curve, introduced in subsection 2.2.6. The alternative to the material PSF is to use the design fatigue factor directly in the design criterion. It depends on the accessibility and the atmospheric conditions around the considered structural detail, such as described in table 4-20 in [63]. In the SLS and the ALS, the material PSF becomes $\gamma_m = 1.0$.

The application of the load related PSF γ_f takes place on the load effect side S_d of Equation 2.10 by either multiplying it with the characteristic loads effects S_k or directly with the corresponding characteristic loads $F_{k,i}$. The second procedure is required for geometrical or material related nonlinearities. On the right hand side of Equation 2.10, the inverses of γ_n and γ_m should be multiplied with the characteristic resistance R_k. Thus, the expanded version of the design criterion may look like

$$\gamma_f \, S_k \leq \frac{1}{\gamma_n \, \gamma_m} R_k \tag{2.29}$$

Note that the IEC 61400-1 [50] and the DNV GL standard [63] demand that characteristic loads $F_{k,i}$ of DLCs with turbulent wind fields should be corrected in terms of their probability of recurrence. A corresponding procedure on how to perform such a statistical extrapolation is given in Annex F of IEC 61400-1 [50].

2.2.5 Ultimate Limit State

Observance of the ULS is necessary to prevent the loss of structural resistance, such as excessive yielding and buckling, brittle fracture of components, overturning of the WT, large deformations, and collapse [60]. The ULS is most related to maximum and minimum load components, such as normal forces, shear forces, bending moments, and torsional moments. ASE timeseries of U labeled DLCs in Tables 2.1 and 2.2 have to be analysed and their results have to be summarized in extreme load tables. An example of a load table is presented in appendix H of IEC 61400-1 [50]. These load tables contain the different extreme load components combined with the simultaneous acting other load components. Each of these combinations must be checked for each structural member in the WT and for the whole global structure to cover the most unfavourable load situation.

Strength Analysis

In terms of material strength the von Mises yield criterion may be applied in the design criterion 2.10:

$$\sqrt{\sigma_{x,d}^2 + \sigma_{y,d}^2 - \sigma_{x,d}\,\sigma_{y,d} + 3\,\tau_d^2} \leq \frac{f_y}{\gamma_n\,\gamma_m}, \tag{2.30}$$

where $\sigma_{x,d}$, $\sigma_{y,d}$, and τ_d are the design normal stresses in x- and y-direction and the shear stress at a certain point of a structural member. Index $_d$ denotes that the respective stress component has already been multiplied by γ_f. DNV GL prescribes that stresses have to be calculated under consideration of net cross sectional properties. However, local exceedance of the yield stress is allowed if adjacent structural parts have enough capacity for the redistribution of stresses. If plastic strains from a linear first order analysis with nonlinear stress-strain relation exceeding 1 ‰, the impact to fatigue must be examined. For safety reasons ductile failure modes should occur before brittle fracture modes [63, p. 51].

Buckling Analysis

Section 6.3 in DIN EN 1993-1-1 [42] provides a simple hand calculation for members column buckling resistance $N_{b,Rd}$ against axial loading. There, it is assumed as

$$N_{b,Rd} = \frac{\chi\,A\,f_y}{\gamma_m} \tag{2.31}$$

where A is the cross sectional area and χ is the reduction factor calculated as

$$\chi = \frac{1}{\Phi + \sqrt{\Phi^2 - \bar{\lambda}^2}}, \quad \text{but } \chi \leq 1.0 \tag{2.32}$$

In Equation 2.32 Φ is

$$\Phi = 0.5\left[1 + \alpha\left(\bar{\lambda} - 0.2\right) + \bar{\lambda}^2\right] \tag{2.33}$$

where α is the buckling curve dependent imperfection coefficient from table 6.2 in DIN EN 1993-1-1 [42] and $\bar{\lambda}$ is the slenderness ratio with

$$\bar{\lambda} = \frac{L_{\mathrm{cr}}}{i\,\lambda_1} \tag{2.34}$$

L_{cr} in Equation 2.34 is the buckling length in the considered load plane, i is the radius of gyration with

$$i = \sqrt{\frac{I}{A}} \tag{2.35}$$

and λ_1 is

$$\lambda_1 = \pi \sqrt{\frac{E}{f_{\mathrm{y}}}} \tag{2.36}$$

where E is the Young's modulus for the material. From Equations 2.31 to 2.35 it is clear that an increased bending stiffness $E\,I$ of a member means increased global buckling resistance. The geometric influence to the bending stiffness is given by the second area moment of inertia I, which increases with more material being distributed as far away from the bending axes as possible.

In numerical global buckling analyses the most unfavourable buckling mode must be considered for validation, while initial imperfections and residual stresses must be accounted for [63, p. 51]. Eurocode 3 is appropriate for buckling analyses [63, p. 52] where first order calculation is allowed if the relation α_{cr} of the design load F_{d} to the ideal buckling load F_{cr} is

$$\alpha_{\mathrm{cr}} = \begin{cases} \dfrac{F_{\mathrm{d}}}{F_{\mathrm{cr}}} \geq 10 & \text{for} \quad \text{elastic analysis} \\ \dfrac{F_{\mathrm{d}}}{F_{\mathrm{cr}}} \geq 15 & \text{for} \quad \text{plastic analysis} \end{cases} \tag{2.37}$$

Otherwise Eurocode 3 [42] demands that the stability analysis of structures must account for second order influences and in any case for imperfections. These effects may be examined within the calculation of the whole structure, the whole structure and partwise equivalent beam approaches, or equivalent beam approach for each member with buckling lengths and buckling shapes according to the whole structures buckling mode shape. If structural members imperfections are not accounted appropriately in the whole structure calculation, additional equivalent beam proofs may be necessary. In each case appropriate equivalent imperfections must be defined to account for residual stresses, geometric imperfections, such as inclination and deviation from straightness, and eccentricities. Thereby, equivalent imperfections have to be applied to the whole structure and to individual members locally. Equivalent imperfection shapes may be derived from the most unfavourable buckling mode shape, while their actual values of initial inclination and curvature are given in DIN EN 1993-1-1 5.3.2(3) [42] or may be provided in the National Annex. It is likewise allowed to use equivalent loads or to use the most unfavourable buckling mode shape with a prescribed amplitude to account for imperfections. Considering the design criterion of Equation 2.10 in terms of buckling, the Equation 2.38 must hold, where f_b is the calculated buckling load of the most unfavourable load combination and imperfections and $f_{b,Rd}$ is the corresponding buckling resistance load.

$$\frac{\gamma_f \, f_b}{\gamma_n \, \gamma_m \, f_{b,Rd}} \geq 1.0 \tag{2.38}$$

Local buckling must be checked for cross sectional parts, which do not meet the criteria of cross section class 3. These criteria are related to the shape, the dimensions, and the normal stress distribution of a cross sectional part and are listed in table 5.2 of DIN EN 1993-1-1 [42] and DNV GL appendix B [63]. Each cross section which does not observe the criteria of type 3 becomes a type 4 and must be treated with respect to plate buckling in DIN EN 1993-1-5 [43] or to shell buckling in DIN EN 1993-1-6 [44], respectively. Furthermore, the DNV GL standard [63] notes that the interaction of local and global buckling modes may be relevant and should be considered especially for tubular towers.

2.2.6 Fatigue Limit State

WTs are dynamically loaded structures, whereby its components experience many damaging load cycles over their lifetime. Therefore, the FLS has high relevance especially for bolted and welded details in WTs. Structural details may fail according to one of the following three crack growth cases [62, p. 15]:

- fatigue crack growth from the weld toe into the base material
- fatigue crack growth from the weld root through the fillet weld
- fatigue crack growth from the weld root into the base material
- fatigue crack growth from a surface irregularity or notch into the base material

Common fatigue strength assessment methods are represented by the nominal stress, structural stress, notch stress, notch strain, and crack growth concepts. Each of these methods are briefly explained by Keindorf [117]. Other references for aspects of fatigue analysis are from Haibach [89] and Radaj and Vormwald [154]. Owing to the mainly focussed tower concept in this work, only base material structural details occur for the main members, wherefore the most simple nominal stress concept is applied. The nominal stress concept uses the stresses extracted by means of fundamental beam analysis with its net cross sections. All lifetime stress cycles at a structural point may be extrapolated from short time timeseries of ASE analyses and counted with respect to a certain cycle count algorithm. A widely used counting algorithm for stress cycles is the Rainflow counting developed among others [154, p. 271] by Matsuishi and Endo [138]. Counted stress cycles are assembled to binned stress range spectra, where a certain stress range corresponds to a certain number of occurring cycles over the lifetime. Hypothetical beam discontinuities are accounted for on the resistance side by corresponding nominal stress S-N curves for a characteristic detail.

The DNV GL guideline [62] proposes different S-N curves with respect to the applied fatigue analysis methodology. S-N curves result from experimental data for certain structural details and are plotted as stress range over its amount of load cycles until failure on logarithmic axes. Figure 2.5 indicates such S-N curves for different structural details, depicted in appendix A of [62].

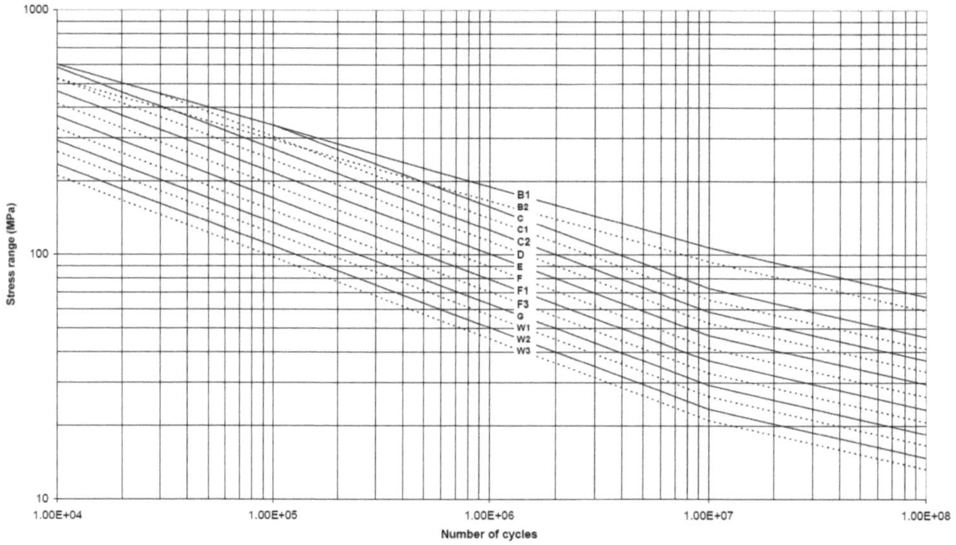

Figure 2.5: S-N curves for different structural details, from DNV GL [62, p. 23]

Miner's rule [141] may be applied to calculate the lifetime damage sum of a certain detail of a structural component [50] [63]. Miner's rule assumes that damage of a detail increases linear and is independent for each load cycle. The resulting characteristic cumulative damage D_c of a considered detail must be multiplied by the design fatigue factor DFF. Thus, the design criterion of Equation 2.10 becomes

$$DFF\ D_c \leq 1.0 \tag{2.39}$$

Owing to computational expense ASE simulated timeseries represent only short time periods compared to a WTs design lifetime of $T_{DL} \geq 20$ years. Therefore, an expected lifetime damage $E\langle D_c \rangle$ must be calculated out of scaled short time damages, which are integrated over each wind speed weighted by their occurrence probability according to Equation 2.40.

$$E\langle D_c \rangle = \frac{T_{DL}}{T} \int_{V_{in}}^{V_{out}} \int_0^{\infty} \frac{n_{ST}(S|V,T)}{N(S)} p(V_H)\ dS\ dV \tag{2.40}$$

$p(V_\mathrm{H})$ of Equation 2.40 is the probability density function for the current wind speed at hub height. A cumulated Weibull distribution

$$P_\mathrm{W}(V_\mathrm{H}) = 1 - e^{-\left(\frac{V_\mathrm{H}}{C}\right)^k} \tag{2.41}$$

is a common assumption to be used to calculate the probability density function $p(V_\mathrm{H})$. Thereby, $P_\mathrm{W}(V_\mathrm{H})$ describes the probability that $V < V_\mathrm{H}$. $p(V_\mathrm{H})$ can now be calculated by derivation according to Equation 2.42.

$$p(V_\mathrm{H}) = \frac{\mathrm{d}P_\mathrm{W}(V_\mathrm{H})}{\mathrm{d}V_\mathrm{H}} = \frac{k}{C}\left(\frac{V_\mathrm{H}}{C}\right)^{k-1} e^{-\left(\frac{V_\mathrm{H}}{C}\right)^k} \tag{2.42}$$

where C is the Weibull scale parameter and k is the Weibull shape parameter. Implying a Rayleigh distribution these parameters become $C = \frac{V_\mathrm{ave}\,2}{\sqrt{\pi}}$ if $k = 2$.

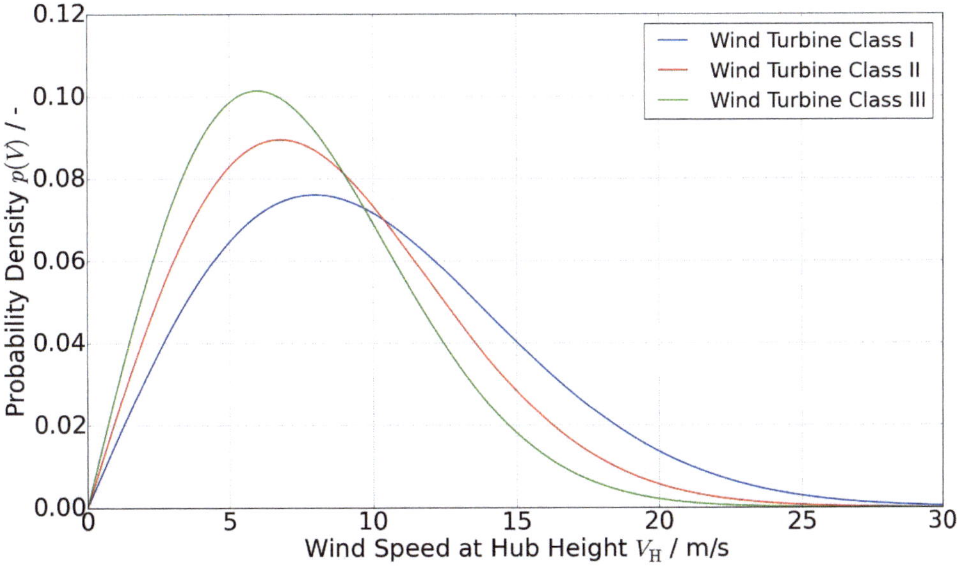

Figure 2.6: Weibull wind speed distribution for different wind turbine classes

Figure 2.6 shows the probability density for the three WT classes, defined in Table 2.3 and a shape factor of $k = 2$. Thereby, higher wind speeds are more likely at a lower WT class, while lower wind speeds occur more likely for a higher WT class. $n_\mathrm{ST}(S|V,T)$ of Equation 2.40 is the amount of short time load cycles for time period T, wind speed V, and load range S. $N(S)$ represents the number of Rainflow counted load cycles until failure occurs for a given S. In case of detail categories provided by DNV GL [62], S-N curves represent a survival probability of 97.7 %. The IEC 61400-1 based fatigue evaluation tool MLife [94] models the S-N curve by

$$N(S) = \left(\frac{S_0 - |\overline{S}_F|}{\frac{1}{2}S} \right)^m \tag{2.43}$$

where S_0 is the theoretical load range at the intersection between the curve and the $\log(S)$ axis, $|\overline{S}_F|$ is an absolute prescribed fixed load mean, and m is the Wöhler exponent. S_0 and m depend on the specific material and detail category. Equation 2.43 is valid only for one load mean, but in reality each load cycle appears over an individual mean value. Therefore, each calculated load range S_i must be corrected according to Goodman

$$S_{G,i} = S_i \left(\frac{S_0 - |\overline{S}_F|}{S_0 - |\overline{S}_i|} \right) \tag{2.44}$$

where a Goodman exponent of one is assumed.

2.2.7 Accidental Limit State

According to DNV GL [63] special accidental events and their consequences must be considered within the ALS. Thereby, accidents, such as

- dropped objects
- collision impact
- explosions
- fire
- change of intended pressure difference

may occur under different environmental and operating conditions, which should be examined likewise. ALSs with impacts of agriculturally machinery for land-based structures or boat collisions for offshore structures demand dynamic nonlinear time simulations. These simulations must represent each collision phase and must account for the impact energy dissipation of the WT support structure. The energy dissipation of the colliding object may also accounted for, but must be fully documented. For fire or explosive ALSs the engineer may consider to reduce the amount of explosive matters and to use fire suppressing equipment. DNV GL [63] lists some possible damages as consequence of an accidental event:

- large permanent deformations
- loss of stability or disconnections, e.g. denting of a monopile shell wall

- buckling failure of a brace member in a jacket structure
- formation of a plastic hinge in a boat bumper
- rupture of a bolted joint

Such damages should occur without complete loss of structural integrity and should not lead to further post accidental damage developments. The post accidental structural behaviour in terms of safety for humans and the environment is likewise important. A post accidental analysis should account for expected gravity loads, specified variable loads, and specified environmental loads, but no additional accidental actions. Detailed information about design against accidental loads are given in the respective recommended practice report of DNV GL, called DNVGL-RP-C204 [65]

2.2.8 Serviceability Limit State

The SLS considers the consequences of extreme characteristic loads and loads with a load duration distribution of 10^{-4} and 10^{-2} of the whole WT lifetime [63]. Relevant consequences are permanent deformations or yielding of structural members, excessive vibrations, differential settlements of foundations, intolerable tilt of a WT, collision of the blade with the support structure, and temperature induced deformations.

Natural Frequency Requirements
Excessive structural vibration should be avoided to prevent rapid fatigue damage accumulation, affecting turbine operation and energy production, and discomfort for people. Therefore, the main excitation frequencies $f_{\mathrm{R},i}$ should occur with certain distance to the structural natural frequencies $f_{0,j}$. DNV GL gives the rule

$$\frac{f_{\mathrm{R},i}}{f_{0,j}} \leq 0.95 \quad \text{or} \quad \frac{f_{\mathrm{R},i}}{f_{0,j}} \geq 1.05 \tag{2.45}$$

for each combination of i and j [63, p. 35]. In practice two main excitation frequencies are relevant for land-based WT support structures: The 1-per-revolution (1p) and n-per-revolution (np) excitation frequency where n denotes the amount of blades. Frequency overlaps of $f_{\mathrm{R},i}$ and $f_{0,j}$ may occur during start-up and shut-down of the WT, but their relevance is small for short time periods. Furthermore, overlapping excitation frequencies with structural natural frequencies are allowed if resonance effects can be omitted by damping devices or vibration monitoring systems and corresponding control manoeuvres [63].

Blade to Tower Clearance
Rotor blades should never collide with the WT support structure. Therefore, DNV GL defines some requirements to the minimum blade to tower clearance under serviceability load conditions in the rotor blades standard DNVGL-ST-0376 [61]. In general blade to tower clearance shall not become less than 30 % of the unloaded state and must be evaluated by ASE simulations. The limitation can be reduced to 25 % by means of additional deflection measurements at

3 full scale test blades under highest tower clearance loads. Additional continuous bending stiffness control during blade series production allows for another limitation reduction to 20 %. These limitations are related to rotating rotor situations. In all cases where the rotor stands still clearances should not be smaller than 5 % of the unloaded state.

2.3 Foundations

2.3.1 Concepts

Land-based WT foundations are used to transfer the loads into the ground, while certain requirements must be fulfilled over its design lifetime. The choice of a foundation concept depends on the expected loads and soil conditions. Most common solutions for multi-megawatt WTs are the concrete slab and concrete slab with piles foundations, which occur in different shapes, such as a circular, polygonal, cross, or square. The concrete slab option is used in firm soils, while the one with piles is more expensive and is used for worse conditions, such as for large clay depths [180]. Figure 2.7 shows how pile foundations work. They use the piles to reach deep and more firm soil layers to transfer loads over friction into them. In case of lattice towers smaller point foundations to the number of legs are used, wherefore less concrete is necessary. Bearing capacity proofs for the tension forces under extreme overturning tower base moments are thereby challenging for the point foundations.

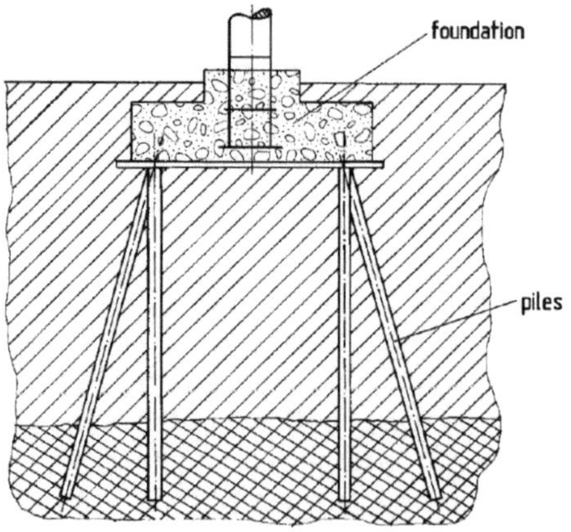

Figure 2.7: Slab foundation with piles from Hau [92]

2.3.2 Assessment and Guidelines

The assessment of WT foundations must ensure at least the following requirements according to Svensson [180]:

- The structure must be stable
- The material strength must not be exceeded
- The function of the structure must be maintained
- The structure should be aesthetically pleasing
- The structure has to be resistant against external factors, such as fire, earthquake, flooding, frost, moisture, temperature differences, traffic accidents, and vermin such as termites and insects etc.

However, in most cases this means that the foundation and the surrounding soil must resist extreme tower base overturning moments, shear forces, and fatigue loadings, while certain crack widths in the foundation should not be exceeded. Especially important guidelines for the foundation assessment are the Eurocodes 0 - Basis of Structural Design (EN 1990), 1 - Actions on Structures (EN 1991), 2 - Design of Concrete Structures (EN 1992), 3 - Design of Steel Structures (EN 1993), 4 - Design of Composite Steel and Concrete Structures (EN 1994), 7 - Geotechnical Design (EN 1997), and 8 - Design of Structures for Earthquake Resistance (EN 1998).

2.4 Manufacturing Technology

This section gives a brief overview over tubular steel tower and lattice tower steel member manufacturing.

2.4.1 Tubular Steel Tower Manufacturing

The most common tubular steel tower concept is manufactured out of rolled and welded steel sheets. In terms of rolling, these steel sheets are assumed to be limited to wall thicknesses of about 75 mm [68]. After rolling a longitudinal weld seam connects both edges of one sheet. Afterwards, several of such cylindrical sheets will be welded together circumferential to one transportable pipe section, such as shown in Figure 2.8.

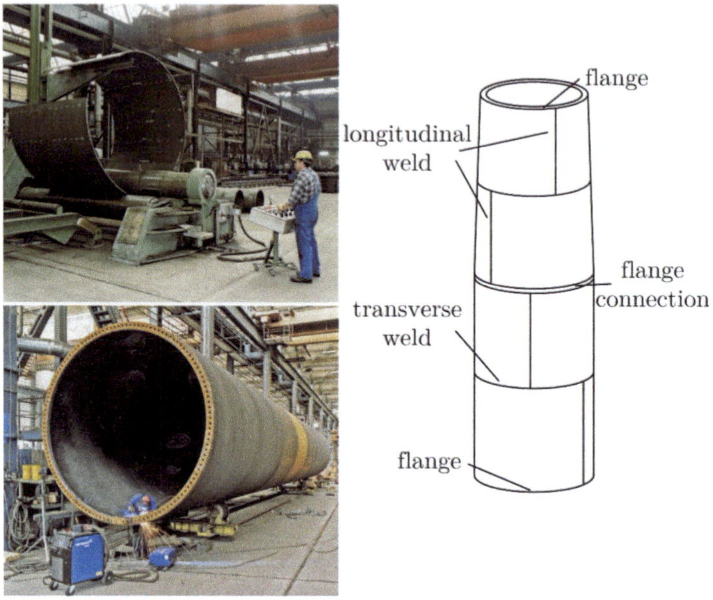

Figure 2.8: Manufacturing of conventional tubular steel towers on the left [93] and the welding scheme of two pipe sections on the right

New on-site manufacturing technologies, such as tapered spiral welded towers, shown in Figure 2.9 have been developed to make tubular steel towers more affordable for larger hub heights. They work by welding different trapezoidal steel sheets together to a continuous plate. Next, the plate will be brought into its tubular shape by rolling and connecting it to the previous rolled section by a continuous transversal weld. This technology can also be beneficial for rotatable tubular tower concepts. However, the temperature and dust sensible welds and high necessary tolerances have likewise to be ensured on-site. Protecting the manufacturing area from the environment by an air-conditioned tent is one way to meet this requirement. An ultrasonic testing inspection behind the manufacturing chain checks for the appropriate quality of the continuous welding seam. A patent of this manufacturing method is hold by Keystone Tower Systems Inc [174].

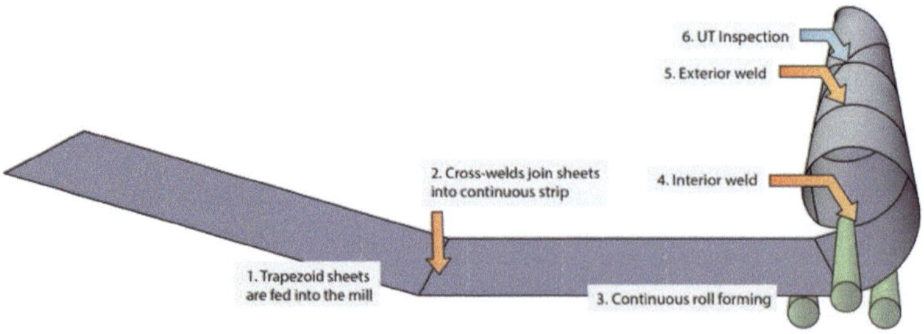

Figure 2.9: Tapered spiral welding technology for WT towers [118]

2.4.2 Lattice Steel Tower Member Manufacturing

Land-based lattice steel towers are assembled from many members with different cross sections. These members are designed to resist especially normal forces in terms of buckling, while good geometrical connectivity to each other should be provided. The later is the reason for circular cross sections to be rare for land-based applications, because of their expensive joint options. These are welding or special joint structures. Welding of circular cross sections is only possible for small lattice structures to ensure transportability and requires cost intensive welding preparations. Polygonal or L-shaped members are more common, because of their better connectivity to each other. In case of cold formed members, higher yield strengths may occur locally, but folding radii are limited. Therefore, extrusion moulding is an alternative way to manufacture these prismatic members instead. Afterwards, holes for bolts will be punched into the material and it will be hot-dip galvanized for corrosion protection. Note that such coatings increase the bolts settlement, wherefore preload forces decrease more than without a coating and this kind of settlement should be considered within the joint proofs [83, p. 566].

2.5 Transport and Erection

While the members of lattice towers are transportable to almost arbitrary sites owing to their small member dimensions, large tubular steel tower segments need to be transported by special heavy trucks. Transport of such WT towers became a problem, since larger hub heights and rotor diameters lead to larger tower diameters to ensure economic and technical feasibility. For German country roads, the minimum bridge and tunnel heights are 4.5 m [74], which is in some cases not enough for tower passage. Figure 2.10 shows the tightly transport of a large tower segment under a bridge. In case of towers with exceeding diameters cost intense detours or uneconomical wall thickness increases are the consequence.

Figure 2.10: Tower section transport under the 1935 Ramford bridge [116]

All tower segments will be mounted together on the construction site by means of a crane. Thereby, the first tower segment has a double flange with inner and outer ring for the bolts at the bottom to get connected with the foundation. For larger hub heights and because of the high heave weights, special cranes are used. In the case of lattice towers, parts of the tower will be assembled at the ground first before they will be heaved onto the already mounted tower parts.

2.6 Yaw System

The yaw system is necessary for the WT to react on wind direction changes. It is used to align the rotor correctly to the mean wind direction to increase the wind energy harvest. Conventional yaw systems are placed at the top of the tower and carry the whole RNA. Figure 2.11 shows an exemplary yaw system where the yaw bearing represents the rotative joint between the tower and the RNA. A ring gear on the outside of the yaw bearing works together with the yaw drives to achieve controlled yaw motions. The yaw drives are mounted on the bedplate, which carries the RNA and is connected to the yaw bearing on its moving ring. For large WTs of ≥ 1 MW 8 or more yaw drives are common practice.

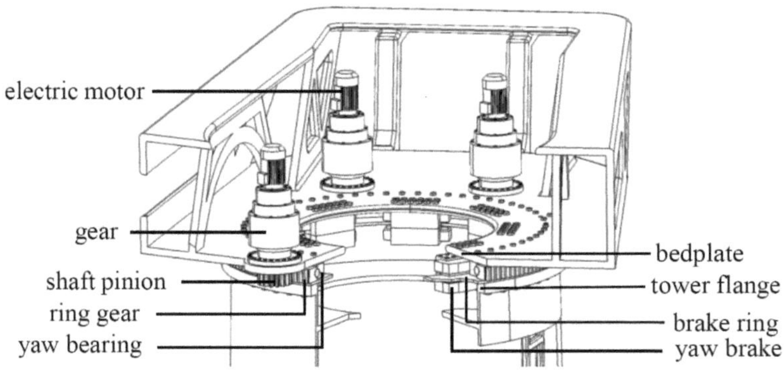

Figure 2.11: Exemplary yaw system composition. The picture was taken from M.-G. Kim and P. H. Dalhoff [119] with permission to use from P. H. Dalhoff.

Since RNA yaw motions cause gyroscopic loads by the acceleration of the high mass and inertia of the RNA, only small yaw accelerations are allowed. Gyroscopic loads increase the fatigue of mechanical and structural components and have to be balanced against a higher energy yield through low yaw misalignments within the control strategy. Common yaw speeds are around 0.5 deg/s [93, p. 470] and the acceleration path to that speed has to be adjusted with respect to the allowed loads. In reality a WT faces always yaw misalignments, but it is a question of their mean value over a certain time, whether the WT should yaw or not. Hau [93, p. 469] states that 10 s mean yaw misalignments of small magnitudes, such as 10 deg will be corrected after 60 s, medium mean yaw misalignments of 20 deg will be corrected after 20 s, and larger yaw misalignments than 50 deg will be corrected immediately. However, yaw misalignments are especially relevant for the WTs efficiency in operational conditions below rated wind speed, because in higher than rated wind speeds the blades pitch can be used to increase the energy yield out of that skewed inflow condition. The later would in turn also increase the fatigue of components. In general, no larger energy losses than 1 to 2 % are expected by yaw misalignments [93, p. 470]. Since this work treats a rotatable tower with yaw system at its bottom, subsection 6.2.4 discusses the effect of yaw motions with respect to the occurring loads in more detail.

2.7 Lattice Towers

Figure 2.12: 2.5 MW rated power WTs with 100 m rotor diameter and 100 m hub height mounted on a Butzkies lattice tower in Bahrenfleth, Germany. Photo: A. Struve 2018

In this work a rotatable lattice tower was developed and investigated, wherefore it is appropriate to provide a short introduction to main aspects of WT lattice towers. Lattice towers were popular in the first years of commercial wind energy usage, because of their relatively material saving and stiff properties for small hub heights [93, p. 517]. They can be seen as plate towers, which consist only out of plate stiffenings where the plate material itself is cut away. This is the reason for their material saving nature. Nevertheless, history shows that tubular steel towers displaced lattice towers as support structure for wind turbines, owing to the trend of soft designs [93, p. 517-518]. Recently, the interest in lattice towers, such as shown in Figure 2.12, has been increased again for hub heights $\geq$ 100 m [93, p. 518]. The increasing interest is caused by the uneconomical wall thicknesses of tubular steel towers with large hub heights under transport limitations, such as mentioned in section 2.5. Figure 2.13 shows the mass and cost development along different hub heights of an exemplary 3 MW WT with 100 m rotor diameter. Unfortunately, Hau [93, p. 529-530] provides no information about the assumed transport boundary conditions, which were applied to estimate the curves. However, not only the absolute masses and costs of lattice towers are smaller than the compared conventional steel towers, but more important the gradients of the curves are likewise smaller. This indicates that lattice towers in general are more appropriate for WT support structures with larger hub heights.

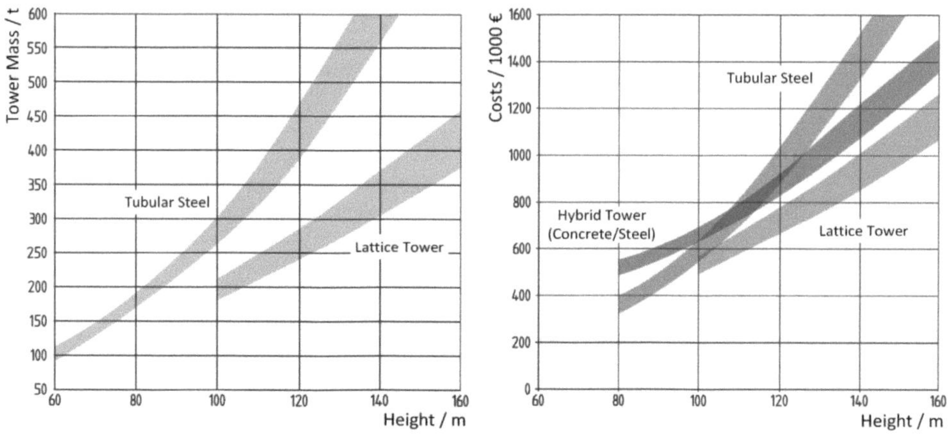

Figure 2.13: Mass and cost increases of different tower concepts, dependent on the hub height of a 3 MW wind turbine with 100 m rotor diameter. Taken from [93, p. 530] and translated by the author.

2.7.1 Bracing Concepts

Bracing members serve the reduction of buckling lengths for legs and other bracing members. They may contribute to the shear and torsional stiffness and to small amounts to the axial stiffness of a lattice tower. Thereby, their bracing angle and their cross sectional area are the main properties of contribution, such as indicated by the equations in table 13.1 and 13.2 in Petersen [149, p. 895-896]. In general lattice towers may have different bracing concepts, such as shown in Figure 2.14 and explained by Petersen [149, p. 890-891] as follows. A variety of bracing member and lateral member arrangements are more or less appropriate for certain load requirements.

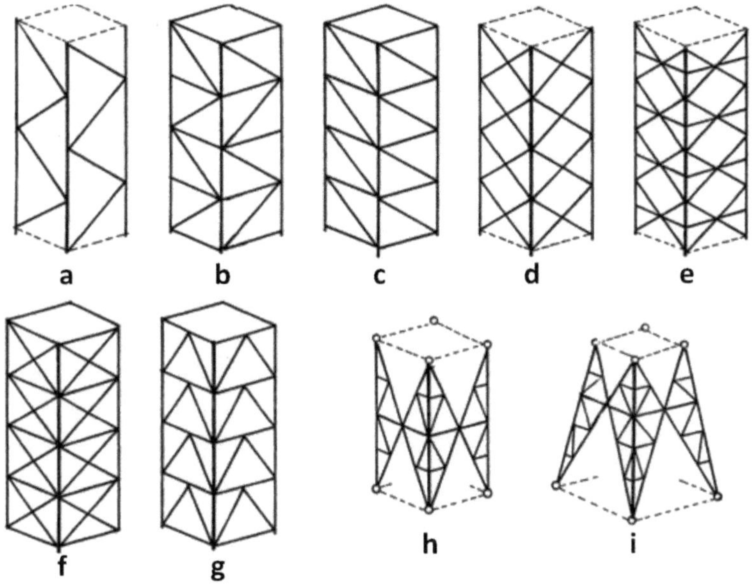

Figure 2.14: Lattice tower bracing concepts by Petersen [149]

Concept **a** is often used for masts with low loadings, because of the large buckling lengths of the bracing members. The lateral member in concept **b** halves the buckling lengths of the leg members compared to **a**. For concept **c** the same statements as for concept **b** are true, but the spiral bracing form tend to drill for huge axial tower loadings, wherefore it is not that popular. In diamond-bracing concept **d** buckling lengths of the bracing members are halved and with concept **e** the leg members experience another halving compared to the previous three concepts. But on the other side, the joint complexity of the bracing members increases as well as the wind attack area. Concept **f** provides no further improvements, but the leg joint complexity increases without any further buckling lengths reductions. The K-bracings in concept **g** are relatively steep, which reduces the torsional stiffness. Diamond- and K-bracing concepts have the advantage of comparatively small secondary stresses for huge axial loadings [149, p. 890]. For small lateral loadings the diamond-bracing may be stretched, such as shown for concepts **h** and **i** and the leg members will be supported by secondary bracing members to reduce the legs buckling lengths. Another key aspect for bracing concepts is the kinematic stability. Meskouris and Hake [139] propose indeterminancy counting criteria for lattice structures with ideal hinged members. A measure for indeterminancy is the integer number n, which is calculated as

$$n = a + s - 3k \tag{2.46}$$

for a three dimensional lattice structure with the amount of possible supporting forces a, the amount of members s, and the amount of nodes k. Thereby, $n < 0$ means that the structure is kinematic unstable and $n = 0$ that the structure is static determined. Incidentally, according to Petersen [149, 898] $n = 0$ is not an adequate criteria for infinitesimal kinematic stability.

Moreover, $n > 0$ means that the structure is statical undetermined. Modern static software is capable to model lattice structures with bending stiff member joints. Meskouris and Hake [139] propose the indeterminancy counting criteria for such structures with

$$n = (a + s\,p) - (g\,k + r), \tag{2.47}$$

where

- a: amount of support reactions
- s: amount of independent load components per member (= 6 for three dimensional structures)
- p: amount of all members
- g: amount of equilibrium conditions per node (= 6 for three dimensional structures)
- k: amount of all nodes inclusive support nodes
- r: sum of all constraints between the members

Consequently, assuming only bending stiff joints for concept **d** in Figure 2.14 sets $r = 0$ and 4 base nodes with 6 reactions each, sets $a = 4 \times 6 = 24$. In one x-segment $k = 12$ nodes and $p = 20$ members occur, wherefore n must be > 0 and the structure is statical undetermined and kinematic stable.

2.7.2 Bracing Member Cross Sections

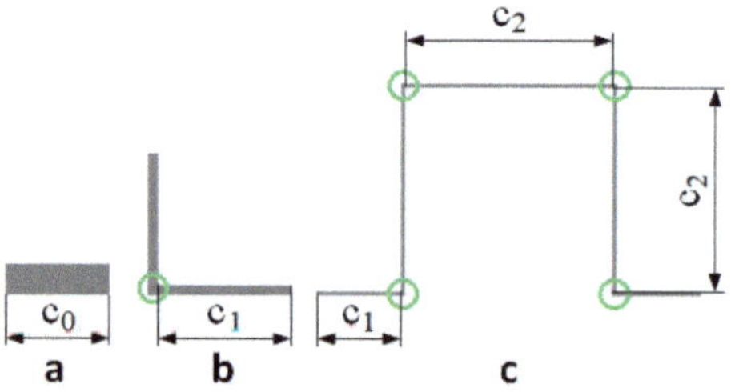

Figure 2.15: Bracing member cross sectional comparison with respect to material efficiency

Bracing members are primarily loaded by axial forces, especially if they are attached by hinges. In reality bracing members experience additional bending moments owing to eccentric attachments and because ideal hinged bracing joints do not occur. However, in most cases bracing members are designed against global and local buckling. Nowadays, buckling analyses are performed with computer software, such as RSTAB/RFEM, ANSYS, or ABAQUS, where bending stiff and eccentric joints are accounted for in the modelling phase. From subsection 2.2.5 it is shown that global buckling resistance of structural members increases with their bending stiffness. In terms of material efficiency the engineer has an interest in bracing members with large second area moments of inertia I, while low amounts of material are used. Figure 2.15 shows some potential lattice member cross sections. Material efficiency in terms of bending stiffness leads to small wall thickness, wherefore the resistance against plate buckling must be checked. From DIN EN 1993-1-1 [42] and DIN EN 1993-1-5 [43] it is shown that the way how plate sections of a cross sections are supported influences the buckling resistance. For instance the L-profile **b** in Figure 2.15 provides only one support for its two legs in its bottom left corner, marked by a green circle. Here, such one-side supported plate sections are declared with c_1, whereas two-side supported plate sections, such as given for the hat-profile **c** in Figure 2.15 are declared with c_2. Other names for the one-side and two-side support are outstand flange and internal compression part. Two-side supported plate sections provide higher plate buckling resistance, wherefore they can be wider as one-side supported ones. This leads to the recognition that in terms of bending stiffness a hat profile with three two-side supported plate sections is more material efficient than a L-profile.

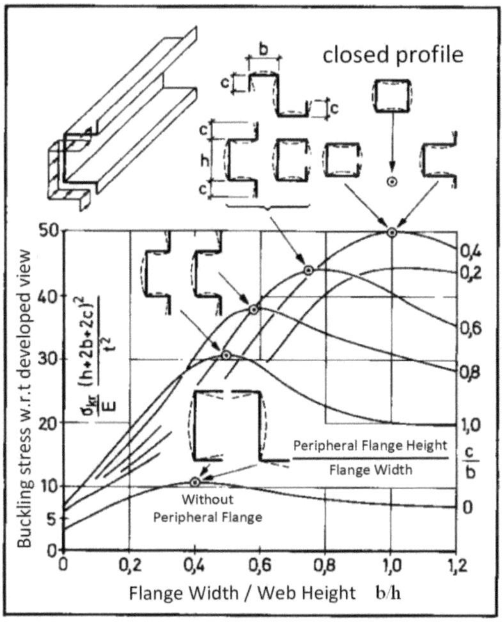

Figure 2.16: Hat-profile dimension ratio optimization with respect to its plate buckling resistance from Wiedemann [196, p. 140], translated by the author

On the other hand more complex profiles cause higher production effort and therefore higher costs, at least for cold formed profiles. Figure 2.16 shows how the buckling resistance of two flanged profiles with additional peripheral flanges, such as hat-profiles changes with respect to its dimension ratios. The resulting buckling stresses are related to the developed view length $(h + 2\,b + 2\,c)^2$ and an optimum can be found at dimension ratios $b/h \approx 1$ and $c/b \approx 0.4$. This result is reasonable, because the flanges with width b and the web with height h have the same wall thicknesses and the same amount of supports, namely two. From table 5.2 in DIN EN 1993-1-1 [42] c/t limits for both support situations are given by

$$\begin{aligned} (c_1/t)_{\max} &\leq 21\ \varepsilon\ \sqrt{k_{\sigma,\min}} && \text{for one support with } 21\ \sqrt{k_{\sigma,\min}} = 21\ \sqrt{0.43} = 13.77 \\ (c_2/t)_{\max} &\leq 42\ \varepsilon && \text{for two supports} \end{aligned} \tag{2.48}$$

where

$$\varepsilon = \sqrt{\frac{235\ \mathrm{N/mm^2}}{f_y}} \tag{2.49}$$

If the c/t ratios respect the limits in Equations 2.48, the cross section is not of class 4 and must not be checked for plate buckling, such as stated in DIN EN 1993-1-5 [43]. Moreover, with $(c_1/t)_{\max}/(c_2/t)_{\max} = 0.328 = (c/b)_{\mathrm{opt}}$ a more exact optimum for the peripheral flange to flange width ratio is given. Z-profiles with peripheral flanges, such as suggested at the top in Figure 2.16 are unfavourable in terms of corrosion, because they tend to hold rainwater. Other profiles such a circular ones, are not common for land-based WT lattice towers, because of their high production costs and problematic joint connections.

2.7.3 Load Derived Leg Inclination

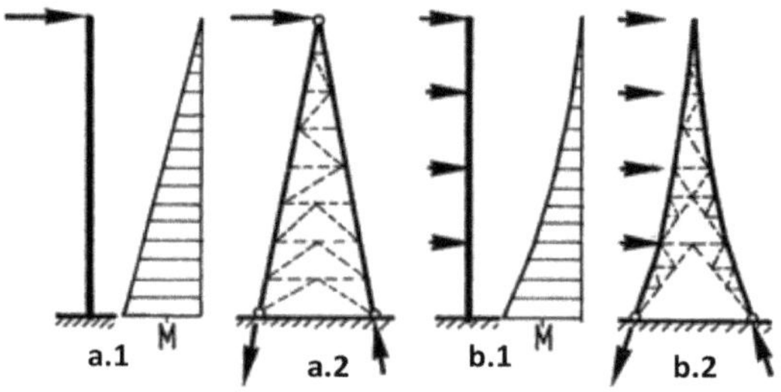

Figure 2.17: Leg forces in lattice towers under different loadings by Petersen [149, p. 891]

Based on Figure 2.17 constant axial leg forces along the whole tower height are achievable by adjusting the leg inclination angle with respect to the bending moment shape. Normal leg forces N_{Leg} in a lattice tower with quadratic cross sectional shape of side length b are estimated by

$$N_{\mathrm{Leg}} = \pm \frac{M_{\mathrm{x,y}}}{2\,b} + \frac{F_{\mathrm{z}}}{4}, \tag{2.50}$$

where $M_{\mathrm{x,y}}$ is the corresponding bending moment about one principal axis x or y and F_{z} is the normal force at the same tower cross sectional cut. By prescribing b changes with $M_{\mathrm{x,y}}$ and F_{z} over z, constant axial leg forces are the consequence.

Neglecting the axial force and assuming a linear bending moment, such as shown in **a.1** a linear leg inclination **a.2** can be used. A polygonal bending moment, such as shown in **b.1** requires consequently a polygonal leg inclination **b.2**. In such situations the bracing members would not experience any loadings and are only required for leg buckling lengths reductions. In reality additional vertical loadings and load variations hinder a perfect load adapted leg inclination design, but it may serve as an orientation.

2.7.4 Leg Cross Sections

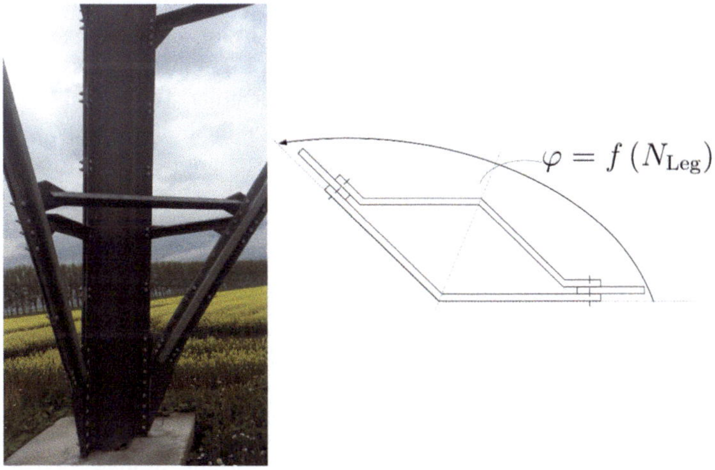

Figure 2.18: Leg to foundation integration of a Butzkies tower (left) and drawing of an analogous leg profile concept in more generalized form (right). Photo: A. Struve 2018

The legs of modern WT lattice towers have special designs to resist the enormous axial loads with a minimum of material usage. Thereby, the general insights of subsection 2.7.2 apply likewise to leg profiles, where additional corners in the profile increase the plate buckling resistance with respect to the amount of used material. In some cases the leg profiles are assembled out of two profiles, such as shown in Figure 2.18 to reach the same goal. The principal sketch at the right position in Figure 2.18 indicates, that the number of lattice tower legs determines the leg cross sectional angle φ. Both parts of the leg are bolted together in certain distances along the length over integrated filler steel sheets. This fulfills the following four purposes: (1) The local package of three steel sheets bolted together can be seen as a support for the angled steel plate, which increases its c/t limit at that member cut, (2) the longitudinal bolt distance determines the buckling field width, (3) the torsional leg stiffness is drastically increased at the longitudinal member cut with filler steel sheets compared to the open cross sections at other cuts, and (4) the leg member can be evaluated as one beam in terms of column buckling.

Another option for the legs is to connect both cross sectional parts by a longitudinal weld seam. This would save a lot of work to drill all the holes as well as the many expensive bolts or lockbolts and their maintenance. On the other hand welding requires also some factory work, although it is easy to automate. Depending on the detail category and accessibility may welded legs require larger wall thicknesses to resist the fatigue loads.

2.7.5 Joints

Joints of land-based WT lattice towers are normally bolted in contrast to offshore support structures, where joints are welded. This is caused by the limited space for pre-fabricated support structure components, because of transport constraints on streets, such as discussed in subsection 2.5. In case of the Butzkies lattice tower in Figure 2.19 gusset plates are used to attach bracing members with legs and to each other over bolts. In general bolted connections are designed such that they force the shear planes between the bolted components to be sliding resistant over lifetime [121, p. 103]. In cases where this is not possible fitted bolts must bear the shear force partially or complete in themselves [ibid.]. Verification requirements for bolted joints are documented in DIN EN 1993-1-8 [45] and its National Annex [46]. In the past, conventional bolts stood out with high maintenance effort to ensure the minimum pre-load in the bolts under dynamical loading, required by the DIBt-Standard [152]. Therefore, the cost of multiple checks and retightening of thousands of bolts over the lifetime of a WT lattice tower should not be underestimated.

Slip resistant connections, such as exemplary shown in Figure 2.20 transmit loads over the shear planes between the components, whereby the tension force $F_{\mathrm{p,C}}$ in the bolts causes the necessary contact loading. Thereby, $F_{\mathrm{p,C}}$ is measured with respect to the applied tightening procedure, defined in the National Annex of DIN EN 1993-1-8 [46]. According to DIN EN 1993-1-8 [45] the slip resistance $F_{\mathrm{s,Rd(,ser)}}$ must be larger than the acting shear force per bolt $F_{\mathrm{v,Ed(,ser)}}$. The corresponding verification equation is

$$F_{\mathrm{v,Ed(,ser)}} \leq F_{\mathrm{s,Rd(,ser)}} = \frac{k_{\mathrm{s}}\, n\, \mu \left(F_{\mathrm{p,C}} - 0.8\, F_{\mathrm{t,Ed(,ser)}}\right)}{\gamma_{\mathrm{M3(,ser)}}} \tag{2.51}$$

where k_{s} is the geometry coefficient to account for the clearance between the hole and the bolt. Its value is defined in Table 3.6 of DIN EN 1993-1-8 [45]. For normal bolts reductions are neglected, wherefore $k_{\mathrm{s}} = 1.0$. n is equal to the amount of load bearing shear planes and μ is the coefficient of static friction. It depends on the coating and pre-treatment of the bolted component surfaces and can be taken from Table 3.7 in DIN EN 1993-1-8 [45] in dependency of the surface class. $F_{\mathrm{t,Ed(,ser)}}$ is the acting tension force in the connection and $\gamma_{\mathrm{M3(,ser)}}$ is a safety factor to account for the kind of connection. For the both possible slip resistant categories B and C, $\gamma_{\mathrm{M3,ser}} = 1.1$ and $\gamma_{\mathrm{M3,ser}} = 1.25$, respectively.

Figure 2.19: Different joints of a Butzkies lattice tower. Photo: A. Struve 2018

Furthermore, the bolted parts must be checked against material yielding in the net cross section and for the bearing of the hole. Depending on the bolt connection category the bolts must be checked against shearing. The fatigue check of the shear planes can be neglected, because of the distributed loads over the contact plane.

To overcome the challenges concerning maintenance, slip resistant lockbolt joints with low variations in the pre-load force, good visual inspection capability, and high possible pre-load forces are a common solution [83, p. 562]. Moreover, Glienke et al. [83] investigated and quantified the pre-load losses for slip resistant connections under dynamical loading in the context of WT lattice towers. They propose a promising verification procedure of slide resistant connections, which accounts for the pre-load losses to reduce or to avoid later maintenance effort. However, further research is required to account for common component surface coatings and pre-treatments and for clamping length to diameter ratios [83, p. 569].

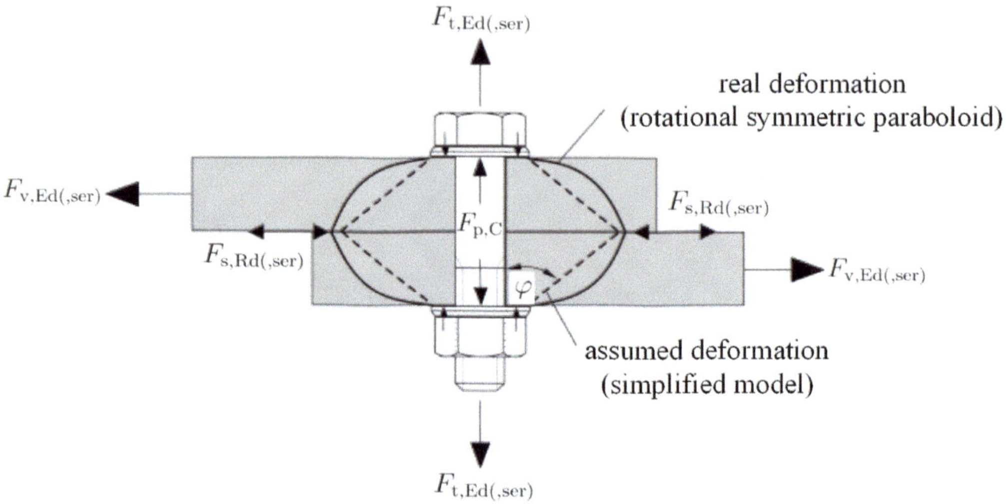

Figure 2.20: Bearing resistance of a slip resistance connection by Glienke et al. [83, p. 562], translated by the author

2.7.6 Aerodynamics

Drag coefficients of sharp edged lattice members, such as L-profiles are higher than of aerodynamically shaped cross sections. However, the front surface of lattice towers with the same height as tubular ones is smaller, except visibility improving coverings are used, such as shown for the space frame tower of GE in Figure 2.21. Aerodynamic load assumptions for uncovered lattice structures are documented within the standard DIN EN 1991-1-4 [67] with National Annex [41] and DIN EN 1993-3-1 [48] with National Annex [49].

Figure 2.21: View into the pentagonal space frame lattice tower of GE [78]

The DIBt standard [152, p. 11] proposes the application of DIN EN 1991-1-4 and its annex for all aerodynamic loads. There, the wind force F_w on a structure becomes

$$F_\mathrm{w} = c_\mathrm{s}c_\mathrm{d}\, c_\mathrm{f}\, q_\mathrm{p}\,(z_\mathrm{e})\, A_\mathrm{ref}, \tag{2.52}$$

with the structural coefficient $c_\mathrm{s}c_\mathrm{d}$, the force coefficient c_f, the gust velocity pressure q_p in reference height z_e, and the reference area A_ref. The c_s part of the structural coefficient $c_\mathrm{s}c_\mathrm{d}$ accounts for the fact that peak wind pressures would not occur over the whole surface simultaneously and the c_d part accounts for structural dynamic interaction with the wind, such as turbulence induced resonance. For more details about the determination of $c_\mathrm{s}c_\mathrm{d}$, a review of chapter 6 in DIN EN 1991-1-4 [67] is encouraged. The force coefficient c_f assembles from the basic force coefficient $c_\mathrm{f,0}$ and the reduction factor ψ_λ according to

$$c_\mathrm{f} = c_\mathrm{f,0}\, \psi_\lambda \tag{2.53}$$

The basic force coefficient $c_\mathrm{f,0}$ is representative for an infinite slender structure and depends on the member density φ and the Reynolds number Re. φ is calculated as

$$\varphi = \frac{A_\mathrm{ref}}{A_\mathrm{c}} = \frac{\sum_i b_i\, l_i + \sum_k A_{\mathrm{g},k}}{d\, l}, \tag{2.54}$$

with width b_i and length l_i of member i, projected area $A_{\mathrm{g},k}$ of joint steel sheet k, width of the lattice structure d, and height of the lattice structure l. Incidentally, the reference area A_ref of Equation 2.52 is defined in Equation 2.54. Re is calculated as

$$Re = \frac{\overline{b}\, v\,(z_\mathrm{e})}{\nu} \tag{2.55}$$

with mean member width $\overline{b}$, current wind speed at the reference height $v\,(z_\mathrm{e})$ and kinematic viscosity ν. The basic force coefficient can finally be read from Figure 2.22.

In case No. 2 of table 7.16 in DIN EN 1991-1-4 [67] and in accordance with the previous defined nomenclature, the slenderness of the structure is

$$\lambda = \begin{cases} \min\left\{1.4\frac{l}{d}, 70\right\} & \text{for} \quad l \geq 50\ \mathrm{m} \\ \min\left\{2\frac{l}{d}, 70\right\} & \text{for} \quad l < 15\ \mathrm{m} \end{cases} \tag{2.56}$$

where it must be linear interpolated for the case between both situations. The reduction factor ψ_λ accounts for the finite slenderness and corresponding lower drag resistance at free ends of a structure. It can be read out of the diagram in Figure 2.23 by means of λ and φ.

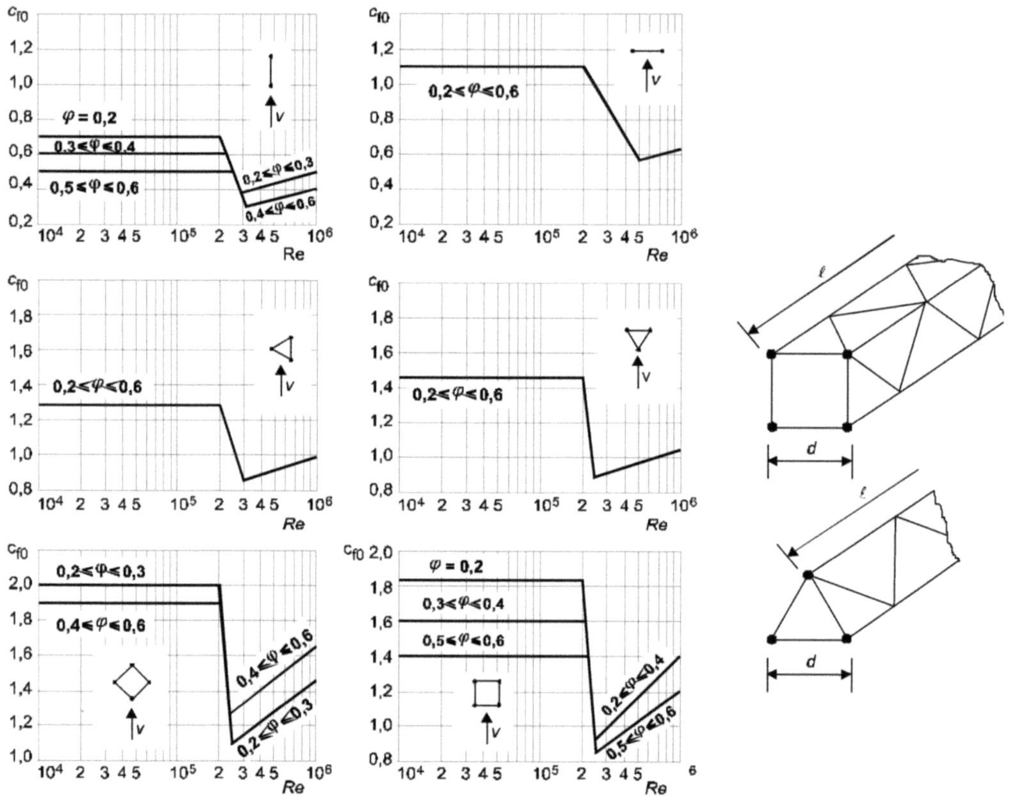

Figure 2.22: Basic force coefficient for different lattice structures, member densities, and Reynolds Numbers, from DIN EN 1991-1-4 [67]

The gust pressure q_p, which accounts for the mean and short term wind speed changes depends on the turbulence intensity I_v and the terrain category. The later prescribes the vertical distribution of the mean wind speed V_m with respect to equation 4.3 in DIN EN 1991-1-4 [67]. Equation 2.57 shows the gust pressure.

$$q_\mathrm{p}(z) = [1 + 7\, I_\mathrm{v}(z)]\, \frac{1}{2}\, \rho_\mathrm{air}\, V_\mathrm{m}^2(z) \tag{2.57}$$

Depending on the structural shape a partwise evaluation of Equation 2.52 and vectorial addition afterwards may be more accurate than an evaluation of the whole structure at once. Incidentally, vortex-induced vibrations should be checked for dense lattice structures with structural densities of $\varphi > 0.6$, such as required by the annotation in figure B.2.3 of DIN EN 1993-3-1 [48].

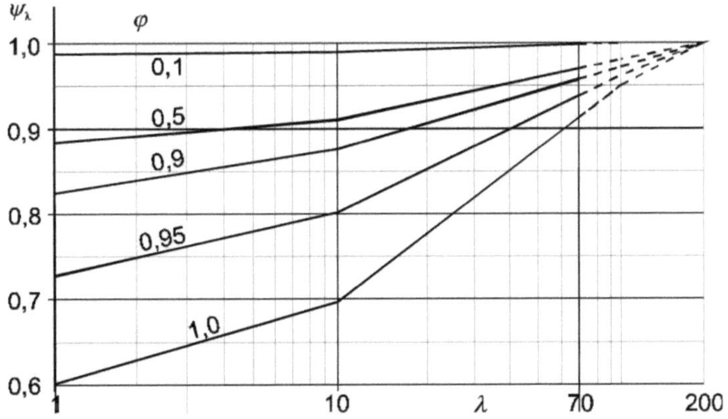

Figure 2.23: Reduction factor for finite slenderness of lattice structures in dependency of its member density, from DIN EN 1991-1-4 [67]

2.7.7 Challenges and Solutions

In practice, lattice towers are not that popular as conventional tubular steel towers. Lattice concepts encounter challenges, such as low torsional stiffness, many member joints, which tend to cause increased erection and maintenance effort, several different member eigenfrequencies, and semi-subjective visibility problems.

The low torsional stiffness applies especially to triangular tower cross sections, owing to their relatively low enclosed tower cross sectional area. To overcome this problem lattice towers with four or more legs have been developed, such as shown in Figure 2.12 and 2.21. In general the more joints a lattice tower has the less material is needed [114]. For example the more legs, and thereby joints, a lattice tower has, the lower become the buckling lengths of the diagonal members, which achieves additional material savings for them. Note that more legs may achieve usage of more simple diagonal member cross sections: While the Butzkies towers in Figure 2.12 have double hat profiles, the pentagonal space frame tower of GE in Figure 2.21 does only have simple L-profiles. On the other hand more cost intensive joints and more point foundations are the consequence of more legs. The argument of high maintenance effort for member joints is refuted through the usage of lockbolts with issued general building approval Z-14.4-591 [38] and new assessment approaches regarding the consideration of remaining pre-load forces, such as investigated by Glienke et al. [83] and discussed in subsection 2.7.5. The almost maintenance free lockbolts are likewise used for the built Butzkies towers, such as shown in Figure 2.24. Maintenance costs for lockbolts only arise for visibility checks of painted control bolts (not shown in Figure 2.24). Relative movements between steel sheets and lockbolts indicate too low pre-loading in the shaft, wherefore the hardened thin painting layer around a lockbolt cracks and the lockbolt must be exchanged.

The visibility problem of lattice towers is controversial discussed, whereby Figure 2.25 indicates that lattice towers must not necessarily look more obtrusive than conventional tubular steel towers do. Visibility is always a question of structural density, painted colour, floor colour, background colour, weather, luminous intensity, distance of view, and light angle of incidence. Therefore, the visibility must be investigated for different support structures and landscape scenarios individually during the realization procedure.

Figure 2.24: Lockbolts of the Butzkies tower and diagonal member with hat cross section. Photo: A. Struve 2018

Figure 2.25: Subjective visibility comparison between conventional tubular steel tower and lattice towers. Photo: Sinning and taken from Hau [93, p. 520]

2.8 Load Evaluation

Knowledge about the nature of WT loads is very important, because a rotatable tower design should be suited to the behaviour of loads to reach the highest possible material efficiency. ASE simulation is the state-of-the-art method to estimate WT loads. Following subsections describe the general loads behaviour of WTs and their modelling in ASE simulations.

2.8.1 Wind Turbine Load Categories

Several loads are acting on WTs and they can be classified by their time history and origin. Time related classes are constant (quasi-steady), such as gravity forces, cyclic (periodic), such as mass unbalance, stochastic (random) from the turbulent wind, and short-time (transient), such as shut-downs or braking events [77, p. 274]. Some exemplary loads are presented in Table 2.5.

The last three time classes in Table 2.5 lead to fatigue, because of the corresponding stress variations within mechanical components and constant loads are responsible for the mean value of stress cycles. Furthermore, unfavourable combinations of time dependent loads lead to ultimate stress situations, which force a component to yield or to buckle. Collision of structural components may also be a resulting failure mode, which has to be prevented. The most tower relevant static loads are the gravity forces caused by nacelle-, rotor-, and tower-weight, the mean rotor thrust, and the mean wind pressure on the tower. Figure 2.26 shows how WT loads can be simplified and applied to a cantilevered beam model. Depicted SS shear forces and moments mainly occur as result of skewed wind inflow, the generator moment and vortex-induced vibrations. Torsional moments are most often negligible for conventional tubular towers, because of their high torsional stiffness, but may be critical for other tower concepts, such as lattice towers. Corresponding torsional moments originate from skewed wind inflow on the rotor disc and yaw events.

Dynamic loads originating from the 1p excitation caused by the mass unbalance within the rotor, and the np excitation frequencies and their harmonics result from the rotor revolution with n blades, which pass the pressure gradient in front of the tower. Excitation frequencies may also be reasoned through the aerodynamic stall behind the tower, mentioned in subsection 4.4.4. Another source for periodic excitation is aerodynamic unbalance through pitch errors, skewed wind inflow, and uneven wind speeds across the rotor disc. If these excitations meet the tower eigenfrequency, resonance occurs and leads to critical fatigue and ultimate loads. Gyroscopic loads caused by yawing and other tower head inducing moments are not listed, but have to be mentioned especially for rotatable WT towers, such as described in section 6.2.4. Some loads are more relevant than others and it is important to know about their significance in terms of component design. This information determines the effort, which should be spent to represent loads with a certain accuracy within a simulation.

Origin / Time	Type of force	Source	Operating condition
Constant (quasi-steady)	Gravity force, Centrifugal force, Mean thrust	Weight, Rotor revolution, Mean wind	Normal operation
Cyclic (periodic)	Mass unbalance, Aerodynamic forces	Unbalance, Tower dam, Oblique flow, Blade passage	Normal operation
Stochastic (random)	Aerodynamic forces	Turbulence of the wind, Earthquake	Normal operation
Short-time (transient)	Frictional and Braking forces, Aerodynamic forces	Shut-down of the wind turbine, Yawing of the nacelle	Manoeuvre, Malfunction, Extreme conditions

Table 2.5: Classification of some exemplary excitating forces according to time history and origin [77, p. 274]

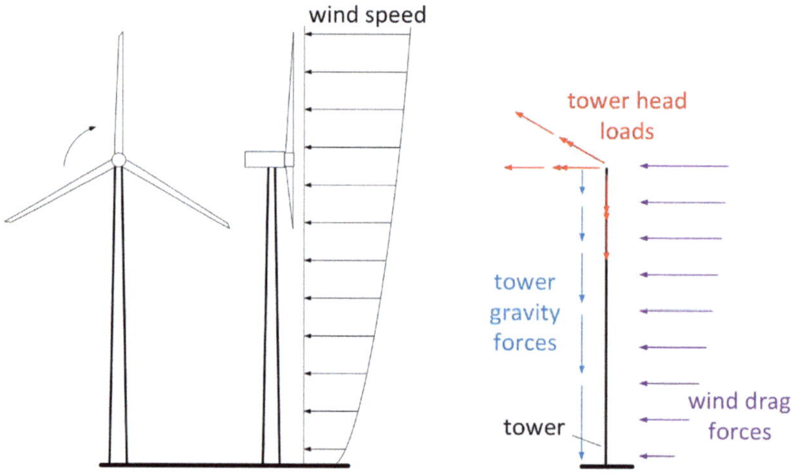

Figure 2.26: Simplified tower loads modelling

2.8.2 Aero-Servo-Elastic Loadsimulations

ASE simulation is a method to represent the WT behavior and loads in a more realistic way than static assumptions do. Main reason for ASE calculations is the account for coupling effects between the aerodynamics, the control system and the mechanical response of components. These nonlinear simulations require iterative procedures at each time step to reach convergence. A good example is the wind ↔ blade interaction, where the blade deforms according to the wind pressure. This deformation creates new aerodynamic behavior along the blade, whereby a new blade deformation follows and so forth. Additional influences through wind turbulences and blade pitch or nacelle yaw motions increase the nonlinearity further. Furthermore, the rotor is mounted on the tower and therefore tower head vibrations are superimposed together with its soil interactions. Most of the computer models are using BEM theory combined to flexible multibody dynamics (FMD) to represent the coupling of blade aerodynamics and the elasticity of the blades. Further flexible components may be the drivetrain, the tower, and the soil.

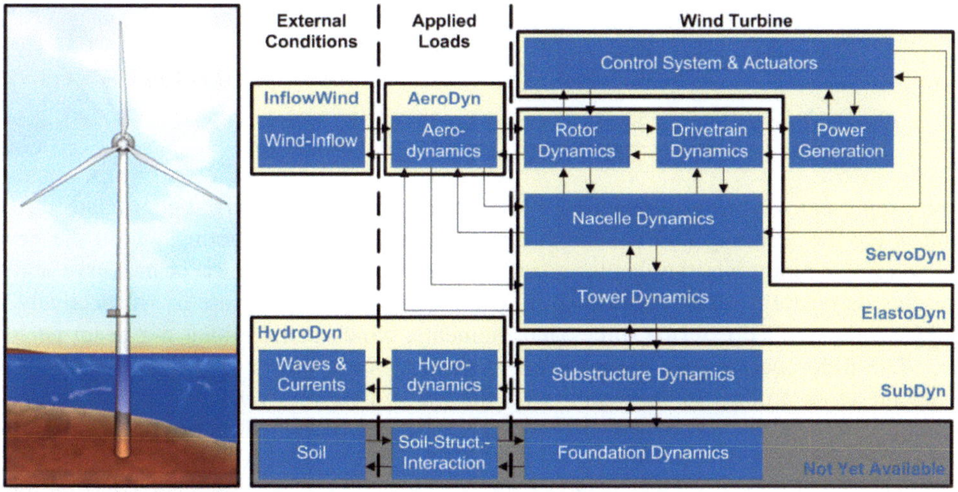

Figure 2.27: FAST module interaction [105]

Several commercial and cost-free ASE codes are available to simulate WT dynamics, such as HAWC2 (Risø National Laboratory, Denmark), DUWECS (TU Delft, Netherlands), FLEX5 (TU Denmark), Bladed (Det Norske Veritas (DNV) Germanischer Lloyd (GL), Norway), QBlade (TU Berlin, Germany) and FAST (National Renewable Energy Laboratory (NREL), USA). A comparison of ASE codes used for certification had been done in year 2006 by NREL and GL [27], where GL issued a statement, that FAST could be used for land-based WT certification. The certified[6] FAST code is cost-free and all source codes are customizable by each user, wherefore it is widely used in the scientific field such as in this work. The FAST

[6] NWTC Information Portal (Certification of FAST and ADAMS®with AeroDyn). `https://nwtc.nrel.gov/SimulatorCertification`. Last modified 18-August-2014; accessed 13-February-2019

glue-code driver connects different modules, such as InflowWind, which chooses the right wind conditions or files, AeroDyn [111], which models the aerodynamics, ServoDyn, which contains the controller system, ElastoDyn, which includes the rotor, drivetrain, nacelle and tower dynamics, the SubDyn module for substructures, and HydroDyn [113], which models hydrodynamics, such as shown in Figure 2.27.

FAST combines a multi-body formulation for the platform, nacelle, generator, gears and hub with a modal-dynamics formulation for the tower [107]. New versions of FAST contain a geometrically exact beam theory (GEBT), which are applied for blades within the new BeamDyn module [192]. The GEBT is able to represent large displacements and rotations of pre curved and pre twisted beams. Cross sectional stiffness matrices, which account for six degrees of freedom (DOF) contain coupling between extension, bending, shear, and torsion. Incidentally, geometrically exact means according to Hodges [97] that no approximations were made to represent the initial and deflected geometries. GEBT concerning equations of motion are given through Bauchau [8, p. 625] with

$$\frac{\partial \underline{h}}{\partial t} - \frac{\partial \underline{f}}{\partial \alpha_1} = \underline{F} \tag{2.58}$$

$$\frac{\partial \underline{g}}{\partial t} + \frac{\partial \underline{\widetilde{u}}}{\partial t}\underline{h} - \frac{\partial \underline{m}}{\partial \alpha_1} - \left(\frac{\partial \widetilde{x}_0}{\partial \alpha_1} + \frac{\partial \widetilde{u}}{\partial \alpha_1}\right)^T \underline{f} = \underline{M}, \tag{2.59}$$

$\underline{h}$ and $\underline{g}$ represent the linear and angular momenta resolved in the inertial coordinate system. $\underline{f}$ and $\underline{m}$ are the resulting internal forces and moments, which are acting within the beam. Furthermore, $\underline{u}$ represents the displacement of a point on the reference beam curve and $\underline{x_0}$ represents its position vector as a function of the curvilinear coordinate α_1, respectively. $\underline{F}$ and $\underline{M}$ are the applied external forces and moments, which acting along the beam per unit span. The tilde operator $(\widetilde{\bullet})$ indicates the skew-symmetric tensor of a vector and can be seen as the cross product of the vector with the following one. Variable t is representative for the time. Multi-body equations of motion in FAST are derived and implemented using Kane's Method [115] and time integration will be done by 4th-order Adams-Bashforth-Adams-Moulton [6] [143] predictor-corrector fixed-step-size explicit integration scheme with 4th-order Runge-Kutta scheme [165] [128] initialization. Modal-dynamics are modeled with Bernoulli-Euler beams in ElastoDyn, which do not represent axial, torsional and shear deformations. Modal-dynamics are implemented by modal superposition of the lowest modes described by polynomial coefficients with small angle approximations and small strains are assumed.

2.8.3 Resolved Support Structures in FAST

The Bernoulli-Euler beam elements mentioned in subsection 2.8.2 are suitable as representative tower elements, even for lattice towers. For fatigue analyses of the detailed members in a lattice structure a transforming procedure for loads between the low computational cost demanding representative beam and the resolved lattice structure was needed. Therefore, the call for resolved support structure modelling in ASE simulations has been answered by an U.S.

Department of Energy Wind and Wave Power Program supported NREL project, where the FAST compatible module SubDyn [32] has been developed. It is a time-domain structural-dynamics module for multimember fixed-bottom substructures, which can either be driven in standalone or coupled with FAST, such as shown in Figure 2.28.

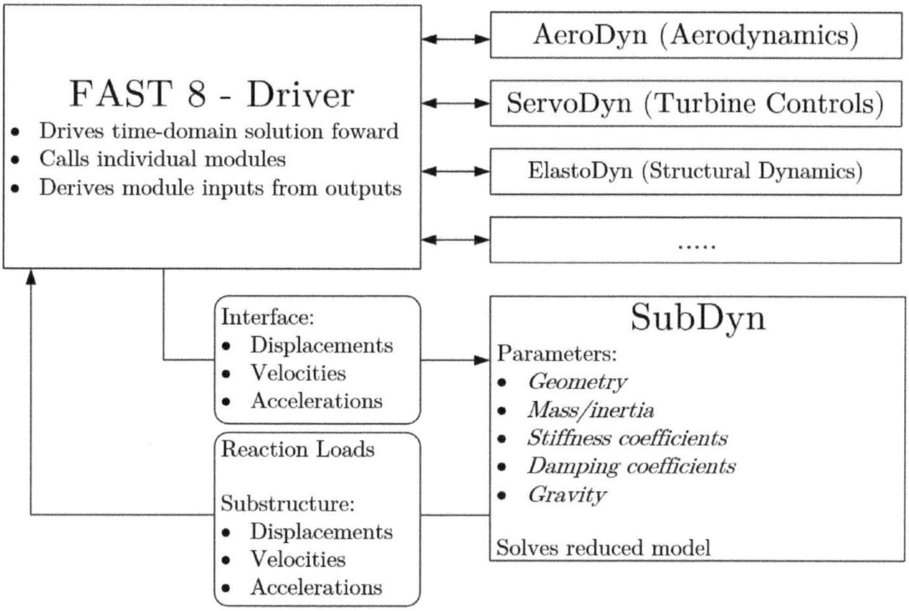

Figure 2.28: FAST to SubDyn coupled interaction, derived from Damiani, Jonkman, and Hayman [32]

In both cases it gets motions, such as displacements, velocities and accelerations at a prescribed interface to the attached structure and responses with reaction loads and structure motions. Typical inputs for the SubDyn module are the substructure's geometry, such as node coordinates and member connectivity, masses, inertias, stiffness, damping, and gravity. Offshore support concepts are additionally realizable with hydrodynamic loading on the members. Since the Craig-Bampton reduction, the necessary state-space formulation, and the static-improvement method are essential parts of SubDyn, the next paragraphs provide a short overview over their governing equations. Further information about how to derive the procedures and matrices in detail are provided by references [32], [33], [34], and [176].

Craig-Bampton Reduction
The main equation of motion for a finite element system, such as in SubDyn is

$$\underline{\underline{M}}\,\underline{\ddot{U}} + \underline{\underline{C}}\,\underline{\dot{U}} + \underline{\underline{K}}\,\underline{U} = \underline{F} \tag{2.60}$$

where $\underline{U}$ and $\underline{F}$ are the nodal displacement and external load vectors [32]. $\underline{\underline{M}}$, $\underline{\underline{C}}$, and $\underline{\underline{K}}$ are the global mass, damping, and stiffness matrices for the whole system. Thereby, $\underline{\underline{M}}$ and $\underline{\underline{K}}$

are assembled from the element mass and stiffness matrices and $\underline{\underline{C}}$ is derived with respect to different mode damping ratios for the whole system. These matrices are of dimensions $N \times N$, which would lead to computational expensive mathematical operations for typical numbers of substructures DOFs with $N > 10^3$. A modal reduction of the linear frame finite-element beam model via Craig-Bampton method, supplemented by a static-improvement method achieves a significant reduction of DOFs, while accurate results are obtained. Thereby, only the boundary DOFs $\underline{U}_{\mathrm{R}}$ from the interface nodes and the reduced interior generalized modal DOFs $\underline{q}_m$ are retained. The Craig-Bampton transformation between the boundary DOFs together with all internal DOFs and the reduced generalized DOFs is performed by

$$\begin{Bmatrix} \underline{U}_{\mathrm{R}} \\ \underline{U}_{\mathrm{L}} \end{Bmatrix} = \begin{bmatrix} \underline{\underline{I}} & \underline{\underline{0}} \\ \underline{\underline{\Phi}}_{\mathrm{R}} & \underline{\underline{\Phi}}_m \end{bmatrix} \begin{Bmatrix} \underline{U}_{\mathrm{R}} \\ \underline{q}_m \end{Bmatrix} \tag{2.61}$$

where $\underline{\underline{\Phi}}_{\mathrm{R}}$ and $\underline{\underline{\Phi}}_m$ are transformation matrices for the interior node DOFs $\underline{U}_{\mathrm{L}}$ with respect to the rigid body interface motions $\underline{U}_{\mathrm{R}}$ and for the retained internal eigenmodes with respect to the generalized modal DOFs $\underline{q}_m$.

Deriving the full Craig-Bampton reduced equation of motion by means of some mathematical steps one arrives at

$$\begin{bmatrix} \underline{\underline{\widetilde{M}}}_{\mathrm{BB}} & \underline{\underline{\widetilde{M}}}_{\mathrm{B}m} \\ \underline{\underline{\widetilde{M}}}_{m\mathrm{B}} & \underline{\underline{I}} \end{bmatrix} \begin{Bmatrix} \underline{\ddot{U}}_{\mathrm{TP}} \\ \underline{\ddot{q}}_m \end{Bmatrix} + \begin{bmatrix} 0 & 0 \\ 0 & 2\,\zeta\,\underline{\underline{\Omega}}_m \end{bmatrix} \begin{Bmatrix} \underline{\dot{U}}_{\mathrm{TP}} \\ \underline{\dot{q}}_m \end{Bmatrix} + \begin{bmatrix} \underline{\underline{\widetilde{K}}}_{\mathrm{BB}} & 0 \\ 0 & \underline{\underline{\Omega}}_m^2 \end{bmatrix} \begin{Bmatrix} \underline{U}_{\mathrm{TP}} \\ \underline{q}_m \end{Bmatrix} = \begin{Bmatrix} \underline{\widetilde{F}}_{\mathrm{TP}} \\ \underline{\widetilde{F}}_m \end{Bmatrix} \tag{2.62}$$

where $\underline{\widetilde{F}}_m$ represents the interior loads, which account for the retained modes, $\underline{\widetilde{F}}_{\mathrm{TP}}$ are the transition piece (TP) loads after Craig-Bampton reduction, $\underline{\underline{\widetilde{M}}}_{\mathrm{BB}}$ is the substructures equivalent mass matrix, $\underline{\underline{\widetilde{M}}}_{\mathrm{B}m}$ is a matrix partition after the Craig-Bampton reduction, $\underline{\underline{I}}$ is the identity matrix, $\underline{\underline{\widetilde{K}}}_{\mathrm{BB}}$ is the substructures equivalent stiffness matrix. In Equation 2.62, the equivalent mass and stiffness matrices are referred to the TP and the damping ratio ζ is defined as percent of the critical damping ratio. $\underline{\underline{\Omega}}_m$ is a $m \times m$ diagonal matrix, which represents the eigenfrequencies of the retained Craig-Bampton eigenmodes. With this reduced system the number of DOFs reduces from $6 \times$ free nodes to $6 + m$. Such a significant DOF reduction makes the SubDyn module suitable for practical analyses of resolved beam structures in computational expensive ASE simulations.

State-Space Formulation

The state-space formulation is used to compute the substructures response at each time step. It connects the outer inputs, such as the transition piece displacements, velocities, accelerations, and substructure hydrodynamic forces called $\underline{u} = \{\underline{U}_{\mathrm{TP}}, \underline{\dot{U}}_{\mathrm{TP}}, \underline{\ddot{U}}_{\mathrm{TP}}, \underline{F}_{\mathrm{L}}, \underline{F}_{\mathrm{HDR}}\}^T$ to the structural response outputs, such as the transition piece loads and the deflections, velocities, and accelerations of the substructure. $\underline{F}_{\mathrm{L}}$ and $\underline{F}_{\mathrm{HDR}}$ are the hydrodynamic forces on every interior and on the boundary nodes, respectively. A review of Figure 2.28 where hydrodynamic interaction is neglected $(\underline{F}_{\mathrm{L}} = \underline{0};\ \underline{F}_{\mathrm{HDR}} = \underline{0})$ may help for clarification.

Transition piece loads $\underline{F}_{\mathrm{TP}}$ are calculated by the state-space formulation

$$\underline{F}_{\mathrm{TP}} = \underline{\underline{C}}_1 \underline{x} + \underline{\underline{D}}_1 \overline{u} + \underline{\underline{F}}_{\mathrm{Y1}} \tag{2.63}$$

where $\underline{\underline{C}}_1$, $\underline{\underline{D}}_1$, and $\underline{\underline{F}}_{\mathrm{Y1}}$ contain input independent and therefore constant model modal parameters, which are likewise used in the Craig-Bampton reduced equation of motion 2.62. In Equation 2.63 $\underline{x} = \left\{\underline{q}_m, \underline{\dot{q}}_m\right\}^T$ and the input vector $\underline{u}$ is slightly modified with respect to the hydrodynamical forces on the boundary nodes, wherefore it is denoted as $\overline{u}$. The later has no influence if no hydrodynamic loads are considered.

Substructure movements are denoted as $\underline{Y}_2 = \left\{\overline{U}_{\mathrm{R}}, \underline{U}_{\mathrm{L}}, \dot{\overline{U}}_{\mathrm{R}}, \underline{\dot{U}}_{\mathrm{L}}, \ddot{\overline{U}}_{\mathrm{R}}, \underline{\ddot{U}}_{\mathrm{L}}\right\}^T$ and distinguished in terms of the interface DOFs $\overline{U}_{\mathrm{R}}$ and the internal DOFs $\underline{U}_{\mathrm{L}}$. A state-space connection between the inputs and movements is obtained by using again substructure describing modal parameters, contained in the Craig-Bampton reduced equation of motion 2.62 and denoted as $\underline{\underline{C}}_2$, $\underline{\underline{D}}_2$, and $\underline{\underline{F}}_{\mathrm{Y2}}$. The state-space formulation for substructure movement becomes the similar form as in Equation 2.63

$$\underline{Y}_2 = \underline{\underline{C}}_2 \underline{x} + \underline{\underline{D}}_2 \underline{u} + \underline{\underline{F}}_{\mathrm{Y2}} \tag{2.64}$$

Static-Improvement Method

Modelling of the important gravity and buoyancy forces requires the inclusion of a huge amount Craig-Bampton modes, because the corresponding frequencies are comparatively high. Thereby, the advantage of having only a few modal DOFs and a resulting improved computational performance would vanish. The static-improvement method is incorporated to overcome this issue. It works through accounting for dynamically not included modes by means of a quasi-statically approach: It adds the difference between two static solutions to the dynamic solution, which was explained before. These two static solutions of the Craig-Bampton modal system lead to another transformation as it was in Equation 2.61, namely

$$\begin{Bmatrix} \underline{U}_{\mathrm{R}} \\ \underline{U}_{\mathrm{L}} \end{Bmatrix} = \begin{bmatrix} \underline{\underline{I}} & \underline{\underline{0}} & \underline{\underline{0}} & \underline{\underline{0}} \\ \underline{\underline{\Phi}}_{\mathrm{R}} & \underline{\underline{\Phi}}_m & \underline{\underline{\Phi}}_L & -\underline{\underline{\Phi}}_m \end{bmatrix} \begin{Bmatrix} \underline{U}_{\mathrm{R}} \\ \underline{q}_m \\ \underline{q}_{L0} \\ \underline{q}_{m0} \end{Bmatrix} \tag{2.65}$$

In Equation 2.65 the additional terms $\underline{\underline{\Phi}}_L \underline{q}_{L0}$ and $-\underline{\underline{\Phi}}_m \underline{q}_{m0}$ are the two static solutions for the full Craig-Bampton system with all L DOFs and the reduced system with m DOFs, respectively. The corresponding necessary slight changes to the state-space formulation is explained together with more detailed derivations by Damiani, Jonkman, and Hayman [32].

2.9 Rotatable Support Structures for Wind Turbines

The general introduction to the state-of-the-art in some aspects of WT technology of the previous sections will be extended by the state-of-the-art for rotatable support structures for WTs in this section. The investigation of these exotic structures are the main focus of this work. The first subsection 2.9.1 provides an overview about the industrial approaches and lists some patents, which are related to rotatable wind turbine support structures and in the next subsection 2.9.2 some scientific considerations about this topic are reviewed.

2.9.1 Industrial Approaches and Patents

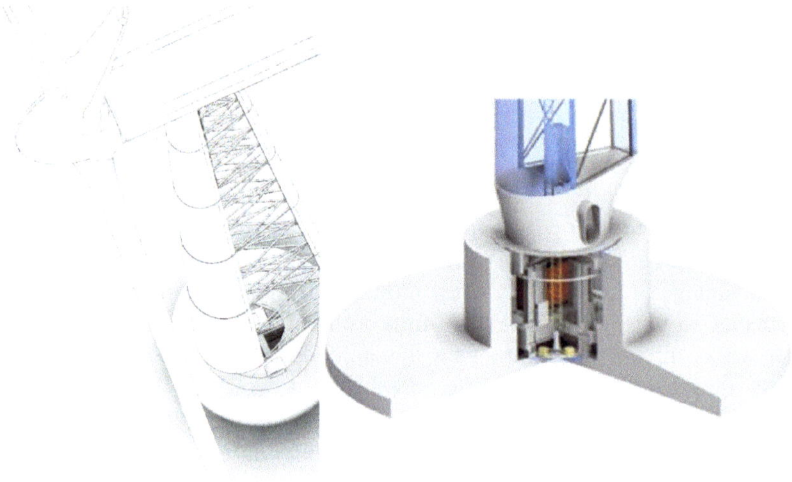

Figure 2.29: Rotatable WT tower concept ASPARAGO® by ©Steel Pro Maschinenbau GmbH [177] and from `http://www.steel-pro-maschinenbau.de/windenergie.html`; accessed 23-January-2019

Industrial approaches for modern WTs with rotatable tower are presented by Steel Pro Maschinenbau GmbH from Berlin, Germany. They are trying to establish their ASPARAGO® tower with yaw bearing at the tower base in the wind energy business. It has a divided tower profile, as shown on the left side in Figure 2.29. The shear connection between both half cylindrical shells is realized by strut members. Steel Pro proposes already a solution for the yaw bearing at the tower base. This foundation integrated approach is depicted on the right side in Figure 2.29, but no detailed informations are published[7].

[7] State from Mai 2018

The Internet page[8] postulates a linear cost growth as function of the hub height as opposed to a polynomial cost growth for conventional steel shell towers. This statement should make the tower more beneficial for larger hub heights. The concept is discussed in more detail in the later section 6.3.1. Up to February 2019, such a tower has not been built.

Clipper Windpower from California, USA holds two patents. One for a rotatable WT tower [95] and one for a corresponding yaw bearing inside the tower [96]. The principal approach is shown in Figure 2.30a and contains a roller bearing at the end of the first quarter of the tower. Therefore, the rest of the tower is capable to rotate and aerodynamically shaped.

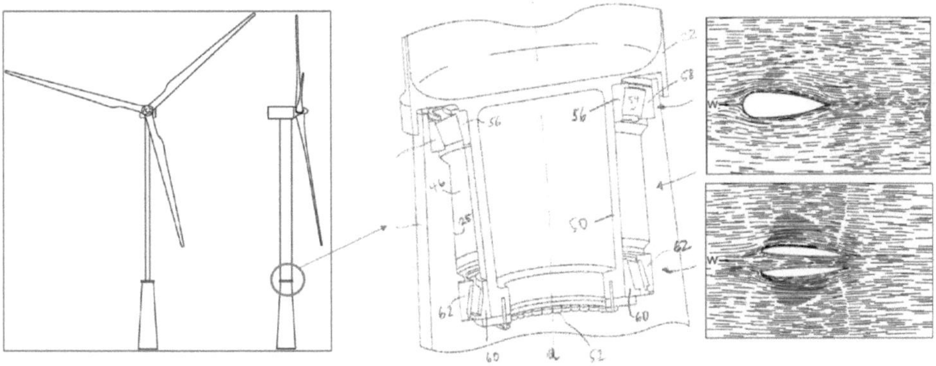

c

d

a b

Figure 2.30: Patent pictures of a rotatable WT tower concept, held by Clipper Windpower [95] [96]

Figure 2.30b shows the yaw bearing construction inside the tower. It consists of two o-arranged roller bearings with a certain distance to minimize the internal radial forces. Figure 2.30c shows the aerodynamicly shaped cross section of the tower within a streamline field. The patent text states:

> *At an average wind speed (free stream velocity) of* 25 m/s*, the minimum wind speed in the aerodynamic wake zone directly behind (downwind) the tower 24 is still about* 20 m/s*, or about* 80 % *of the free stream velocity.* [95]

For the divided aerodynamic profile in Figure 2.30d, the patent text states that the wind speed behind the tower is about 94 % of the free stream velocity. A conventional tower profile under the same condition slows the wind velocity behind the tower down to 1.31 m/s, which is 4.52 % of the free stream [95]. This indicates a high potential in terms of tower shadow reduction. The tower shadow is responsible for periodic exitations from vortex induced vibrations. Note that this topic is treated later in subsection 4.4.2. The patents are focused on the aerodynamic tower properties and neglect the potential of material saving at the bottom part of the tower, because it is not capable to align with the wind. Clipper Windpower does not spend effort to this subject anymore, such as indicated by the mail conversation in Appendix 14.1.

[8] `http://www.steel-pro-maschinenbau.de/windenergie.html`; accessed 23-January-2019

Houly Co. Ltd. from Taipei City, Taiwan owns a patent [132] where the principal of a rotatable tower with yaw bearing within the foundation is applied, such as shown in Figure 2.31. The bearing consists of a hollow cylinder as body and two steel rings (17 A and B in Figure 2.31) to lead the forces into the ground. To make the cylinder more stable, it can be filled with concrete mortar or other materials. Furthermore, the patent does make a proposition to build a frame on a floating structure to contain this kind of yaw bearing in the tower base for offshore WTs.

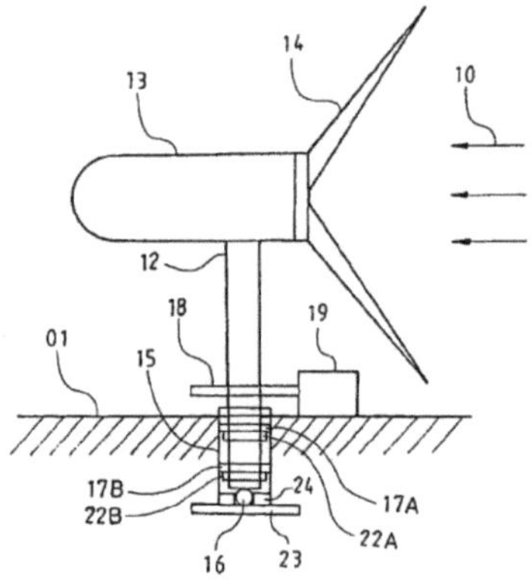

Figure 2.31: Schematic sketch of a rotatable tower with yaw bearing in the tower base by Houly Co. Ltd. [132]

General Electric Company from New York, USA holds a patent with the peculiarity of an inclined rotatable tower section at the top of the support structure, such as shown in Figure 2.32. Regarding the patent the rotatable section should have a length of 2 to 15 times the tower diameter at the top and an inclination angle between 1° and 20°. Remarkably, the rotor axis of the hub (26 in Figure 2.32) should keep the inclination of 2° to 5° and therefore gives away the opportunity to increase the energy harvest through a vertical rotor alignment. This aspect is more discussed in the later subsections 6.2.3 and 10.2. The approach in Figure 2.32a induces an additional bending moment according to the increased horizontal distance of the RNA center of mass, relative to the vertical axis of the fixed tower part. The other option in Figure 2.32b prevents this for the fixed tower section, but leads to larger rotatable section lengths on the other side. Moreover, it induces a large imperfection, which has to be considered within the stability analysis.

Siemag Tecberg Group GmbH from Haiger, Germany does have a patent [10] where the generator should be provided at the tower base instead of the nacelle. To realize the proposed belt drive along the whole tower, it has to yaw together with the rotor and therefore the tower has to have a rotatable degree of freedom. This patent sets its focus on the drive train instead of the concept of a rotatable tower. A similar patent is provided by Lutz [134]. The patents [99], hold by Siemag Tecberg Group Gmbh and [81] by Gernandt are also related to rotatable WT towers.

Aerodyn Engineering GmbH from Büdelsdorf, Germany works on a project called SCD® nezzy. Their goal is to develop an 8 MW offshore WT with a guided leaning profiled tower and a self-adjusting, 2 bladed downwind rotor. Three stabilizing floaters are the main parts of the floating structure. The company is postulating cost savings of about 40 % compared to conventional offshore solutions [166]. The concept does not need a yaw bearing and is aligned to the wind by the thrust force on the downwind rotor, the wind pressure on the eccentric tower, and the water current. Mooring lines at the joint of one of the floaters enables the structure to rotate around a fixed point.

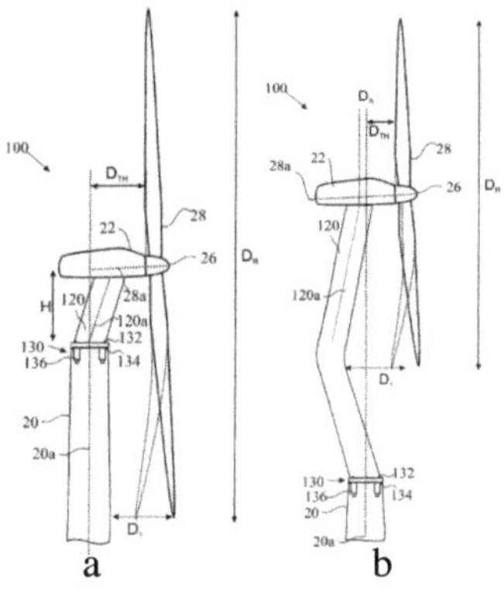

Figure 2.32: Inclined tower section with yaw bearing inside the tower by General Electric Company [144]

All these approaches show, that other people and companies see the potential of rotatable WT towers. Therefore, scientific investigations should bring more consistency and show how rotatable towers contribute to sustainable and efficient support structures for WTs.

2.9.2 Scientific State of the Art

One of the first scientific papers concerning rotatable WT towers was written at the Reichsarbeitsgemeinschaft Windkraft [158] community. It was an economic independent community of scholars under the NSDAP regime of Germany with the objective to explore the utilization of wind as energy source. One ambitious project was the development of the MAN-Kleinhenz WT. It should have a rated power of 10 MW with a lee sided rotor. The diameter of the rotor should amount to 130 m, the tip speed ratio was 5 and a hub height of 250 m was intended [120], such as shown in Figure 2.33.

The support structure is characterized by a fixed section and a rotatable aerodynamicly shaped section. One of the proposed fixed substructures was an eight leg lattice concept with the largest width of 110 m at the bottom and 30 m at the top, shown on the left side in Figure 2.33. The right side in Figure 2.33 shows a guyed tower section with the consequence of low bending moments at the tower base and therefore a small diameter of 8 m is sufficient. The anchor position diameter of the right concept is proposed to be 385 m. The aerodynamic shape of the rotatable tower section reduces the wind loads at the WT top and the connected aerodynamic nacelle leads the air stream around itself to accelerate the air at the blades root area. Furthermore, the nacelle overhang of 27 m behind the tower centerline and the rotor are used as a wind vane, wherefore the necessary yaw drive power can be reduced. The MAN-Kleinhenz WT was never built owing to the Second World War, but the conceptual considerations of the Reichsarbeitsgemeinschaft Windkraft show that rotatable towers did play a role for large WTs many decades ago.

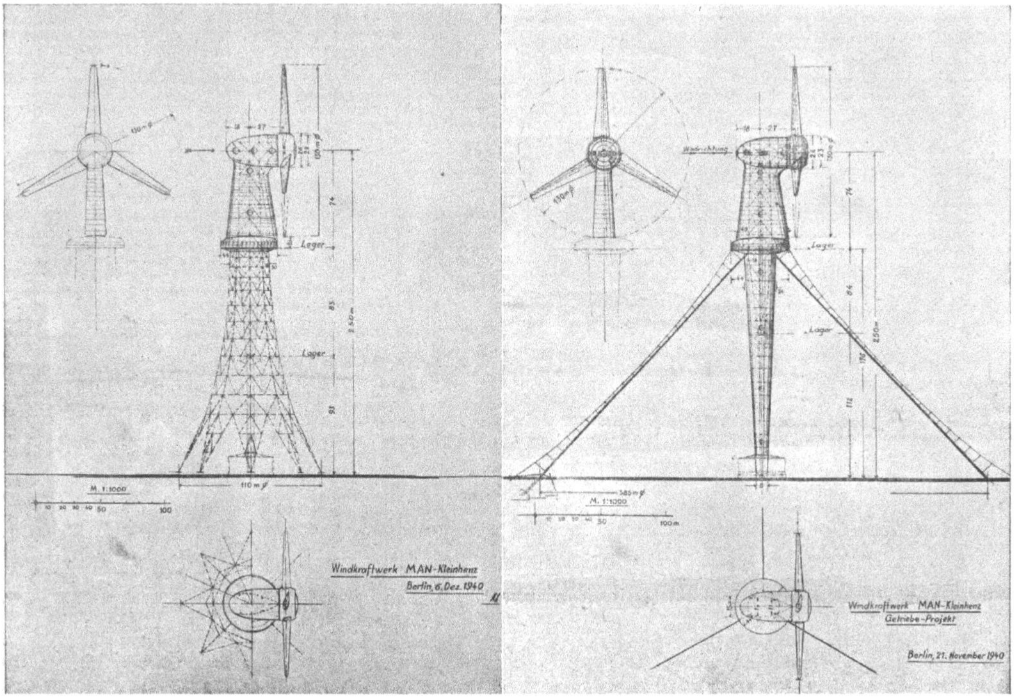

Figure 2.33: Sketch of the WT MAN-Kleinhenz with two different support structure concepts [120]

An article from Beye, Rudolf, and Faber of the Wind Energy Technology Institute [14] is focused on the relocation of the yaw bearing to the tower base of modern WTs and was based on Beye's bachelor thesis [13] about the same topic. In these works, smaller tower profile heights and larger profile widths are proposed, which facilitate the transport of welded steel shell tower sections through bridges and tunnels. Thereby tower material is saved. The confirmation of this statement should be brought within a comparison of two different tower designs for a 2.5 MW reference WT with a hub height of 79.43 m. The conventional one has a conventional circular cross section and the new tower design has an elliptical cross section. The simplifying assumptions are to hold the area moment of inertia around the main bending axis at each cross section cut equal to the conventional one. Elementary dimensioning without strength, fatigue, and buckling analysis postulates material savings of 41.6 % or 61,234 t for the elliptical tower compared to the conventional one. Stiffness requirements are the main design criteria for most of the modern wind turbine towers [92, p. 427], wherefore this rough calculation gives an indication to the high material saving potential through the concept. Material savings at the tower lead to cost savings, but standing against cost increases for the yaw bearing at the tower base. The loads at the tower base were assumed to be ten times larger than at the tower top. Furthermore, only small bearing slackness is allowed to avoid additional loads, induced by second order displacements at the tower top. A secure tower base yaw drive operation requires much effort with respect to high tolerance manufacturing, wherefore costs will increase further. Three different yaw bearing concepts are provided within the article, as shown in Figure 2.34.

The double row four point bearing in Figure 2.34a is a conventional solution for yaw bearings at the tower top. It has a low rolling friction resistance, but it tends to have a wear caused bearing slackness, wherefore additional second order loads will be induced after certain operation times. Furthermore, damaged double row four point bearings have to be replaced by dismantling the whole WT. A dry slide bearing, such as in Figure 2.34b is sometimes used in conventional yaw systems. It can carry high loads regarding its required space and the adjustment screw can be used in combination with the spiral spring to reduce bearing slackness and to adjust the frictional torque. In case of damage the bearing housing has to be removed whereby all plastic slip bodies can be exchanged without dismantling the tower. The yaw bearing solution in Figure 2.34c reminds of the approach in Figure 2.30b with two o-arranged roller bearings. This concept needs an inner circular steel tube as an adapter between the rotational asymmetric tower profile and the circular bearing block. The connection between adapter and the outer tower side is realized by welding and requires high manufacturing effort to realize the required small tolerances. Bearing slackness can be adjusted by a nut, but in cases of bearing damage the whole WT has to be dismantled. Two bearing manufacturers [2], [126] were commissioned to dimension a roller bearing at the base of the reference WT tower by the assumption of simplified loads. The results in Table 2.6 show the wide price range of $180,000$ € to $450,000$ € for roller bearing solutions.

Furthermore, a relation between the mass and the price is provided where the half weight of the Liebherr-Components bearing should cost less than half of the price from IMO Energy. Moreover, the mass seems to have a relation to the outer diameter, wherefore large diameters should be preferred. This small study shows the technical feasibility for yaw bearings at the tower base. Economic feasibility under consideration of the high yaw bearing prices from Table 2.34 is not given in the article [14], but the important factor of quantity discount is neglected and can have a positive influence. Note that Table 2.6 shows only the bearing costs, but not the whole yaw system costs.

Manufacturer	Liebherr-Components	IMO Energy
Type	–	three row roller bearing
Weight / t	15	30
Inner Diameter / mm	–	4450
Outer Diameter / mm	6000	5500
Running Thread Diameter / mm	–	5000
Price / 1000 €	180 to 220	450

Table 2.6: Basic manufacturer cost estimates for yaw bearing at tower base [2], [126]

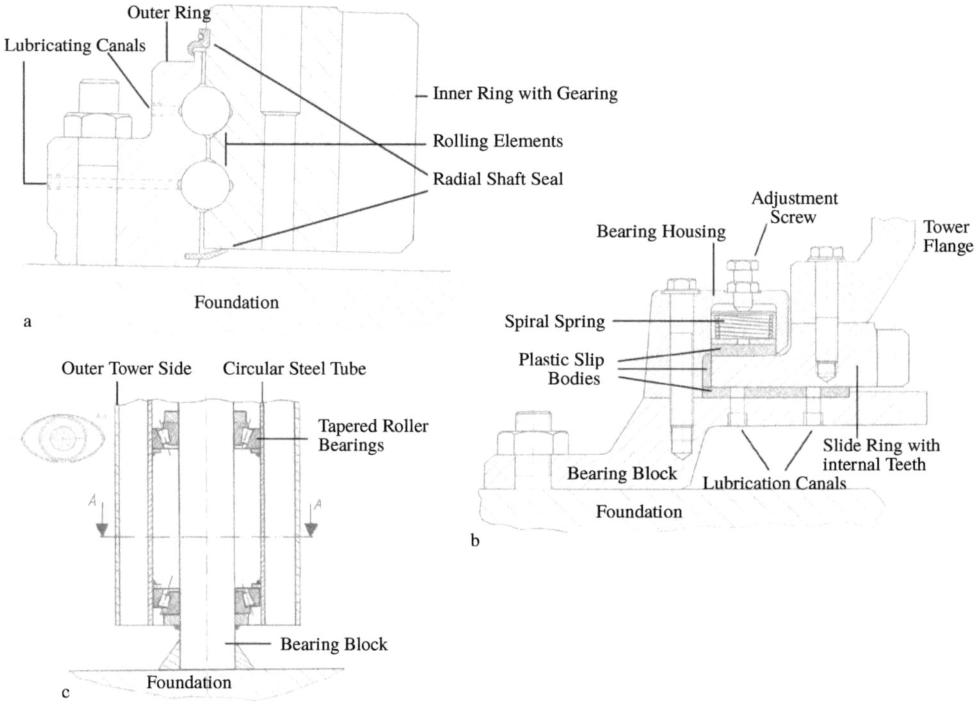

Figure 2.34: Yaw bearing concepts for rotatable wind turbine towers from Beye, Rudolf, and Faber [14]. Translated and rearranged by the author

2.10 Other Related Structural Systems

To develop a sustainable rotatable tower concept with yaw bearing at the tower base, it is appropriate to consider other systems with the ability to resist huge loads while rotating. Therefore, railway turntables, rotatable buildings, mobile cranes, bucket wheel excavators and rotatable shortwave antennas are analysed with respect to possible tower and bearing solutions throughout the following subsections.

2.10.1 Railway Turntables

Railway turntables are used to turn locomotives in such a way that they are able to drive in the opposing direction or be transferred to another track. They are likewise used to maneuver locomotives into roundhouse depots for maintenance or parking. Several different principles were already built near the mid-nineteenth century. Concepts with bearing surface at the center, at the circumference or at both positions were realized. Roller bearings and pivot wheels were used to minimize the rotational resistance [15], such as shown in Figure 2.35.

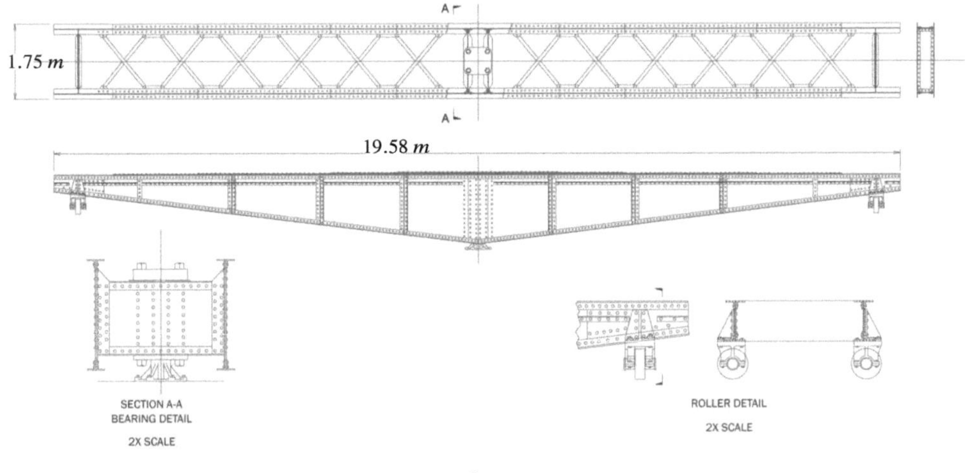

Figure 2.35: Drawing of a railway turntable[9]- dimensions converted to meters by the author

Modern turntables are able to rotate vehicle masses of 100 t to 250 t and have diameters of 24 m to 30 m [31]. Railway turntables have to transfer large and along the rail uniformly distributed gravity forces into the ground. These loads produce bending moments within the rail, but not within the monovalent and divalent bearings at the circumference and center position. Therefore, railway turntables do not fit in the load characteristic of large bending moments within the wind turbine tower base, but they show that pivot wheels can be an appropriate solution for large bearing diameters.

[9] http://www.jdkds.com/blog/wp-content/uploads/2013/07/illustration_evmcrr_turntable.jpg; accessed 13-February-2019 - original reference: [197]

2.10.2 Rotatable Buildings

Often rotatable buildings serve architectural design purposes. Other applications aim for practical improvements, such as a better energy efficiency through higher energy yield from the sun [86]. In either case a reliable support together with the ability to rotate must be ensured for the designed lifetime of those buildings, which may weigh hundreds of tons. An example for a rotatable passive energy house is given in Figure 2.36. Owing to its ability to rotate the house wall with large windows and solar panels together with the solar panels on the roof can be adjusted to the sun over the whole day. Thereby, sun energy is efficiently captured by the house throughout the year. Since the house produces more energy than it consumes it is not only a passive house, but also an active house [86]. On the right hand side in Figure 2.36 one can see the inner timber frame construction. Its ability to rotate is achieved through pivot wheels at the outer diameter right under the ground surface and the slewing ring in the non-rotatable basement of the house [86]. Thus, the concept is similar to the previous mentioned railway turntables in subsection 2.10.1 and has its own patent, owned by Rinn [162].

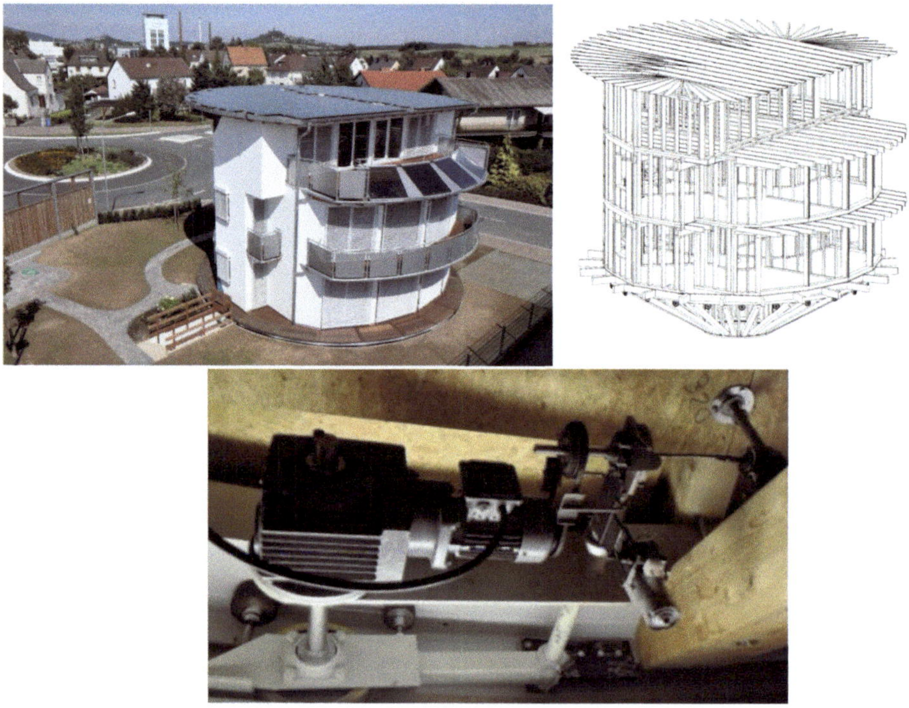

Figure 2.36: Rotatable passive energy house, developed by RINN XI. GmbH. Pictures from Gröger [86]

2.10.3 Mobile Cranes

Mobile cranes have the ability to drive and build themselves up on the building site. The norm DIN EN 13000 contains the main informations about construction, manufacturing, mounting and safety devices [40, p. 7]. Considering mobile cranes mobility it is appropriate for them to have the bearing at the base of the boom, such as shown in Figure 2.37.

The telescopic crane boom of the LTM 1100-4.2, shown on the left side in Figure 2.37 has a maximum hoist height of 91 m and is supported by a triple-roller slewing rim [130, p. 20] with an approximately diameter of 1,742 mm (derived from the drawing in [130, p. 2]). The word triple refers to three rowed rolling elements within the bearing. More bearing rows increase the ability to support heavier loads [29, p. 431] and are necessary to lead the maximum load capacity of 100 t to the frame. The frame is equipped with four hydraulic telescopable outriggers to pass the loads on to the ground. To heave large weights of 750 t the LG 1750 can be used with its guyed lattice concept and a maximum hoist height of 193 m. High overturning moments are counterbalanced by counterweights. The technical data sheet shows the use of a roller slewing bearing [131, p. 6] with an approximately diameter of 3,468 mm (derived from the drawing in [131, p. 11]). This information shows that roller bearings are preferred for high axial loads and applications with higher production quantity. Another important aspect is the load direction oriented design of the telescopic crane booms, owing to the constant position of heaved loads with respect to the boom, such as shown in Figure 2.38a.

Figure 2.37: Mobile crane LTM 1100-4.2 (left) [129] and LG 1750 (right) [131] by Liebherr

The paper [91] documents the historical development of their cross sections. The first cross sections had simple box shapes without any curvature. This was possible without plate buckling, because of the low steel strengths, such as S460Q. Consequently low allowable stresses led to a large wall thickness. As a consequence of the modern higher steel strengths, such as S1100Q, corresponding smaller wall thicknesses provoke local plate buckling. One step within the development was to provide more foldings and to weld reinforcements into the profile, such as shown in Figure 2.38b. This led to great manufacturing expenditure and could be avoided by adding more foldings in such a way that the plate buckling problem became a shell buckling problem with higher buckling resistance, such as presented in Figure 2.38c.

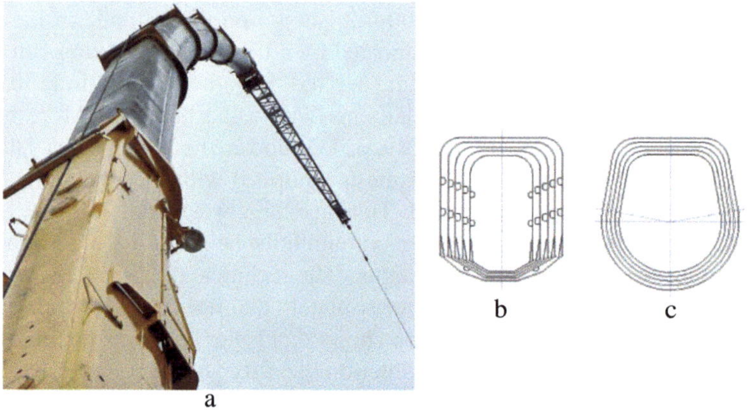

Figure 2.38: Deformation of a telescopic boom under heavy loads (a) and their old and new cross section shapes (b and c) [91]

2.10.4 Bucket Wheel Excavators

Bucket wheel excavators gained the widest application in the area of surface mining [101] and their engineering principals are mentioned in DIN 22261-2 [39]. They consist of a slewing superstructure, which contains the bucket wheel boom (1) and the counterweight boom (2) with winches (3) to position the bucket wheel (4), such as shown in Figure 2.39.

The bucket wheel is used to mine mineral volume flows of up to 14,000 m^3/h and can be raised to a height of up to 50 m in case of a SR6300.[10] Superstructure booms can be realized by solid wall or lattice concepts and will be made most commonly out of the steel grade S355J2+N [101]. Exact counterweight adjustments with respect to different bucket wheel boom positions are very important to guarantee static stability of the excavator [23]. Therefore, bucket wheel boom positions have an influence to the acting moment at the slewing bearing. Many different roller bearing types can be used to make the superstructure rotatable. The main bearing load is axial owing to the high mass of the superstructure, wherefore cylindrical roller bearings are preferred. Excavators, such as the 288 giant-type reach bearing diameters of 20 m.[11] This affirms the use of roller bearings even for large diameters, but is restricted to special manufactures, such as

[10] `http://bucket-wheel-excavator.fam.de/`; accessed 13-February-2019

[11] the same homepage as the previous footnote

bucket wheel excavators and not necessarily applicable for mass production applications, such as WT yaw bearings.

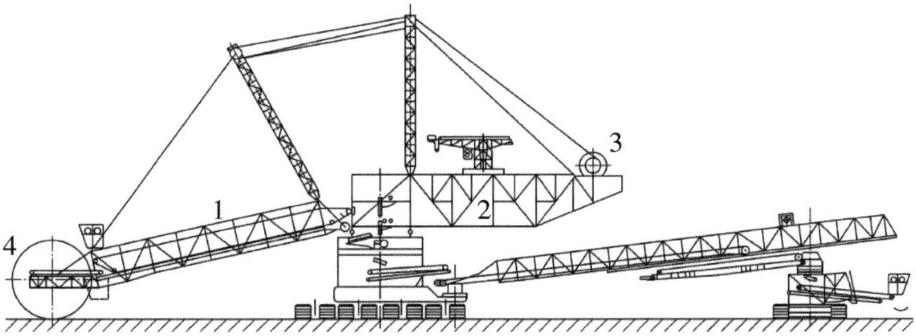

Figure 2.39: Schematic drawing of the bucket wheel excavator RBW 262 [39, p. 112]

2.10.5 Rotatable Shortwave Antennas

Shortwave antennas are used for wide range radio signal transmission of frequencies between 6 MHz and 26 MHz [125]. Concepts of the past need a huge amount of material and field area, because their radiation elements (dipoles) had not the ability of self alignment. Recent developments overcame these problems through a rotatable structure, which bears all dipoles and reflectors, such as shown on the left side in Figure 2.40.

It consists of a hollow concrete foundation, which contains the short wave radio station. A short non-rotatable circular shell section is on its bottom flanged to the foundation and bears the rotary joint on its top. This rotatable joint, depicted on the right side in Figure 2.40, is a three row roller bearing known from the discipline of mechanical engineering.

High requests to the circular and planar shape accuracy of the fixed shell flange, the roller bearing, and the rotatable circular shell is a challenge in terms of manufacturing. These requests are necessary to guarantee constraint free rotation. Corresponding ovalization of the shells is prevented by three triangular ordered stiffening members within the tubes. A light weight lattice structure bears secondary lattice arms, which support all dipoles and reflectors. It is placed on top of the rotatable tubular shell section. The lattice arms are connected rigidly to the load-direction-derived rectangular main lattice shaft. Its dimensions are 2.1 m × 3.4 m and therefore a different bending stiffness along the main load direction and its orthogonal direction is achieved. The outer dimensions of this rotatable shortwave antenna are 80 m height and 76 m width. The resulting area is filled with many members and antenna ropes, wherefore high wind speeds combined with icing conditions are most critical for the structure. However, fatigue loads are not as high as in the case of wind turbines, because the later experience huge changing thrust forces from the rotor disc additionally to the comparatively low aerodynamic drag forces on the support structure.

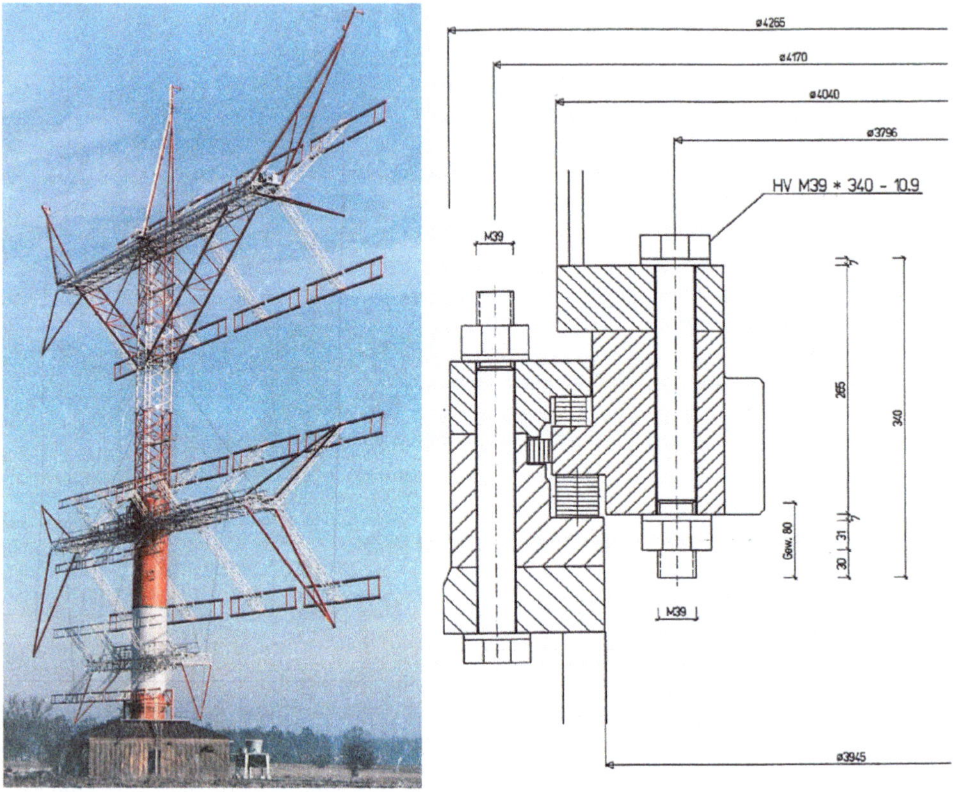

Figure 2.40: Rotatable short wave antenna, built in 1996 for Deutsche Bundespost (left) and its detailed roller bearing joint (right) [125]

Low fatigue loads make the used tubular members with its resulting welding joints convenient for a rotatable shortwave antenna. Another shortwave antenna concept described in [25] and [168] has guyed dipole arms instead of rigid connections, wherefore less material is required. The exchange of the roller bearing at the antenna base is possible without dismounting in the case of failure, as given in [168]. A lifting device is used to heave the whole antenna, but previous guying is necessary to reduce the acting overturning moments.

2.10.6 Summary of Findings from Related Systems

Mobile crane booms, bucket wheel excavator booms, and rotatable shortwave antenna support shafts are mainly loaded from one direction, wherefore some of their general concepts are applicable for rotatable WT towers. Boom profile shapes are suited for their load direction and account for the difference between a pressure and a tension side. Therefore, higher curvature is provided on the pressure side of solid wall cross sections. Lattice structures account for the main load direction through different side lengths to increase the bending stiffness around the mainly loaded axis. Modern mobile crane booms are out of high-strength fine-grained steel, because of static loads. In opposition, bucket wheel excavator booms are more designed with respect to fatigue and therefore consisting of lower steel grades, such as the hot-rolled structural steel S355J2+N. This achieves lower steel costs, because the current Eurocode 3 norm DIN EN 1993-1-9:2010-12 for fatigue does not distinguish between steel strengths regarding their fatigue resistance [47]. In contrast to this, research activities of Ummenhofer et al. [188], [187] show that relations between steel strengths and their fatigue resistance exist especially if welding seems are treated after welding by a High Frequency Impact Treatment (HiFIT) or by an Ultrasonic Impact Treatment (UIT).

Different bearing solutions are proposed by the previous industry applications. The most of them are working with roller bearings in between a large diameter range from less than 1 m up to 20 m. Production quantity has to be considered if the former findings should be applied to WT towers. WTs are produced in higher quantity as bucket wheel excavators and shortwave antennas, wherefore the economical benefit of mass production of large roller bearings can be exploited. Another way to realize large bearing diameters is obtained by pivot wheels, rolling on a rail and optionally combined with an inner slewing ring, such as railway turntables and rotatable houses have. All the findings in related technical fields show that huge bearing diameters are possible, while large axial loads have to be carried. At the same time, rotatable shortwave antennas show that large bending moments are likewise supportable by a roller bearing with comparatively small diameter. These findings show already that technical feasibility for rotatable WT towers is given.

3 Load Analysis of Conventional Designs

3.1 Methodological Approach

This work focuses on the load analysis of rotatable WT support structures, but it was appropriate to analyse the loads of a conventional WT first. Resulting findings helped to understand the driving influences and the loads analysis methodology could be validated. After an appropriate reference WT was chosen in section 3.2, simplified analytical equations show the influence of WT dimensions regarding the tower base bending moment in section 3.3. Next, a literature research for ASE calculated loads of scientific relevant reference WTs in section 3.4 is presented. In the end, the analysis of the reference turbine in section 3.5 achieved a systematic comparison between the conventional reference and the later presented rotatable support structure. Moreover, it delivered a validation for the ASE load simulation procedure. Note that the state-of-the-art framework, FAST, mentioned in subsection 2.8.2, was used for all ASE load simulations.

3.2 Reference Wind Turbine

Reference WTs serve the purpose of comparability between different research and development projects and to set new standards. Therefore, it was appropriate to chose one reference WT as basis for the development of a rotatable support structure. Some published reference WTs are summarized with fundamental specifications within Table 3.1.

In this work, the NREL 5 MW Baseline WT and its conventional tubular steel tower was chosen as reference, due to following reasons:

- Practical relevant land-based WT power class
- Representative tower concept for the most land-based WTs
- Properties are based on the real existing REpower 5M
- Comprehensive parameter documentation
- Already implemented in the ASE FAST framework

- Detailed ASE load simulation results are published
- Most used reference WT in research projects

Its properties are listed in Table 3.2.

Institution	Rated Power / MW	Rotor Diameter / m	Hub Height / m	Tower Concept	Ref.
UT Dallas, U.S. UM Michigan, U.S. TU Delft, Netherlands	20.0	276.0	160.2	Tubular Steel	[4]
DTU Risø, Denmark	10.0	178.3	119.0	Tubular Steel	[5]
UCC Cork, Ireland DNV GL Bærum, Norway	8.0	164.0	110	Tubular Steel	[37]
NREL Golden, U.S.	5.0	126.0	90.0	Tubular Steel	[109]
WETI Flensburg, Germany	3.5	140.0	150.0	Steel/Concrete Hybrid	[183]

Table 3.1: Reference wind turbines

Property	Unit	Value
Rated power	MW	5.0
Rotor orientation	–	Upwind
Blades	–	3
Control	–	Variable Speed, Collective Pitch
Gearbox	–	Multiple-Stage
Rotor diameter	m	126.0
Hub height	m	90.0
V_{in}, V_{r}, V_{out}	m/s	3.0, 11.4, 25
Rotor speed in, rated	1/min	6.9, 12.1
Rotor mass	t	110.0
Nacelle mass	t	240.0
Tower mass	t	347.46
1. Tower FA freq.	Hz	0.324
1. Tower SS freq.	Hz	0.312
2. Tower FA freq.	Hz	2.900
2. Tower SS freq.	Hz	2.936

Table 3.2: Property of the 5 MW NREL reference wind turbine from [109]

3.3 Quasi-Steady Loads Sensitivity to Structural Dimensions

Quasi-steady loads, such as defined in Table 2.5 of subsection 2.8.1, are dependent on the structural properties of a WT. This subsection treats analytical load estimations for different WTs with conventional tubular towers to show these dependencies. Tubular tower design driving load components are the internal normal forces and the acting bending moments. While normal forces are due to the structural self-weight, bending moments result form the mean rotor thrust and the aerodynamic drag forces on the tower with their respective lever arms. Thereby, the mean rotor thrust is a function of the rotor diameter and the hub height. Tower drag forces are due to the tower diameter and the tower height. Mean rotor thrust T and tower drag W estimates can be carried out according to Gasch [77] by the relations

$$T = \frac{1}{2}\rho_{\text{air}}\, A_{\text{R}}\, c_{\text{T}}\, V_{\text{H}}^{2} \tag{3.1}$$

$$W = \frac{1}{2}\, \rho_{\text{air}}\, A_{\text{T}}\, c_{\text{D}}\, V_{\text{eq}}^{2}, \tag{3.2}$$

where ρ_{air} is the air density, A_{R} and A_{T} are the rotor and projected tower area, c_{T} and c_{D} are the assumed rotor thrust and tower drag coefficients, V_{H} equals to the wind speed at hub height, and V_{eq} equals to the equivalent wind speed at the weighted wind attack point on the tower, respectively. Furthermore, an exponential wind profile according to IEC 61400-1 [50] was assumed to calculate the wind speed distribution $V_{\text{w}}(z)$ along the WT height. It followed from the relation, given in Equation 2.11. The thrust force of the rotor in Equation 3.1 was calculated by the mean wind speed of the exponential wind profile in Equation 2.11: $\langle V_{\text{w}}(z)\rangle\{z_{\text{H}} - \frac{1}{2}D_{\text{rot}} \leq z \leq z_{\text{H}} + \frac{1}{2}D_{\text{rot}}\}$. Furthermore, the wind drag force per unit length was given through normalization of Equation 3.2 by the hub height z_{H}.

$$q_{\text{W,x}}(z) = \frac{1}{2}\, \rho_{\text{air}}\, D_{\text{T}}(z)\, c_{\text{D}}\, (V_{\text{w}}(z))^{2} \tag{3.3}$$

For the purpose of this loads study, the outer tower diameter $D_{\text{T}}(z)$ is represented by the linear Equation 3.4

$$D_{\text{T}}(z) = D_{\text{T,b}} + \frac{D_{\text{T,t}} - D_{\text{T,b}}}{z_{\text{H}}} z, \tag{3.4}$$

where $D_T(z = 0) = D_{T,b}$ is the tower base diameter and $D_T(z = z_H) = D_{T,t}$ is the tower top diameter, respectively. Usage of Equation 3.4 and integration of the wind line load of Equation 3.3 delivers the accumulated shear force distribution according to

$$
\begin{aligned}
Q_{W,x}(z) &= \int_0^{z_H - z} q_{W,x}(z)\, dz \\
&= \int_0^{z_H - z} \left[\frac{1}{2}\, \rho_{air} \left(D_{T,b} + \frac{D_{T,t} - D_{T,b}}{z_H}\, z \right) c_D \left(V_H \left(\frac{z}{z_H} \right)^{\alpha} \right)^2 \right] dz \\
&= \underbrace{\frac{1}{2}\, \rho_{air}\, c_D \left(\frac{V_H}{z_H^{\alpha}} \right)^2}_{C} (z_H - z)^{2\,\alpha+1} \left[\frac{D_{T,b}}{2\,\alpha + 1} + \frac{D_{T,t} - D_{T,b}}{z_H\,(2\,\alpha + 2)} (z_H - z) \right]
\end{aligned}
\tag{3.5}
$$

Another integration of $Q_{W,x}(z)$ and T delivers the bending moment due to tower wind loads

$$
\begin{aligned}
M_{W,y}(z) &= \int_0^{z_H - z} Q_{W,x}(z)\, dz \\
&= \int_0^{z_H - z} C\,(z_H - z)^{2\,\alpha+1} \left[\frac{D_{T,b}}{2\,\alpha + 1} + \frac{D_{T,t} - D_{T,b}}{z_H\,(2\,\alpha + 2)} (z_H - z) \right] dz \\
&= -C\,(z_H - z)^{2\,\alpha+2} \left[\frac{D_{T,b}}{4\,\alpha^2 + 6\,\alpha + 2} + \frac{D_{T,t} - D_{T,b}}{z_H\,(4\,\alpha^2 + 10\alpha + 6)} (z_H - z) \right] \Bigg|_0^{z_H - z} \\
&= C \left[\frac{D_{T,b}}{4\,\alpha^2 + 6\,\alpha + 2} \left(z_H^{2\,\alpha+2} - z^{2\,\alpha+2} \right) + \frac{D_{T,t} - D_{T,b}}{z_H\,(4\,\alpha^2 + 10\alpha + 6)} \left(z_H^{2\,\alpha+3} - z^{2\,\alpha+3} \right) \right].
\end{aligned}
\tag{3.6}
$$

and due to rotor thrust

$$
\begin{aligned}
M_{T,y}(z) &= \int_0^{z_H - z} T\, dz \\
&= T\,(z_H - z)
\end{aligned}
\tag{3.7}
$$

Application of these equations together with the parameter set from Table 3.3 reveals the magnitude of FA bending moments in a rated steady wind speed situation for different hub heights and rotor diameters.

The thrust coefficient for rated wind speed was calculated according to the 5 MW NREL reference WT [109] and outer tower dimensions were taken from it, likewise. An exponential wind profile with $\alpha = 0.2$ was chosen, with respect to IEC 61400-1 [50].

Figure 3.1 shows the resulting bending moments from Equations 3.6 and 3.7 at $z = 0$ for the tower base in different relations and WT configurations. The top diagram depicts the absolute tower base bending moments for the rated wind speed of $V_r = 11.4$ m/s for different tower heights and rotor diameters. The moments increase with increasing dimensions, as expected. In the middle diagram, tower base bending moments caused by the tower drag are related to moments caused by the rotor thrust. For this parameter set, the tower drag load contribution of 2 % to 11 % is relatively small with respect to the rotor thrust contribution. The bottom diagram relates the bending moments, caused by tower drag from a parking extreme wind situation with $V_{\mathrm{ref}} = 50.0\, m/s$ to the same rated rotor thrust induced moments as in the middle diagram. It stands out, that tower drag in such situations is no more negligible and may cause multiple higher bending moments than in rated production states. For rotatable WT support structures, SS bending moments and their magnitude with respect to FA bending moments should be considered, likewise. This study indicated that a rotatable WT support structure with yaw failure may experience significant side loads in parking extreme wind situations. In case of the 5 MW NREL reference WT parameters with a hub height of $z_{\mathrm{H}} = 87.6$ m and a rotor diameter of $D_{\mathrm{rot}} = 126$ m, a ratio between the extreme wind tower drag induced bending moment and the rated rotor thrust induced moment of $\frac{M_{\mathrm{W,b,y}}(\mathrm{extreme})}{M_{\mathrm{T,b,y}}(\mathrm{rated})} = 0.38$ resulted. Another moment, which acts in the SS plane is the rated generator moment, but it amounted to merely 0.041 MNm and is therefore small compared to the FA moments, presented in Figure 3.1. Other significant SS bending moments may result from ASE load simulations, where skewed wind inflow and dynamic effects can be included.

Symbol	**Value**	**Unit**
V_{r} normal	11.4	$\frac{\mathrm{m}}{\mathrm{s}}$
V_{ref} extreme	50.0	$\frac{\mathrm{m}}{\mathrm{s}}$
ρ_{air}	1.225	$\frac{\mathrm{kg}}{\mathrm{m}^3}$
c_{T}	0.856	–
c_{D}	1.1	–
α	0.2	–
$D_{\mathrm{T,b}}$	6	m
$D_{\mathrm{T,t}}$	3.87	m

Table 3.3: Assumed parameters for tower quasi-steady loads estimation

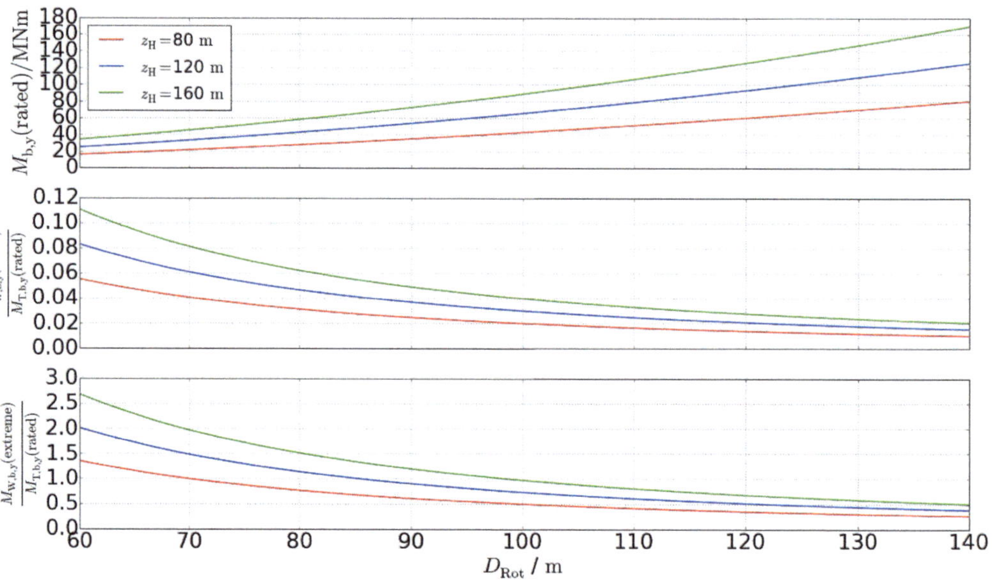

Figure 3.1: Absolute tower base bending moments for different hub heights and rotor diameter (top), tower base bending moments from tower drag relative to bending moments from rotor thrust for rated wind speed (middle), and bending moments from tower drag in extreme wind parking situation relative to bending moments from rotor thrust for rated wind speed (bottom)

3.4 Loads from the Literature

Two of the most relevant scientific reference WTs are the 5 MW NREL [109] and the 10 MW DTU [5]. The performance class of the 5 MW NREL WT is closer to already built WTs, wherefore it has more practical relevance and was selected as reference WT for this work. Unfortunately, the public literature provides not that much information about detailed ASE calculated extreme and fatigue loads for modern WT configurations with the exception of NREL's 5 MW onshore and offshore WT [110], NREL's WindPACT studies about 0.75 MW, 1.5 MW, 3.0 MW WTs [136] and DTU's 10 MW offshore turbine [5]. The most load direction relevant factor for rotatable WT support structures is the bending moment ratio of the moment around the rotor plane parallel axis (FA) divided by the moment around the rotor plane orthogonal axis (SS). Previous listed references provide extreme and fatigue bending moment ratios, whereby fatigue loads are calculated according to Miner's cumulative damage rule [141].

Figure 3.2 shows the summary of bending moment ratios dependent on the rotor diameter for the different WT configurations. Abbreviations ULS and FLS in the legend are representative for the ultimate limit state and fatigue limit state, respectively. Formula symbols P_r and H_{Hub} are equal to the rated power and the hub height of the WT. Unfortunately, the 10 MW DTU documentation has some inconsistencies regarding the tower base bending moments, wherefore the tower top bending moment ratio was entered instead.[1] Life time equivalent fatigue bending moment ratios of the DTU WT are insecure likewise, but they were entered in the case, that the inconsistency exists within a unit conversion fault, which would not affect the ratio itself. Furthermore, each study considered a different set of DLCs, which are summarized within Appendix 14.2. All DLC meanings are presented within subsection 2.2.2. The examined DLCs by DTU are the most comprehensive. In contrast to that, the WindPACT study does only respect a few DLCs. The analysis of the DTU and the analysis of the 5 MW NREL WT show, that the most critical DLC for ultimate loads is the DLC 1.3 with the extreme turbulent wind model. This DLC has not been examined within the WindPACT study, whereby the lower tower base bending moment ratios are explainable.

All presented ratios are > 1, which means that the general tendency for higher loads according to the FA bending axis compared to the SS bending axis can be assumed irrespectively of modern WT configurations. Thus, material saving potential of a rotatable support structure compared to a non-rotatable one is given for different WT configurations.

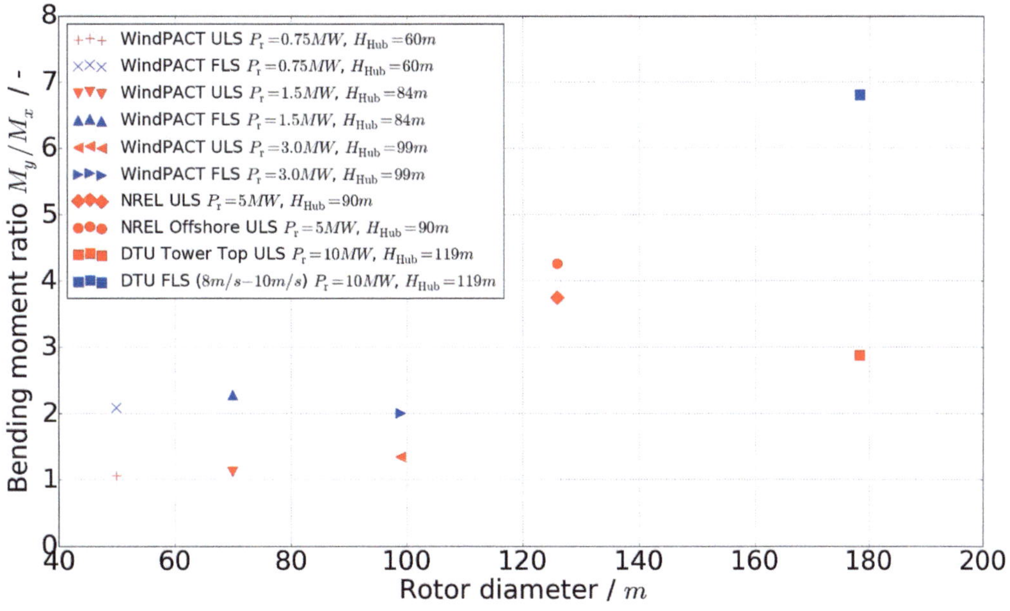

Figure 3.2: Tower base bending moment ratios M_y/M_x according to different references and WT configurations

[1] The tower base extreme bending moments of the 10 MW DTU reference WT are in the same magnitude as the tower top bending moments. Own tower base load estimates according to the given rotor thrust revealed them as being too low. Nevertheless they would result an bending moment ratio of M_y/M_xx = 2.87, which is equal to the ratio of the tower top bending moment ratio.

3.5 ASE Load Analysis of the Reference Wind Turbine

While the previous section presents loads from the literature, in this section the FAST framework, introduced in subsection 2.8.2, was used to perform own loads studies for the non-rotatable reference tower. This first analysis served as validation of the general ASE load simulation procedure and as reference to the later ASE load simulation of the rotatable support structure. Python[2] scripts were written to automatically generate all required input files for the TurbSim turbulent wind field generator, for the IECWind[3] steady wind field generator, and for the FAST environment. Thereby, the parameter settings for the different DLCs could be set by minimizing the probability of typing errors. Moreover, to shorten the ASE simulation time, the analysis was parallelized by distributing the start command of the different simulation jobs over several batch files. Thereby, one batch file for each thread of the computer was created.

3.5.1 Validation of ASE Load Analysis Procedure with DLCs 1.x

ASE load simulations are sensitive to many parameters, whereby inexperienced engineers tend to make mistakes in setting the correct values and to pick the wrong underlying models. For this reason, an existing study was used as basis for further investigations. The previous mentioned study of Jonkman [110] presents load comparisons between the land-based 5 MW NREL reference WT and an offshore floating version of it. The following own loads analysis used all the parameters, given in this study [110] to reproduce the same loads analysis results. This step ensured the correct application of this complex ASE tool. Nevertheless, a brief summary of the most important applied parameter settings, simplifying assumptions, and considered DLCs of the comparative study [110] are presented beforehand.

Modelling Assumptions in the ASE Load Analysis of Jonkman [110]
The study [110] uses the NREL 5 MW Baseline WT, presented in 3.2, as a land-based and as an offshore variant with floating barge. For this work, only the land-based version was of interest. For comparison between the land-based and the offshore version, Jonkman [110] chose an offshore reference site in the northeast of the Shetland Islands of Scotland in collaboration with ITI Energy. Since both versions should experience the same wind conditions for comparability, a power law exponent of $\alpha = 0.14$ was used for the land-based and for the offshore variant. The power law exponent is applied in the underlying power law wind profile, introduced in subsection 2.2.2. Note that extreme wind conditions with a recurrence period of 1 year and 50 years were modelled with $\alpha = 0.11$. The reference 10 min average wind speed V_{ref} and the expected turbulence intensity I_{ref} were assumed according to WT class BI. Values for V_{ref} and I_{ref} may be reviewed from Table 2.3. The air density was chosen with respect to IEC 61400-

[2] The Python(x,y)-2.7.10.0 distribution was used. It is available at `https://python-xy.github.io/`; accessed 21-July-2020

[3] IECWind version 4.0 was used in this work

1 [50] to be $\rho_{\text{air}} = 1.225 \frac{\text{kg}}{\text{m}^3}$. Simplifications in this study were made by the exclusion of earthquakes, wake effects from neighbouring wind turbines, atmospheric pressure variations, and the soil conditions. Thus, the WT was clamped to the ground rigidly. Following DOFs were enabled in the analysis: two flapwise and one edgewise bending-mode for each blade, one drivetrain torsion, one variable generator speed, one nacelle yaw, and two FA and two SS tower bending modes. Scaling of two blade masses, one with +0.5 % and one with −0.5 % of the reference mass was incorporated to mimic a rotor imbalance. Furthermore, blade structural damping ratios of each blade mode were set to 2.5 % of the critical value. Aerodynamic tower drag was neglected, because it was not available in AeroDyn, at that time. The ASE simulations were carried out with FAST glue-code version 6.10a-jmj and AeroDyn version v12.60i-pjm.

DLC Specifications in the ASE Load Analysis of Jonkman [110]
Due to the preliminary nature of the study [110], not all DLCs, demanded by the IEC 61400-1 [50], were analysed. Although it is stated, that fatigue behaviour governs the design of WTs in many cases, fatigue DLCs were neglected. This step is justified, because the study claims to cover only a first step of an iterative design procedure where an initial concept is designed with respect to ultimate loads, before it is checked for fatigue.[4] Start-up and shutdown DLCs (3.x, 4.x, 5.x) as well as the transport, assembly, maintenance, and repair DLCs (8.x) were neglected. Jonkman [110] stated that in his experience with land-based turbines, the omitted DLCs did not dominate the ultimate loads. Consequently, the land-based relevant DLC specifications in Table 3.4 were considered.

In case of statistical extrapolated loads in DLC 1.1, the PSF for loads is 1.25. Jonkman stated that in his and others experience loads increase about 20 % after extrapolation. Therefore, he replaced the extrapolation procedure by scaling the loads of DLC 1.1 with a factor of 1.2 [110, p. 100]. In power production DLCs, the quasi-steady BEM axial-induction model with the Beddoes-Leishman dynamic-stall model were used in AeroDyn. For parking DLCs, these models were disabled because they were unsuitable for high post-stall angles of attack [110, p.97]. Moreover, parking DLCs were modelled with an idling rotor and 90° pitched rotor blades. According to IEC 61400-1 7.4.2 [50], fault situations with the most unfavourable consequences have to be considered. Jonkman [110] decided to simulate a fault in DLC 2.1 where one blade is running suddenly with full pitch rate of 8 °/s to 0°. It was assumed that the WT protection system detects the fault and that the control system performs a shutdown with 0.2 s delay. The shutdown happened through feathering the other two blades to 90°.

[4] see also van der Tempel [189].

Design situation	DLC	Wind condition	Controls / Events	PSF
1) Power production	1.1	NTM $V_{\mathrm{in}} < V_{\mathrm{H}} < V_{\mathrm{out}}$	Normal operation	1.25; 1.2
	1.3	ETM $V_{\mathrm{in}} < V_{\mathrm{H}} < V_{\mathrm{out}}$	Normal operation	1.35
	1.4	ECD $V_{\mathrm{H}} = V_{\mathrm{r}} - 2\ \mathrm{m/s}, V_{\mathrm{r}},$ $V_{\mathrm{r}} + 2\ \mathrm{m/s}$	Normal operation; $\pm\Delta$wind dir	1.35
	1.5	EWS $V_{\mathrm{in}} < V_{\mathrm{H}} < V_{\mathrm{out}}$	Normal operation; ±ver./hor.shr.	1.35
2) Power production plus occurrence of fault	2.1	NTM $V_{\mathrm{H}} = V_{\mathrm{r}}, V_{\mathrm{out}}$	Pitch runaway → Shutdown	1.35
	2.3	EOG $V_{\mathrm{H}} = V_{\mathrm{r}} \pm 2\ \mathrm{m/s}$ and V_{out}	Loss of load → Shutdown	1.10
6) Parked (idling)	6.1	0.95 EWM 50-year recurrence period	Yaw=0°, ±8°	1.35
	6.2	0.95 EWM 50-year recurrence period	Loss of grid $-180° < \mathrm{Yaw} < 180°$	1.10
	6.3	0.95 EWM 1-year recurrence period	Yaw=0°, ±20°	1.35
7) Parked and fault conditions	7.1	0.95 EWM 1-year recurrence period	Seized blade Yaw=0°, ±8°	1.10

Table 3.4: By comparison study [110] considered land-based relevant design load case specifications

In DLC 2.3, loss of the generator load was subordinated and the WT control reacted with a shutdown after 0.2 s delay. This shutdown feathered all 3 blades at full pitch rate to 90°. The generator loss happened at different times during the gust, for each wind speed. The parking DLC 7.1 with fault was simulated with one blade at 0° pitch position and the other two blades at 90° pitch position. Prescribed wind speed ranges, such as $V_{\mathrm{in}} < V_{\mathrm{H}} < V_{\mathrm{out}}$, were discretized by means of 2 m/s steps, starting at 4 m/s and ending with 24 m/s.[5] This wind speed range is likewise required for DLC 2.1, but Jonkman [110] decided to simulate only wind speeds V_{r} and V_{out} because they led to the highest loads in his experience. Yaw misalignments were introduced by prescribing non-zero yaw angles where DLC 6.2 was simulated with a 20° yaw angle discretization step size. All turbulent 3D wind fields were generated with TurbSim [106] v1.20 and an assumed Kaimal wind spectrum. Deterministic wind conditions were generated

[5] See also IEC 61400-1 7.4 [50].

with a customized version of IECWind v5.00 to account for a power law exponent of $\alpha = 0.14$ instead of $\alpha = 0.2$. IEC 61400-1 prescribes in 7.5, that the spatial resolution of the wind field grid, such as shown in Figure 3.3, should be smaller than min $\{0.25\ \Lambda_1, 0.15\ D_{\mathrm{rot}}\} = 10.5$ m. Finally, Appendix 14.4 summarizes the required simulation time periods and its corresponding random seeds to be used for each wind speed in turbulent wind fields. The rotor speed and the pitch angles had to be initialized with respect to their steady state values for a given mean wind speed. Despite such initialization precautions, complete suppression of transient behaviour in the beginning of a computational analysis is difficult. Therefore, each simulation time period was extended by another 30 s. Thus, the first 30 s of an analysis would not be accounted for in the post processing step. Moreover, transient deterministic wind events were initiated at 60 s after simulation start, due to the same reason.

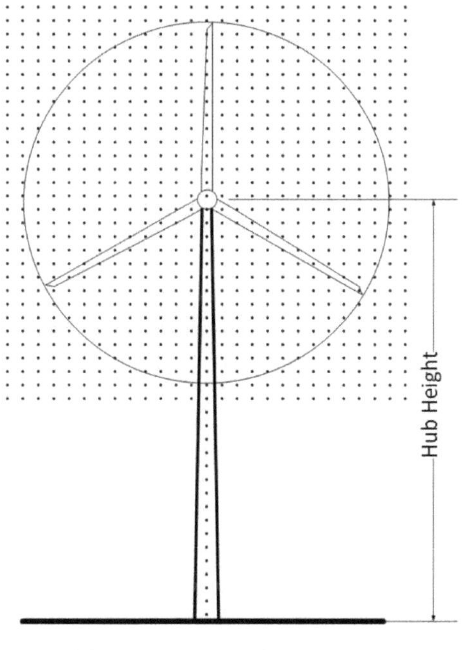

Figure 3.3: Exemplary wind field discretization grid

Validation of own ASE Load Results

Jonkman [110] applied the PSFs in Table 3.4 to the load results of each DLC to weight them properly for comparison. Due to modelling instability problems in [110], not all DLCs of Table 3.4 led to realistic results. Subsection 3.5.2 treats the instability problems in more detail. Jonkman [110] decided to divide the DLCs into two groups: DLCs 1.x and all other DLCs. In the end, he listed only the extreme load tables of the first group because of the uncertainties with the other DLCs. In this validation, the rotor thrust (RotThrust), the rotor torque (RotTorq), the blade root bending moments (RootMxc1 (in-plane), RootMyc1 (out-of-plane)) of blade 1, and the tower bending moments (YawBrMxp,YawBrMyp (at the top about x and y axes), TwrBsMxt,TwrBsMyt (at the bottom about x and y axes)) were considered. Note that the names for the different loads correspond to the load sensor names, defined in the FAST environment. Due to the stochastic nature of ASE load analyses with turbulent wind fields, deviations from the comparison study of Jonkman [110] could be expected. Unfortunately, no further information about the statistical properties, such as the expected value μ or the standard deviation σ of his results with six random seeds per wind speed were available. Haid et al. [90] published information about expected load deviations with respect to the amount of selected random seeds in ASE load analyses of offshore floating WTs. They [90] provide an investigation for the absolute maximum blade root out-of-plane bending moment at 12 m/s mean wind speed in the NTM wind model. This bending moment diverged with 10 random seeds by 3.5 % from the mean of 36 random seeds with 95 % confidence. Similar results could likewise be observed for other loads.

In the present comparison of loads another method was applied for the purpose of validation. The goal was to simulate a larger amount of stochastic sensitive timeseries to show the probability with which Jonkman's [110] results would occur in own load analyses. For this purpose, all 1.x DLCs were simulated with the same settings, as in the comparison study of Jonkman [110]. The only difference was the amount of chosen random seeds for the extreme load relevant wind speeds. For DLC 1.1 with wind speed $V_{\mathrm{H}} = 24$ m/s and DLC 1.3 with wind speeds $V_{\mathrm{H}} = 20$ m/s and $V_{\mathrm{H}} = 24$ m/s, 500 random seeds were simulated instead of 6. These wind speeds were more relevant than the others because according to appendix F in [110], the extreme loads of the most considered sensors occurred within them. Table 3.5 summarizes the results of the comparison study [110] and the own load simulations. Thereby, the first column lists the considered load sensors, the second and third columns contain the calculated extreme loads of each load sensor together with the corresponding DLC and mean hub height wind speed where they occur. Column two contains the results of the comparison study [110] and column three the results of the own simulations. The last column gives information about the unit of the load sensor and its value deviation with respect to the comparison study [110]. A positive deviation indicates, that the own result is higher than the comparison result. Table 3.5 shows, that this was true for each considered load sensor, except for the FA tower base bending moment (TwrBsMyt). The reason for that is discussed later in this subsection. For some sensors, such as the in-plane bending moment of blade 1 (RootMxc1), the FA bending moment at the top of the tower (YawBrMyp), and the SS bending moment at the bottom of the tower (TwrBsMxt) deviations of > 10 % occurred. Furthermore, the driving DLC for the out-of-plane blade bending moment (RootMyc1) changed from 1.4 r+2 to 1.3 V24. Note that r+2 denotes rated wind speed at hub height plus 2 m/s and V24 denotes 24 m/s wind speed at hub height, respectively. All these result differences are justified through the huge difference in the number of considered random seeds. Due to probability, more random seeds in certain DLCs increase their chance to be load drivers.

Therefore, a closer look into the statistical evaluation of extreme loads helped to explain the differences. Figure 3.4 highlights the statistical results for the rotor thrust, the rotor torque, and the blade 1 bending moments. The results are presented in form of the probability density function (PDF)

$$\mathrm{PDF}\left(x, \mu, \sigma^2\right) = \varphi_{\mu,\sigma^2}(x) = \frac{1}{\sqrt{2\,\pi\,\sigma^2}} e^{-\dfrac{(x-\mu)^2}{2\,\sigma^2}}, \tag{3.8}$$

where μ and σ^2 are the expected value and the variance of the extreme loads of all considered 10 min timeseries for load sensor x. In this case the arithmetic mean is the expected value, thus $\mu = \overline{x}$.

Load Sensor	Jonkman [110] DLC	Value	Present DLC	Value	Unit	Deviation %
RotThrust	1.3 V20	1.53e3	1.3 V20	1.57e3	kN	2.62 ± 0.66
RotTorq	1.1 V24	7.84e3	1.1 V24	8.35e3	kNm	6.53 ± 0.13
RootMxc1	1.1 V24	1.11e4	1.3 V24	1.29e4	kNm	16.22 ± 0.97
RootMyc1	1.4 r+2	2.22e4	1.3 V24	2.37e4	kNm	6.66 ± 0.46
YawBrMxp	1.1 V24	8.42e3	1.1 V24	8.89e3	kNm	5.60 ± 0.12
YawBrMyp	1.3 V24	1.51e4	1.3 V24	1.73e4	kNm	14.77 ± 0.71
TwrBsMxt	1.3 V24	4.09e4	1.3 V24	5.73e4	kNm	40.08 ± 0.29
TwrBsMyt	1.3 V20	1.53e5	1.3 V20	1.48e5	kNm	−3.27 ± 0.64

Table 3.5: Extreme loads of the comparison study [110], compared to own results

Each chart in Figure 3.4 shows the result of one load sensor, explained at their ordinate. In the background of the curve the underlying probability distribution is plotted with discrete bars. This is helpful to compare the real distribution with the assumed PDF. Furthermore, μ and the corresponding maximum value of the comparison study are plotted to the PDF curve to show their relative positions. The legend of each chart contains the coefficient of variance and information about the statistical standard deviation distance of the comparison study result with respect to the expected value μ. The PDF curves are cut at the minimum and maximum extreme values, that occur for all random seeds in that DLC and wind speed.

Meaningfulness of the results is given through the high amount of random seeds. This includes some statistical outliers, which occurred in both directions of the blade 1 bending moments (RootMxc1, (RootMyc1)) and in the positive direction of the rotor torque (RotTorq). Especially the maximum out-of-plane blade bending moment (RootMyc1) of the comparison study is close to such an outlier with a distance of 5.8 σ to μ. It was appropriate to investigate the respective timeseries to make sure, that no numerical error caused the outlier. Appendix 14.3 shows the out-of-plane blade bending moment driving timeseries of some important load sensors. Thereby, a dip in the generator power and the pitch angle occurred simultaneous with the extreme load of the out-of-plane bending moment, but no numerical instability was observed. The dip was due to an abrupt drop in the wind speed, which forced the control region to change from 3 to 2. The fast pitch angle adjustment should improve the wind energy conversion efficiency, while higher loads were tolerated. Therefore, this timeseries was considered to be a valid case.

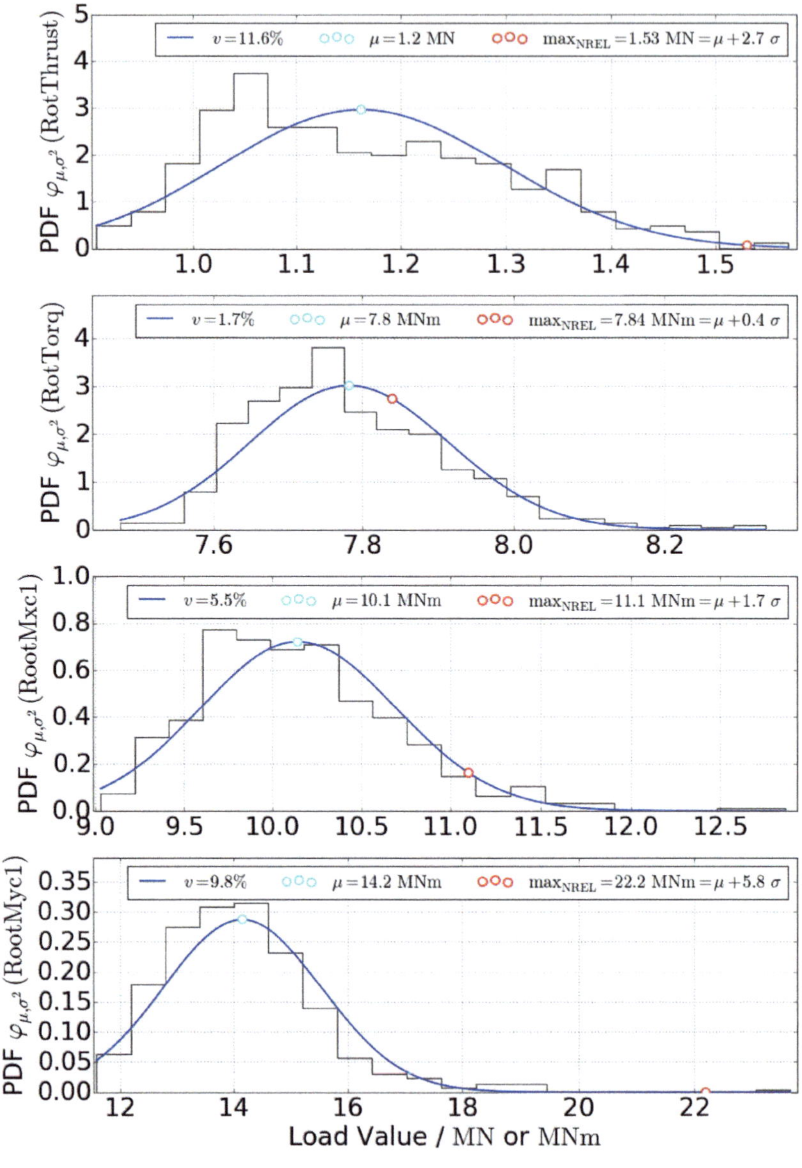

Figure 3.4: Probability density of extreme rotor thrust, rotor torque, and blade 1 bending moments of DLC 1.1 and DLC 1.3, compared to extreme loads of Jonkman [110]. Each PDF chart is based on the simulation of 500 random seeds. Each seed represents a simulation time period of 10 min.

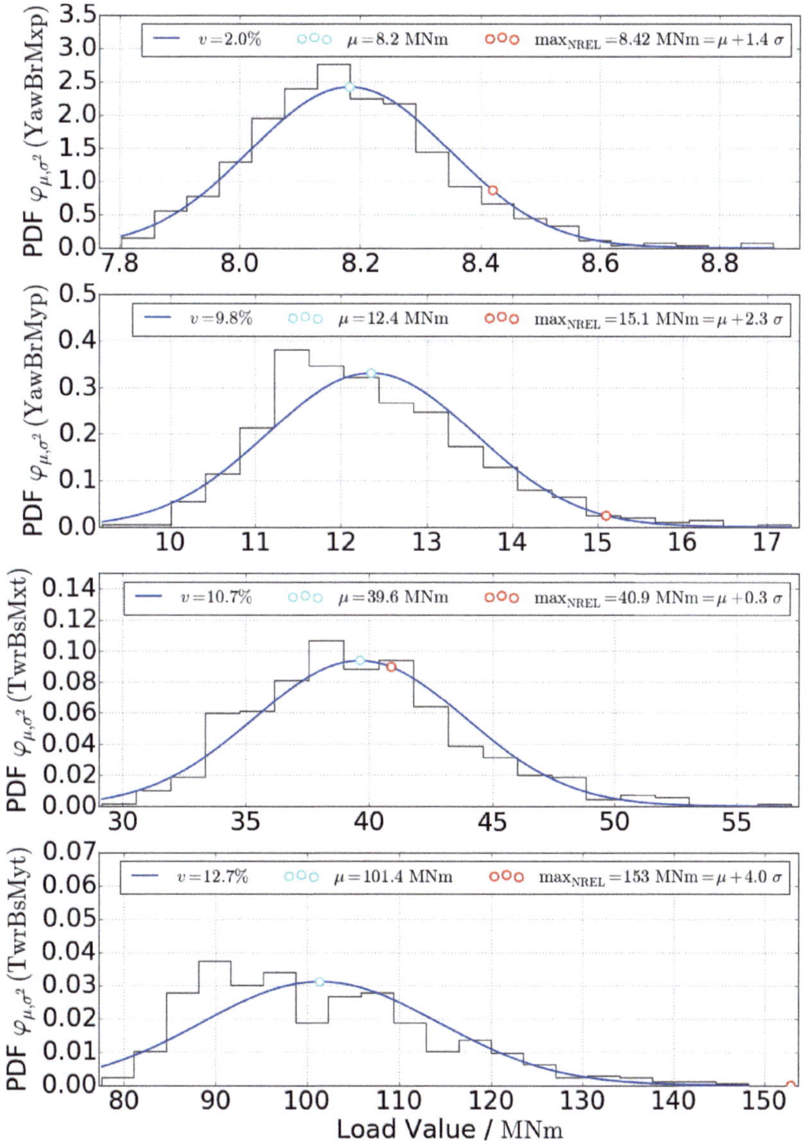

Figure 3.5: Probability density of extreme tower top and tower bottom bending moments of DLC 1.1 and DLC 1.3, compared to extreme loads of Jonkman [110]. Each PDF chart is based on the simulation of 500 random seeds. Each seed represents a simulation time period of 10 min.

Analogous, the statistical results of the extreme tower bending moments are shown in Figure 3.5. Here, only the FA tower base bending moment (TwrBsMyt) shows some significant outliers and the reference result of the NREL comparison study [110] had a distance of 4.0 σ from μ. The maximum value of TwrBsMyt in the own analysis was 1.48e5 kNm and was therefore 3.3 %

lower than the result of the comparison study [110]. Unfortunately, the study [110] gives no information about the numerical stability of the interesting timeseries nor shows the timeseries themselves. This is due to the different scope of Jonkman [110], who compared the land-based and offshore extreme loads with focus on the complex offshore floating barge support. Thus, there are four possible scenarios, which could explain that the own results for the FA tower base bending moment are slightly lower than the results from the comparison study:

1. The present own ASE loads analysis had somewhere different settings than the comparison study.
2. The more recent software versions of FAST, AeroDyn, and TurbSim, used in the own analyses, led to incomparability between the results due to underlying changes in the physical models.
3. The result of the comparison study [110] was a valid statistical outlier.
4. The result of the comparison study [110] could be an invalid statistical outlier due to numerical instability.

To treat scenario 1., Appendix 14.5 shows the main input files, from which the new valid maximum FA tower base bending moment resulted. Showing the corresponding input files is necessary for transparency anyway. The answer to scenario 2. is, that according to the AeroDyn change log [26], no changes to the physical models were done between the relevant software versions, which would lead to the load differences. If more recent options were available, the settings in AeroDyn and TurbSim were set such that they mimic the settings of the comparison study. For example the aerodynamic tower drag CalcTwrAero was disabled and the ScaleIEC parameter in TurbSim was set to 0. Jonkman provided access to the timeseries, which was responsible for his extreme FA tower base bending moment. It is shown together with own results in Appendix 14.6 and reveals, that no numerical instability caused the extreme load in the comparison study [110] or in the own extreme load timeseries. The extreme FA tower base bending moments were rather caused by the change of the control region, such as for the extreme out-of-plane blade bending moment (RootMyc1), mentioned earlier in this subsection. Thus, scenario 4. could be rejected. Furthermore, the FA tower base bending moment arithmetic means and their standard deviation suggests that the own load simulation parameters were set correctly. Although the comparison studies extreme load was about 3.3 % larger than the corresponding extreme of the own ASE analysis, the arithmetic mean and the standard deviation of the comparison studies timeseries were about 3.6 % smaller. These comparable small differences would not play a crucial role for the goal of this work and all important input files are disclosed in Appendix 14.5 for the review of following researchers. The similarity of own results compared to the comparison study supports that the important model settings were chosen correctly. Therefore, the own load simulation procedure is considered to be validated.

3.5.2 Blade-Stall-Induced Vibrations in Parking DLCs

As explained in subsection 3.5.1, Jonkman [110] avoided the presentation of extreme loads for other DLCs than 1.x, but especially for rotatable support structures, other DLCs than 1.x may be relevant. In the comparison study, idling DLC 6.2 showed an unreasonable extreme load driving SS tower instability between certain yaw misalignments. Note that DLC 6.2 treats a loss of electrical power grid in extreme wind speed parking situation, wherefore each yaw misalignment $-180° <$ Yaw Error $< 180°$ was possible. The mentioned instability could likewise be reproduced in own ASE simulations. In DLC 6.2 of IEC 61400-1 [103], an extreme wind speed $V_{e50} = 50$ m/s with recurrence period of 50 years and in DLC 6.3 an extreme wind speed $V_{e1} = 40$ m/s with recurrence period of 1 year is proposed for wind turbine class I. In these cases, the WT rotor idles freely with all blades being feathered to 90 deg. Own experience indicated, that idling of the WT rotor led to smaller loads, than a fixed parked rotor.

Jonkman [110, p. 118] stated in the comparison study that it was difficult to determine, whether the instability was of physical nature or was an artefact of modelling assumptions. He supposed that the amount of structural damping in the first SS tower bending mode was exceeded by the rotors energy absorption. Skrzypinski confirmed this behaviour in his dissertation [172, p. 22] and related it to negative damping for the first edgewise blade mode in parking DLCs with specific yaw misalignments. Figure 3.6 shows the maximum, mean, and the minimum rotor thrusts (RotThrust) and edgewise bending moments (RootMxb1) of DLC 6.2 for different yaw misalignments between -180 deg $\leq$ Yaw Error $\leq$ 180 deg of the own analysis. This time, turbulent wind fields were used. The mean of the standard deviations of six simulations with different random seeds is likewise plotted and indicates on which yaw misalignments instabilities occur.

In this analysis, yaw misalignments of -40 deg $\leq$ Yaw Error ≤ -10 deg and 10 deg $\leq$ Yaw Error $\leq$ 30 deg highlighted significant instabilities. Moreover, Skrzypinski [172, p. 22] associated the negative damping in the edgewise blade mode to stall-induced vibrations. Bertagnolio et al. [11] stated that airfoils under stall conditions experience large fluctuations of aerodynamic forces, caused by trailing edge shear flow instabilities, airfoil surface turbulent boundary layer vortices, vortex shedding, and unsteady detached flow over the airfoil. This phenomenon can be observed within the FAST and the HAWC2 ASE codes, but Skrzypinski [172, p. 54] concluded that in state-of-the-arts ASE codes, blade standstill vibrations in deep stall are probably modelled inaccurately. He suggested that low amounts of temporal lag in the aerodynamic response would already lead to significantly increased aerodynamic damping, which would prevent the unstable behaviour. Based on these and other experts experience[6], it was not sure if the instabilities would occur in the physical world. Therefore, ASE calculations of critical yaw misalignments between -40 deg $\leq$ Yaw Error ≤ -10 deg and 10 deg $\leq$ Yaw Error $\leq$ 40 deg in parking DLCs were discarded in this work. In case that future findings confirm the instabilities as being real in a physical sense, the rotatable WT support structure, developed in this work, must be checked again for these yaw misalignments in extreme wind speed conditions.

[6] See a discussion about this topic by the experts internet community forum of the National Wind Technology Center (NWTC) of NREL: `https://wind.nrel.gov/forum/wind/viewtopic.php?f=3&t=1219&p=5270`; last access on 5. March 2018

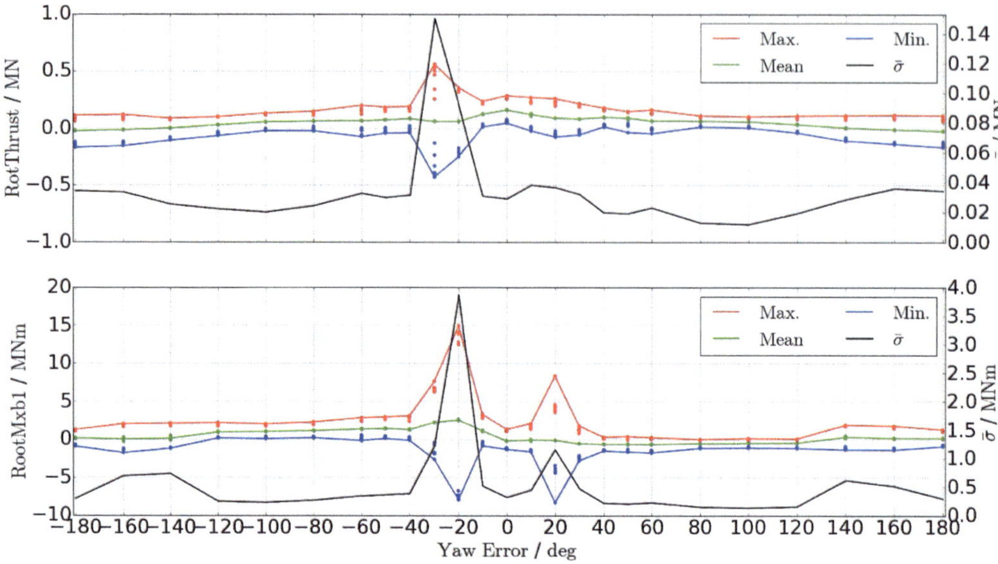

Figure 3.6: DLC 6.2 maximum, mean, minimum, and standard deviation of the rotor thrust and blade one edgewise bending moment over -180 deg $\leq$ Yaw Error $\leq$ 180 deg

3.5.3 Loads Analysis of all Considered DLCs

The design of a rotatable WT support structure required a load assumption for the first design iteration. To get a guess of these loads, an ASE load simulation of all ultimate limit state DLCs, listed in Table 3.4, was performed with the non-rotatable reference WT with conventional tubular steel tower, such as described in subsection 3.2. Thereby, Table 3.4 lists all DLCs, which were considered for the following design of the rotatable WT support structure. In general, the same model settings and DLCs as in the comparison study [110] were applied, but with the following differences.

Additional considered DLCs in this work were fatigue DLCs 1.2, 3.1, 4.1, and 6.4, start-up DLCs 3.2 and 3.3, shut-down DLC 4.2, and emergency shut-down DLC 5.1. Note that fatigue DLCs, marked with F in the PSF column, are contained in Table 3.4, but not accounted for in this first iteration step extreme load analysis. Fatigue DLCs were considered in the final rotatable tower design instead. Different initiation times of the WT actions (start-up or shut-down) at the transient wind events in DLC 3.2, 3.3, and 4.2 helped to find the most unfavourable combination. In EOG wind events, the WT action starts at $\left\{0, \frac{1}{5}\,T_{\mathrm{EOG}}, \frac{2}{5}\,T_{\mathrm{EOG}}, \frac{3}{5}\,T_{\mathrm{EOG}}, \frac{4}{5}\,T_{\mathrm{EOG}}, T_{\mathrm{EOG}}\right\}$ and in EDC it starts at $\left\{0, \frac{1}{3}\,T_{\mathrm{EDC}}, \frac{2}{3}\,T_{\mathrm{EDC}}, T_{\mathrm{EDC}}\right\}$. According to subsection 2.2.2, $T_{\mathrm{EOG}} = 10.5$ s and $T_{\mathrm{EDC}} = 6$ s. Start-up actions were modelled such that the fully feathered blades (90°) pitched to a pitch angle, which corresponded to the current mean wind speed at full pitch rate. The pitch angle to wind speed relation is listed in Table 7-1 of [109, p. 23]. Note that this relation was likewise used to initialize the collective pitch angles for other DLCs.

Design situation	**DLC**	**Wind condition**	**Controls / Events**	**PSF**
1) Power production	1.1	NTM $V_{in} < V_H < V_{out}$	Normal operation, Yaw Error = 0°, ±8°	1.25 1.2
	1.2	NTM $V_{in} < V_H < V_{out}$	Normal operation	F
	1.3	ETM $V_{in} < V_H < V_{out}$	Normal operation, Yaw Error = 0°, ±8°	1.35
	1.4	ECD $V_H = V_r - 2$ m/s, V_r, $V_r + 2$ m/s	Normal operation, ±Δwind dir, Yaw Error = 0°, ±8°	1.35
	1.5	EWS $V_{in} < V_H < V_{out}$	Normal operation, ±ver./hor.shr., Yaw Error = 0°, ±8°	1.35
2) Power production plus occurrence of fault	2.1	NTM $V_H = V_r, V_{out}$	Pitch runaway → Shutdown	1.35
	2.3	EOG $V_H = V_r \pm 2$ m/s and V_{out}	Loss of load → Shutdown	1.10
3) Start	3.1	NWP $V_{in} < V_H < V_{out}$		F
	3.2	EOG $V_H = V_r \pm 2$ m/s, V_{out}		1.35
	3.3	EDC $V_H = V_r \pm 2$ m/s, V_{out}		1.35
4) Stop	4.1	NWP $V_{in} < V_H < V_{out}$		F
	4.2	EOG $V_H = V_r \pm 2$ m/s, V_{out}		1.35
5) Emergency Stop	5.1	NTM $V_H = V_r \pm 2$ m/s, V_{out}	Yaw Error = 0°, ±8°	1.35
6) Parked (idling)	6.1	EWM 50-year recurrence period	Yaw Error = 0°, ±8°	1.35
	6.2	EWM 50-year recurrence period	Loss of grid −180° < Yaw Error < 180°	1.10
	6.3	EWM 1-year recurrence period	Yaw Error = 0°	1.35
	6.4	NTM $V_H < 0.7\ V_{ref}$ recurrence period		F
7) Parked and fault conditions	7.1	EWM 1-year recurrence period	Seized blade Yaw Error = 0°, ±8°	1.10

Table 3.6: Considered design load cases in this work, derived from IEC 61400-1 [50]

For shut-down actions, all blades pitched to the fully feathered position (90°) at full pitch rate and in the emergency case (DLC 5.1), the high speed shaft brake deployed additionally. Furthermore, additional yaw misalignments in DLC 1.1, 1.3, and 5.1 of ±8° were accounted for. For parking DLCs 6.2 and 6.3, the critical yaw misalignments of ±20° (6.2), ±30° (6.3), and ±40° (6.2) were discarded. This was due to the modelling uncertainty, discussed in subsection 3.5.2. From then on, tower drag was included in the ASE simulations, by enabling the AeroDyn switch CalcTwrAero and by using the new tower aerodynamics input file, shown in Appendix 14.7.

This time, some adjustments to the parameters should shift the ASE load analysis to a more realistic land-based one, such as required by the IEC 61400-1 [50]. The amount of random seeds per each wind speed in normal operation conditions was set to six. For the 6.x and 7.x DLCs of offshore WTs, six 1 h simulations per random seed are required by the IEC 61400-3 [102]. The design standard for land-based WTs, IEC 61400-1 [50], requires six random seeds, each with only 10 min simulation time. Therefore, the factor of 0.95 for DLCs 6.x and 7.x in Table 3.4 could be set to 1.00. Furthermore, the power law exponent was set to the land-based value of $\alpha = 0.2$ for all DLCs, except for the parking ones, where $\alpha = 0.11$. All extreme load relevant DLCs summed up to 1082 different ASE simulation files.

Figure 3.7 compares the extreme tower base SS (TwrBsMxt), FA (TwrBsMyt), and torsional (TwrBsMzt) moment results of all DLCs with tower drag (CalcTwrAero = True) and without (CalcTwrAero = False). Additionally, the extreme rotor torque (RotTorq) values are provided for the following discussion about different load influencing control settings. However, the red and blue bars are considered first. In the most cases, tower drag had no significant impact to the extreme tower base bending moments, except for the parking DLCs 6.x and 7.x. This was expected due to the high wind pressure on the tower for the extreme wind speed with 1 year and 50 year recurrence period probability. Considering the SS tower base bending moment of its driving DLC 6.2, a load increase of 20.1 % resulted from enabling the tower drag. On the other hand, this made the blade aerodynamics responsible for ≈ 79.9 % of the extreme SS tower base bending moment in this case. As expected, the tower drag had no visible influence to the extreme torsional moment at the tower base, but the moment was noticeable high in DLC 2.1.

To reduce this outstanding extreme torsional moment, the pitch rate was changed from 8 deg/s to 2 deg/s for any transient pitch maneuvers in DLCs 2.x, 3.x, 4.x, and 5.1. Neither IEC 61400-1 [50], DNV GL [64] nor DIBt [152] prescribe minimum pitch rates or maximum start-up and shut-down times. Note that this valid control adjustment did not affect the normal production pitch rates. Another control adjustment was introduced to reduce the extreme SS tower base bending moment in extreme wind parking situations. In this study, the worst case for the SS tower base bending moment was electrical power grid loss in extreme wind conditions in DLC 6.2, because it was assumed that the yaw system fails and the WT is attacked by the wind potentially from all directions. Therefore, instead of an idling rotor, the rotor was parked with one blade pointing vertically upwards in DLCs 6.x and 7.1 in this new control approach. Furthermore, it was assumed, that a hydraulic aggregate provides enough energy to adjust the pitch angle of the single blade at top position with respect to the wind direction. Note that the participating measurement and control systems must also have enough energy, potentially provided by a battery for the storm hours. The demanded pitch angle $\Theta_{\mathrm{b},1}$ for blade one at the top position calculates as function of the wind direction WndDir as follows

$$\begin{aligned} \Theta_{\mathrm{b},1} &= 90\ \mathrm{deg} - \mathrm{WndDir} && \text{for}\quad \mathrm{WndDir} >= -90\ \mathrm{deg} \\ \Theta_{\mathrm{b},1} &= -270\ \mathrm{deg} - \mathrm{WndDir} && \text{for}\quad \mathrm{WndDir} < -90\ \mathrm{deg} \end{aligned} \tag{3.9}$$

where the other two blades stay at feathered 90 deg position for all wind directions.

The magenta and cyan bars in Figure 3.7 show, that the intended extreme tower base SS bending moment and torsional moment reductions were achieved by the previous discussed control adjustments. Thereby, the extreme SS tower base bending moment with tower drag in DLC 6.2 was reduced about 57.6 %, so that DLC 2.1 became its new driving DLC. Here, the extreme SS tower base bending moment increased slightly about 1 % due to the reduced pitch rate. On the other hand, the reduced pitch rate caused a significant extreme tower base torsional moment reduction of 24 %. Note that the rotor torque in parking situations increased for parked rotors compared to idling ones. This did not affect the drive train loads, because of the locked rotor, but requires a closer look to the hub and lock system design to ensure their sufficient strength. Appendix 14.8 shows the corresponding extreme blade root moments, where the changed controls had no significant impact. For transient start-up and shut-down events, more sophisticated control algorithms may help to reduce the resulting extreme loads further. Nevertheless, such control system improvements were out of scope for this work.

Table 3.7 summarizes the corresponding extreme load results for the load sensors, mentioned in subsection 3.5.1. Thereby, column 1 contains the load sensor name, column 2 declares if it is a minimum or maximum, column 3 lists the file name of the timeseries where the corresponding extreme load occurred, column 4 contains the load value, and column 5 its unit. The first numbers in the file names declare the respective load case and an attached T marks files with turbulent wind field. Following numbers give information about the wind direction with respect to the WT rotor or the turbines action time at transient wind events, the mean wind speed, and in case of stochastic sensitive DLCs, the random seed number.

The resulting extreme loads of each load sensor from all considered DLCs were larger compared to the validation case in Table 3.5, where only DLCs 1.x were accounted for. Thus, the 1.x DLCs seemed to have negligible influence to the extreme loads, but WT shut-down events and parking DLCs dominated the extreme loads of this load sensors, instead. The analysis of all DLCs revealed, that the difference between the absolute extreme SS and FA tower base bending moment was smaller than in the reference result. Figure 3.2 and Table 3.5 may be reviewed for comparison. The absolute TwrBsMyt/TwrBsMxt = 1.72 ratio was small and indicated, that there was not as much material saving potential as suggested by the literature research in Figure 3.2. Even if the new SS bending moment driving DLC 2.1 results could be questioned in terms of the applied control settings, they provided an assumption on the safe side for the design of the rotatable support structure. This assumption was due to the material saving potential, which results from different tower cross sectional stiffness with respect to its load direction, such as mentioned in subsection 4.3.1. A comprehensive summary of important load sensor extreme loads is given in Appendix 14.9.

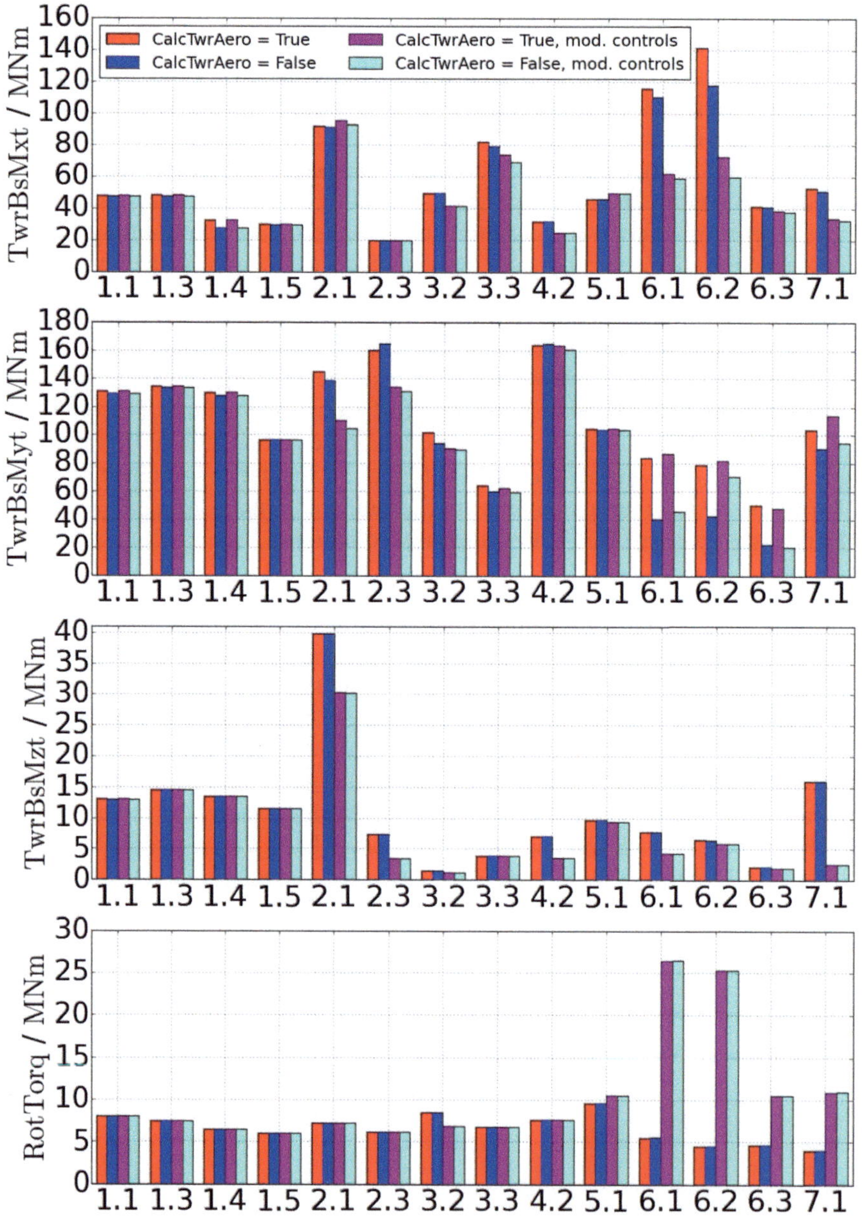

Figure 3.7: Absolute extreme tower base moment and rotor torque comparison between all ultimate limit state design load cases from the ASE analysis of the NREL reference wind turbine with conventional tower. Results are presented with and without tower aerodynamics and with applied changes to the wind turbine controls in certain DLCs.

Parameter	**Type**	**File Name**	**Load**	**Unit**
RotThrust	Min	Dlc_23_6.3_Vout.out	−7.598e2	kN
RotThrust	Max	Dlc_42_6.3_Vr.out	1.721e3	kN
RotTorq	Min	Dlc_61T_-08deg_S5.out	−2.646e4	kNm
RotTorq	Max	Dlc_62T_+160deg_S2.out	2.533e4	kNm
RootMxc1	Min	Dlc_61T_-08deg_S3.out	−8.897e3	kNm
RootMxc1	Max	Dlc_21T_+08deg_Vout_S6.out	1.580e4	kNm
RootMyc1	Min	Dlc_23_6.3_Vout.out	−1.091e4	kNm
RootMyc1	Max	Dlc_14_-08deg_-_Vr+2.out	2.588e4	kNm
YawBrMxp	Min	Dlc_61T_-08deg_S5.out	−2.583e4	kNm
YawBrMxp	Max	Dlc_62T_+160deg_S2.out	2.473e4	kNm
YawBrMyp	Min	Dlc_21T_+08deg_Vout_S6.out	−2.984e4	kNm
YawBrMyp	Max	Dlc_21T_-08deg_Vout_S6.out	2.899e4	kNm
TwrBsMxt	Min	Dlc_21T_+08deg_Vout_S6.out	−9.522e4	kNm
TwrBsMxt	Max	Dlc_33_6.0_+_Vout.out	7.378e4	kNm
TwrBsMyt	Min	Dlc_42_8.4_Vout.out	−8.591e4	kNm
TwrBsMyt	Max	Dlc_42_6.3_Vr+2.out	1.636e5	kNm

Table 3.7: Extreme loads of the NREL reference wind turbine for all considered DLCs

4 Load-Direction-Derived Support Structures

4.1 Definition of Load-Direction-Derived

In the past, WT tower design has been reduced to the optimization of tower diameters and their wall thickness. The expression load-direction-derived (LDD) is a general designation for the extension of the existing design procedures by not only considering the magnitudes, but likewise the directions of loads. Due to the ability of towers to rotate, a LDD design is achievable for WT support structures, such as mentioned in section 2.9. This statement is backed from the different load magnitudes with respect to their direction, such as attested in subsection 3.4. Industrial examples are presented in subsections 2.10.3 and 2.10.5 where mobile crane booms are designed according to a pressure and a tensile side of the beam. This special design stems from the fact, that mobile crane booms experience bending moments mainly about one axis. In the case of rotatable shortwave antennas, a LDD is given through different leg distances of the lattice main shaft, which lead to different bending stiffness in each direction. The term LDD design is not sufficient to provide a comprehensive description for the design of rotatable WT support structures, because other aspects, such as transport, manufacturing, blade tip to tower clearance or even aesthetics of the external appearance may also play crucial roles in finding the optimal tower concept. For convenience, the term LDD should serve as a synonym for rotatable WT support structures in this work.

4.2 Yaw System and Foundation

Since the development of a yaw system and a corresponding foundation for a rotatable WT tower was out of scope for this work, only a few thoughts and ideas about it are documented in this section to provide a starting point for following research activities in this area. Thereby, the first subsection discusses the general aspects and requirements to such a yaw system and the second proposes some reasonable concepts.

4.2.1 Challenges and Requirements

The first obvious difference of a yaw system at the bottom of a WT compared to its conventional position are the occurring loads. The base of a WT tower experiences huge bending moments due to the rotor thrust force and tower wind drag forces with their long lever arms. Additionally, larger shear forces and axial forces can be expected, whereby the later result from the self-weight of the tower which adds to the RNA weight. Furthermore, such a high loaded yaw system must not only withstand the loads, but it must simultaneously ensure its function for all required operational conditions over the lifetime. Functionalities are the ability to rotate, transfer of all loads to the foundation, and ensuring the defined tower alignment. Since the friction moment in the yaw bearing depends on the tower base loads, it is challenging to control the occurring yaw accelerations such, that gyroscopic loads become not too big[1]. Another difficulty arises with respect to the manufacturing tolerances of the components. Assuming a connection between the WT foundation with wide civil engineering tolerances and the yaw system with its narrow mechanical engineering tolerances requires a good transition strategy between both worlds. This aspect is likewise given for the conventional tower top flange to yaw bearing connection, but for concrete foundations wider tolerances are used than for a tower flange.

However, a yaw system at the tower base has also advantages compared to the conventional solution. It has a good accessibility for inspection, maintenance, and the exchanging of defect parts. The later requires an appropriate concept where the dismantling of yaw system parts is possible without taking down the whole WT. A review of the rotatable shortwave antenna in subsection 2.10.5 shows, that such concepts already work for a failed roller bearing beneath a huge rotatable structure. Therefore, it is likewise possible for WTs. This maintenance aspect is very important, because it reduces costs over the lifetime of a WT. Furthermore, a yaw system at the base of a tower can be mounted within the foundation, save thereby foundation material, and provide a good supporting environment, such as suggested in the next subsection.

4.2.2 Concepts

A good starting point for WT tower base yaw systems are the concepts, which are shown in Figure 2.34 and discussed in subsection 2.9.2. For huge bending moments, other concepts with distributed radial forces may become more appropriate. This means that a global static system, such as shown in Figure 4.1, is introduced where the radial forces F_B and F_C decrease with increasing axial support distance L_1. A simple static equilibrium, such as stated in Equation 4.1, shows that the radial forces can be decreased by two orders of magnitudes with $L_1 \leq 2$ m for the example in Figure 4.1. Note the logarithmic scalc of thc ordinate axis.

$$\begin{aligned} \sum_{F} &= 0 = T + F_B - F_C \\ \sum_{M_B} &= 0 = F_C\, L_1 - T\,(L_1 + z_H) \end{aligned} \tag{4.1}$$

[1] Gyroscopic loads and their dependency to yaw accelerations are explained in subsection 6.2.4

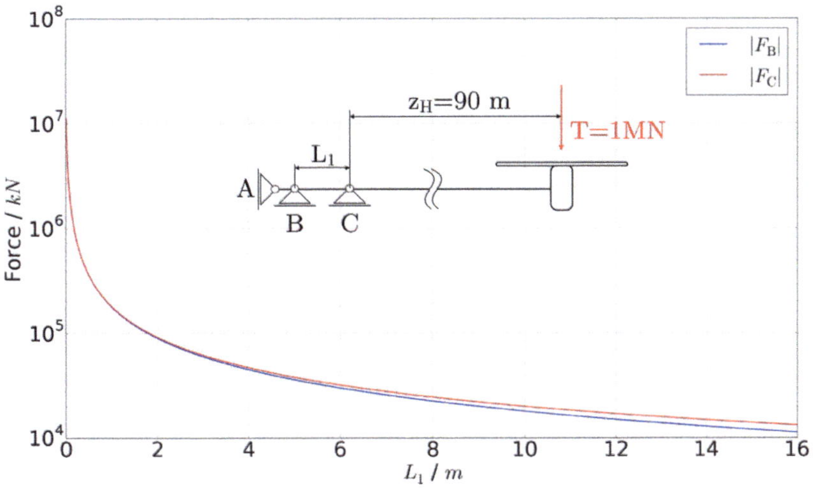

Figure 4.1: Tower base yaw bearing concept with distributed radial forces

This idea is already mentioned within concept c in Figure 2.34 and by Steel Pro Maschinenbau GmbH in Figure 2.29. Furthermore, an expertise survey about the concept of a yaw bearing with distributed radial forces and its feasibility is provided in Appendix 14.10. Two other concepts are shown in Figure 4.2. The left is representative for towers with shell structures, such as circular, elliptical, or airfoil cross sections and the right shows a similar solution for rotatable lattice structures. In both cases, two carrier plates[2] hold radial slide pads to lead the radial forces into the circular sidewall of the foundation. Additionally, axial slide pads are mounted on the bottom to lead the axial forces into the foundation floor. In case of the left variant, the divided carrier plates are connected to each other by bolted flanges and are attached to the tower by a grouted joint connection with shear keys. In both cases in Figure 4.2, slide pads can be exchanged individually. The required yaw drives for yaw motion may be placed at different positions, but one appropriate position would be next to the upper carrier plate. Thereby, yaw drives can be mounted on the foundation and provide good accessibility.

Another approach tries to avoid the axial slide pads through buoyancy. Thereby, the foundation hole contains a liquid and the tower is axially supported by a swimming balloon, such as shown in Figure 4.3. The balloon is held within a cage, which has enough structural integrity to lead the radial forces over slide pads into the foundation, such as it was suggested for the carrier plates in the previous concepts. A swimming concept saves the difficult to reach axial slide pads and provides enough lubrication at all sliding parts. On the other hand, it is challenging to find an appropriate liquid. It should not freeze at common cold temperatures, corrosion of structural and mechanical parts must be avoided, low friction coefficients at the sliding contact surfaces should be obtained, it should be cheap, and environmental friendly. Furthermore, evaporation of the liquid should be stopped or at least being controlled, whereby missing liquid must be refilled. Figure 4.3 shows two possible directions for the balloon: (1) with smaller

[2] Note that other ideas with large diameter cylindrical tubes instead of carrier plates tend to buckle and to ovalize.

diameter and larger draft and (2) with larger diameter and smaller draft. As shown in the previous concept, higher drafts would lead to reduced radial loads, but in case of very large tower base dimensions, a smaller balloon draft may replace sufficient liquid to carry the WT and must hold a lower gas pressure.

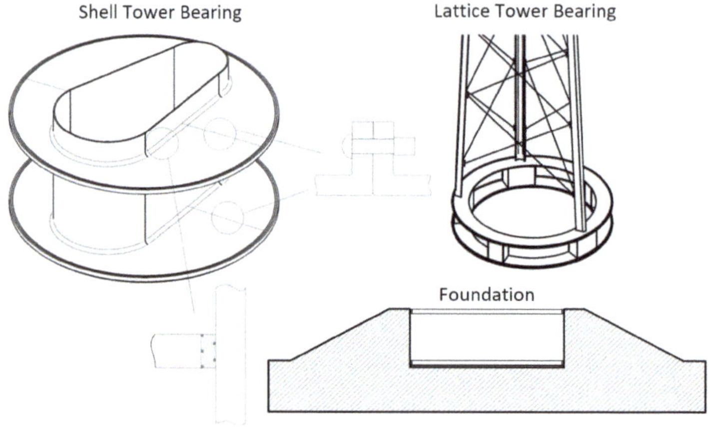

Figure 4.2: Tower base yaw bearing concepts - one concept for towers with shell structures (left) and one concept for towers with lattice structures (right)

A swimming yaw bearing concept mimics a floating offshore foundation, but with better radial support conditions and without currents and high waves. However, additional dynamics through buoyancy together with temperature driven changes in the liquid viscosity and the gas pressure have to be investigated carefully with respect to the operational conditions and requirements of the WT.

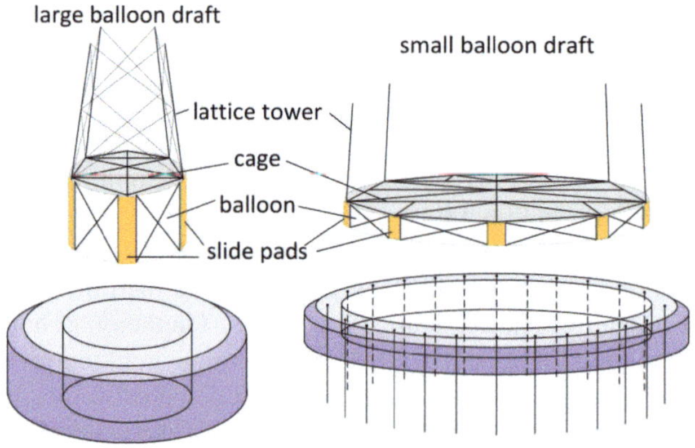

Figure 4.3: Swimming yaw bearing concept with large balloon draft (left) and small balloon draft (right)

4.3 Requirements and Structural Aspects

The favoured concept should stand out with causing low CO_2-emissions throughout the whole production chain and with cost efficiency. CO_2-emissions can be influenced by the kind and amount of used material. Considering $2.8\frac{\text{t CO}_2}{\text{t steel}}$ [148] shows the tremendous contribution to carbon dioxide emissions, which hundreds of tons weighing steel WT support structures have. Cost efficiency should be considered in terms of its fixed and variable costs over lifetime. Fixed costs for towers are highly related to the amount of material, the complexity of used structural members, and its joints. Variable costs are mainly subject to maintenance costs. Tower maintenance costs depend on the kind and amount of joints, corrosion resistance and fatigue behaviour. At this point the structural mass stands out as major factor to ecological and economical properties of a WT support structure. Unfortunately, it is difficult to get reliable cost values, which account for the complexity of structural members and joints. Nevertheless, complexity of members and joints must be considered besides the major interest of developing a material efficient concept. Beyond that, it must account for the ultimate, fatigue, and serviceability limit states, such as described in subsections 2.2.5, 2.2.6, and 2.2.8. Before the discussion about different LDD support structure concepts starts, the following subsections summarize some influencing aspects in general, such as the cross sectional strength, the global stability, and the dynamic response.

4.3.1 Cross Sectional Strength

Most relevant tower load components are the global axial force in combination with the bending moment, because they lead to extreme normal stresses within the wall of tubular towers or the legs and bracing members of lattice towers. In terms of fatigue, the global vibrating bending moments cause the main stress amplitudes, while the self-weight-caused axial force and bending moments, induced by the mean rotor thrust are responsible for the mean stress level in the structural details. As shown within section 3.4, extreme and damage equivalent fatigue bending moments differ according to the FA and SS axis. In terms of tubular shell towers, this behaviour can be exploited for material savings regarding bending stresses. Reduced SS bending moments M_x, compared to the FA bending moments M_y require smaller section moduli $W_\mathrm{x} = \frac{I_\mathrm{x}}{a_\mathrm{max}}$ than $W_\mathrm{y} = \frac{I_\mathrm{y}}{a_\mathrm{max}}$ to satisfy the same bending normal stresses, whereby material can be saved. Consequently a smaller cross sectional area A will increase the normal stresses, caused by the axial force. This effect is assumed to be moderate, because reduced material mass will likewise lead to smaller axial forces.

4.3.2 Global Stability

Stability failure modes of a structure can be distinguished between the global bending and flexural-torsional buckling, and local shell and plate buckling, respectively. Each of these phenomena has to be considered for rotatable WT support structures. Regarding subsection 4.3.1, a LDD tower can have different area moments of inertia for each bending axis. This means, that the ideal Euler buckling load corresponds to the decreased (SS) bending stiffness and global stability becomes lower compared to a conventional steel tower. Another failure mode is the flexural-torsional buckling. It describes the stability collapse of a beam, which tries to evade bending caused by a shear force. Therefore, the beam profile rotates around the torsional axis especially when the torsional stiffness is very low. Richard and Sander [161] propose an equation to determine the maximum shear force F_{FTB} for a beam until flexural-torsional buckling begins. Thereby, it considers the interaction between bending and torsional stiffness according to

$$F_{\mathrm{FTB}} = \beta_{\mathrm{FTB}} \frac{\sqrt{E\, I_{\min}\, G\, I_{\mathrm{T}}}}{L^2}, \tag{4.2}$$

where $\beta_{\mathrm{FTB}} = 4.013$ and $I_{\min} = \min\left[I_{\mathrm{x}}, I_{\mathrm{y}}\right]$ for cantilever beams with a constant cross section along its length L. The maximum sustainable shear force in equation 4.2 is sensitive to the expression $\sqrt{E\, I_{\min}\, G\, I_{\mathrm{T}}}$, which shows that the SS bending stiffness together with the torsional stiffness drive the flexural-torsional buckling resistance for rotatable support structures. More information about analytical and numerical buckling analyses of column like structures is provided in subsection 2.2.5.

4.3.3 Structural Dynamics

Structural dynamics treat the behaviour of a structure in the time or frequency domain. This dynamic behaviour is primarily determined by the structure's natural frequencies[3] and mode shapes. Natural frequencies and their mode shapes depend on the stiffness distribution, mass distribution and the damping of a structure. They are expressed in a coupled form for each degree of freedom (DOF) within the system stiffness matrix $\underline{\underline{K}}_{\mathrm{sys}}^{N\times N}$, the system damping matrix $\underline{\underline{D}}_{\mathrm{sys}}^{N\times N}$ and the system mass matrix $\underline{\underline{M}}_{\mathrm{sys}}^{N\times N}$ with N as the number of DOFs. The equation of motion

$$\underline{\underline{M}}_{\mathrm{sys}}\, \underline{\ddot{x}}(t) + \underline{\underline{D}}_{\mathrm{sys}}\, \underline{\dot{x}}(t) + \underline{\underline{K}}_{\mathrm{sys}}\, \underline{x}(t) = \underline{F}(t) \tag{4.3}$$

describes the equilibrium between the system reaction on the left hand side and the excitation forces $\underline{F}(t)^{N\times 1}$ on the right hand side. System reactions are the node displacements $\underline{x}(t)^{N\times 1}$ and their derivatives with respect to time, which are the node velocities $\frac{\mathrm{d}\, \underline{x}(t)^{N\times 1}}{\mathrm{d}t} = \underline{\dot{x}}(t)^{N\times 1}$ and the node accelerations $\frac{\mathrm{d}^2\, \underline{x}(t)^{N\times 1}}{\mathrm{d}t^2} = \underline{\ddot{x}}(t)^{N\times 1}$, respectively. For practical applications it is difficult to estimate the damping matrix, wherefore the undamped $\left(\underline{\underline{D}}_{\mathrm{sys}} = 0\right)$ system will be

[3] In this work, the term natural frequency is interchangeable with the word eigenfrequency.

solved first. After usage of the right trial functions, the eigenvalue problem for the undamped natural frequencies ω and its eigenvectors $\underline{\varphi}$ can be solved according to the real form [87]

$$\left(\underline{\underline{K}}_{\text{sys}} - \omega^2\, \underline{\underline{M}}_{\text{sys}}\right) \underline{\varphi} = 0 \tag{4.4}$$

where the determinant $\det \left[\underline{\underline{K}}_{\text{sys}} - \omega^2\, \underline{\underline{M}}_{\text{sys}}\right] = 0$ delivers a polynomial of Nth degree and the square roots of ω^2 are the natural frequencies [153, p. 312]. Equation 4.4 connects the global stiffness and mass distribution to the natural frequencies and eigenvectors of the system. One way to determine the damping matrix is given by Rayleigh [156] with

$$\underline{\underline{D}}_{\text{sys}} = \alpha\, \underline{\underline{M}}_{\text{sys}} + \beta\, \underline{\underline{K}}_{\text{sys}} \tag{4.5}$$

where coefficients α and β will be chosen according to the first or the first two natural frequencies ω_1 and ω_2 and its damping ratios ζ_1 and ζ_2, such as shown by Strømmen [178]:

$$\begin{aligned} \alpha &= \frac{2\,\omega_1\,\omega_2\,(\omega_2\,\zeta_1 - \omega_1\,\zeta_2)}{\omega_2^2 - \omega_1^2} \\ \beta &= \frac{2\,(\omega_2\,\zeta_2 - \omega_1\,\zeta_1)}{\omega_2^2 - \omega_1^2} \end{aligned} \tag{4.6}$$

By providing a yaw bearing at the tower base, the cross section of the tower must not remain circular and can have lower stiffness regarding its secondary load direction (SS). Such one sided stiffness reduction would lead to more diverging bending mode eigenfrequencies. As this work is using the 5 MW NREL WT as a reference, a review of its Campbell diagram helps to identify possible support structure frequency ranges. Figure 4.4 shows the major one-per-revolution (1p) and three-per-revolution (3p) excitation frequencies.

The 1p excitation is caused by rotor mass unbalance and the 3p excitation stems from aerodynamic unbalance due to inclined inflow and tower dam effects of a three bladed WT, mentioned in subsection 4.4.2. Cut-in and cut-out rotor speeds of 6.9 rpm and 12.1 rpm are displayed with vertical lines and mark the operational range of the turbine. Furthermore, horizontal lines indicate the first bending eigenfrequencies of the tubular reference tower in FA and SS directions. The dashed lines represent the ±5 % safety margins for all frequencies and the green areas, soft-soft, soft-stiff, and stiff-stiff mark the frequency ranges appropriate for support structure eigenfrequencies to avoid resonance. For the NREL reference WT, it can be seen that the first two bending natural frequencies of the NREL reference tower diverge about 3.8 %. This divergence is caused by the mass distribution of the RNA. A LDD tower may have even larger differences between these eigenfrequencies, because of a less SS stiffness. Such a configuration will shift the SS bending eigenfrequency farther away from resonance and therefore to less dynamic loads, in case of the present WT configuration. In other situations with other hub heights, or other WT properties the Campbell diagram changes, wherefore reduced SS bending stiffness of a LDD tower may also increase the dynamic response. Thus, the dynamic behaviour of the support structure has to be investigated for each new support structure configuration and especially for LDD towers.

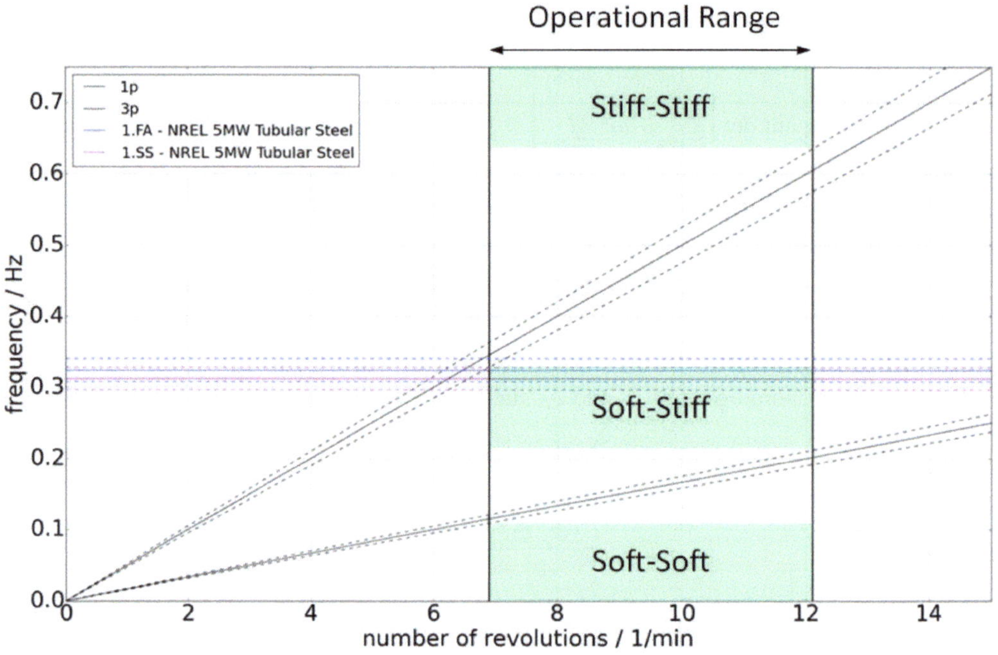

Figure 4.4: Campbell diagram of the NREL 5 MW reference WT. Data from [109]

4.4 Aerodynamics

WT support structure aerodynamics is a wide field and contains amongst others, the concepts of lift, drag, tower dam, tower shadow, gust induced vibrations, vortex-induced vibration and flutter. The following subsections focus on each of these fields with respect to rotatable WT towers. Rotatability of the support structure achieves the ability of the tower to align along the wind direction. Thereby, aerodynamic shaped tower profiles are possible which could be axes asymmetric to raise the stiffness regarding its main load direction, such as explained in subsection 4.3.1. Aerodynamic investigations of arbitrary profiles are possible through experiments and computational fluid dynamics (CFD). Analytical solutions are limited to certain basic geometries, flow conditions, and simplifying assumptions. The potential flow around an obstacle can be described through the inviscid or viscid theory, where the second takes into account viscous forces (friction) between the fluid particles according to the incompressible Navier-Stokes Equations 4.7, documented likewise by White [194].

$$\begin{aligned}
\rho_{\text{air}}\, g_x - \frac{\partial q}{\partial x} + \mu\left(\frac{\partial^2 u}{\partial x^2} + \frac{\partial^2 u}{\partial y^2} + \frac{\partial^2 u}{\partial z^2}\right) &= \rho_{\text{air}}\frac{\mathrm{d}u}{\mathrm{d}t} \\
\rho_{\text{air}}\, g_y - \frac{\partial q}{\partial y} + \mu\left(\frac{\partial^2 v}{\partial x^2} + \frac{\partial^2 v}{\partial y^2} + \frac{\partial^2 v}{\partial z^2}\right) &= \rho_{\text{air}}\frac{\mathrm{d}v}{\mathrm{d}t} \\
\rho_{\text{air}}\, g_z - \frac{\partial pq}{\partial z} + \mu\left(\frac{\partial^2 w}{\partial x^2} + \frac{\partial^2 w}{\partial y^2} + \frac{\partial^2 w}{\partial z^2}\right) &= \rho_{\text{air}}\frac{\mathrm{d}w}{\mathrm{d}t},
\end{aligned} \tag{4.7}$$

where $g_x,\ g_y,\ g_z$ are the gravity components, q is the pressure, μ is the viscosity coefficient, $u,\ v,\ w$ are the flow velocities in $x,\ y,\ z$ direction and t equals to the time. Assuming inviscid, incompressible and irrotational flow,

$$\nabla \times \underline{V} = \nabla \times (\nabla\phi) = 0 \tag{4.8}$$

Laplace's equation applies. The condition for irrotational flow in equation 4.8 contains the nabla operator

$$\nabla = \begin{pmatrix} \partial/\partial x \\ \partial/\partial y \\ \partial/\partial z \end{pmatrix}, \tag{4.9}$$

the velocity potential function $\phi(x, y, z, t)$, which reduces the three unknowns $u,\ v,\ w$ to one unknown $\phi(x, y, z, t)$ and the velocity vector $\underline{V}$. Solving Laplace's equation

$$\nabla^2\phi = 0 = \frac{\partial^2\phi}{\partial x^2} + \frac{\partial^2\phi}{\partial y^2} + \frac{\partial^2\phi}{\partial z^2} \tag{4.10}$$

for $\phi(x, y, z, t)$ provides the searched velocities through its derivatives

$$u = \frac{\partial\phi}{\partial x}, \quad v = \frac{\partial\phi}{\partial y}, \quad w = \frac{\partial\phi}{\partial z}. \tag{4.11}$$

Analytical solving techniques for Laplace's equation are amongst others conformal mapping [155], numerical finite differences [146], numerical finite elements [157] and numerical boundary elements [24]. The inviscid assumption is sufficient for laminar flow without separation from the profile (stall) and for a certain distance away from the profile surface, because the no-slip condition at the wall induces viscous stresses. This region is also called boundary layer. Lift, drag, the tower dam, and the tower shadow can be treated with this simplifying assumption for laminar flows. Investigation of vortex-induced vibrations and flutter requires solving of the viscous problem, such as presented in equation 4.7. Several CFD codes, such as the commercial ANSYS Fluent[4] and the open source OpenFOAM[5] are capable to solve the

[4] `http://www.ansys.com/Products/Fluids/ANSYS-Fluent`; Accessed 03-February-2019

[5] `http://www.openfoam.com/`; Accessed 03-February-2019

Navier-Stokes equations. Furthermore, the open source code XFOIL[6] is specialized to 2D streamline profile aerodynamics and widely used within the wind energy sector. The following sections mention the different aerodynamic fields of WT towers and propose some empirical approaches to handle them. In the end, each is discussed with respect to rotatable aerodynamic shaped WT tower profiles.

4.4.1 Forces and Moments Caused by the Wind

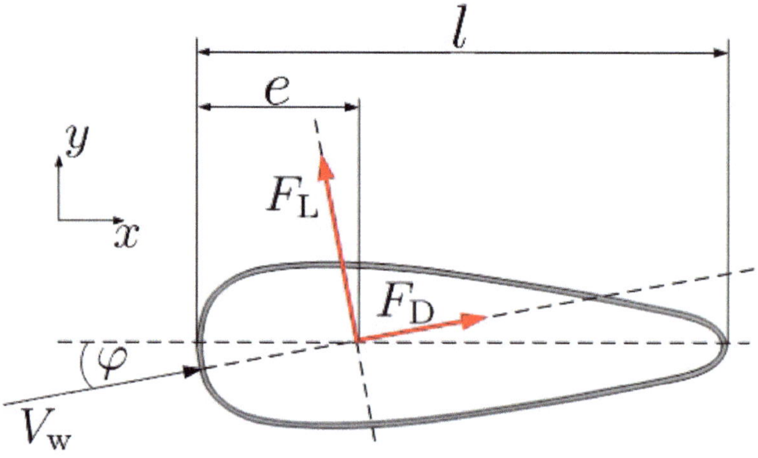

Figure 4.5: Lift and drag forces at aerodynamic shaped axis-symmetric tower profile

Figure 4.5 shows an aerodynamically shaped axis-symmetric tower profile, which experiences a wind flow of wind speed V_w. The wind is inclined by φ with respect to the profile chord. Such conditions increase the local flow velocity at the top and decrease it on the bottom, wherefore corresponding pressure decreases at the top and increases at the bottom occur. Integrating the resulting pressure distribution around the perimeter leads to aerodynamic forces. Another reason for aerodynamic forces is the conversion of momentum. As the wind flow experiences redirection by the profile, a reaction force at the profile itself is the consequence. Both forces, the one from the integrated pressure and the one from conversion of momentum can be divided into one component orthogonal to the flow F_L and one component F_D parallel to it, designated as lift and drag force. The forces attack at the so called pressure point, which represents the locus of the pressure resultant. Incidentally, point symmetric profiles, such as conventional circular cross sections will not experience any lift force due to equal flow conditions on the top and the bottom of the profile.

[6] http://web.mit.edu/drela/Public/web/xfoil/; Accessed 03-February-2019

According to Sockel [175], each body within a flow does have a dimensionless lift and drag coefficient c_{L} and c_{D}, which depends on the impounded pressure q, the related profile shape area A and the corresponding force due to

$$c_{\mathrm{L}}(\varphi, Re, k/l) = \frac{F_{\mathrm{L}}(\varphi, Re, k/l)}{q\,A} = 2\frac{F_{\mathrm{L}}(\varphi, Re, k/l)}{\rho_{\mathrm{air}}\,V_{\mathrm{w}}^2\,A} \tag{4.12}$$

$$c_{\mathrm{D}}(\varphi, Re, k/l) = \frac{F_{\mathrm{D}}(\varphi, Re, k/l)}{q\,A} = 2\frac{F_{\mathrm{D}}(\varphi, Re, k/l)}{\rho_{\mathrm{air}}\,V_{\mathrm{w}}^2\,A}. \tag{4.13}$$

The forces are a function of the angle of attack φ, the Reynolds number Re and the roughness ratio k/l, where k equals to the wall roughness and l is the related body dimension, such as shown in Figure 4.5. Due to Eurocode 1 DIN EN 1991-1-4 [67], wall roughness values are for instance $k = 0.2$ for galvanized and $k = 0.006$ for fine spray paint on steel. Eurocode 1 DIN EN 1991-1-4 [67] and the literature [175], [3], [194] provide aerodynamic coefficients for simple profile shapes, such as circular or rectangular ones. Several streamline profile coefficients are summarized by Abbott and Doenhoff [1]. In most cases, the reference area A corresponds to the projected profile area at the profiles cut with the largest width. Furthermore, the lift and drag forces do not necessarily attack at the shear center. Therefore, they can induce an additional torsional moment around the torsional axis. Sockel [175] relates the moment coefficient c_{M} to the tip of the profile according to

$$c_{\mathrm{M}} = \frac{e}{l}\left(c_{\mathrm{L}}\,\cos(\varphi) + c_{\mathrm{D}}\,\sin(\varphi)\right), \tag{4.14}$$

where e is the distance from the profile tip to the pressure point. Rotor blade shapes are often optimised towards maximum lift to drag ratios under structural constraints. An aerodynamic tower shape should be optimised with respect to low SS aerodynamic lift loads for relevant Reynolds numbers and angle of attacks. This changed focus compared to blades is due to the fact that blades should have an aerodynamic lift to generate torque at the rotor shaft. For rotatable WT towers, the lift causes additional loads in its weak direction and can become a disadvantage.

4.4.2 Tower Dam and Shadow

Blade excitation, caused by its passage through the decreased wind speed area in front of the tower, is called tower dam effect. This effect is only relevant for upwind turbines and should be as low as possible. Accurate investigations regarding blade excitation due to the tower dam effect require experiments, connected with CFD to include blade aerodynamic effects, such as tip losses, stall and the cross-flow along the blade. Such investigations were performed for example by Shkara et al. [171]. For ASE load simulations, expensive computational calculations should be avoided, wherefore several simplified more or less empirical approaches have been introduced within the literature. Most of them are focused on the description of the wind field behind the tower, which is called tower shadow. Its relevance is limited to downwind WTs, because of the changed local wind behaviour, which encounters the passing blades. An exception occurs if periodic effects, such as vortex-induced vibrations result from the tower shadow. This effect is mentioned in the later text. Reiso [160] gives a comprehensive enumeration of different empirical tower shadow models, where some of them are valid for the region in front of the tower to estimate the tower dam effect. The analytical solution of the potential flow around a cylinder is the basis of most of the approaches. According to Reiso [160] it delivers the velocity components

$$V_x = V_\infty \left(1 - \frac{x^2 - y^2}{(x^2 + y^2)^2} \left(\frac{D}{2}\right)^2\right) \tag{4.15}$$

$$V_y = V_\infty \frac{-2\, x\, y}{(x^2 + y^2)^2} \left(\frac{D}{2}\right)^2, \tag{4.16}$$

where V_∞ is the free stream velocity and D equals to the obstacle's diameter. x and y represent the Cartesian position of the velocity components within the 2D flow field. Powles model [150] describes the tower shadow with a cosine squared function and NREL [142] combines it with the potential flow according to

$$V_x = (u - u_{\text{wake}})\, V_\infty \tag{4.17}$$

$$V_y = (v - u_{\text{wake}})\, V_\infty \tag{4.18}$$

with

$$u = 1 - \frac{(x + 0.1)^2 - y^2}{\left((x + 0.1)^2 + y^2\right)^2} + \frac{c_D\,(x + 0.1)}{2\,\pi\,(x + 0.1)^2 + y^2} \tag{4.19}$$

$$v = 2 \frac{(x + 0.1)\, y}{\left((x + 0.1)^2 + y^2\right)^2} + \frac{c_D\, y}{2\,\pi\,(x + 0.1)^2 + y^2} \tag{4.20}$$

and

$$\begin{aligned} u_{\text{wake}} &= \frac{c_D}{\sqrt{d}} \cos^2\left(\frac{\pi\, y}{2\,\sqrt{d}}\right) && \text{for } |y| \leq \sqrt{d} \\ u_{\text{wake}} &= 0 && \text{for } |y| > \sqrt{d}, \end{aligned} \tag{4.21}$$

where

$$d = \sqrt{x^2 + y^2}. \tag{4.22}$$

It is important to note that x and y in Equations 4.19 to 4.22 are the Cartesian coordinates, normalized by the cylinder radius. Further models are developed by Blevin [20], where the wind speed deficit and the wake width depends on c_D and Schlichting and Gersten [167], who consider the Reynolds number additionally to the drag coefficient. This would enable other profiles than circular ones, but the model is restricted to regions $x > 3\,l$ on the lee side, where l represents the characteristic length. Introducing a jet into the stream is proposed by Madsen et al. [135] within the JET wake model and can also be applied as tower shadow approach. The tower shadow of lattice structures is implemented within Bladed [22]. It uses a combination of potential flow for the free stream area, Powles model and a correction for the influence of each cylinder with respect to each other. Unfortunately, none of the models is directly applicable for arbitrary shaped profiles in the context of the tower dam effect. Therefore, advanced investigations have to be carried out to predict the velocity deficit in front of the tower. However, according to Blevin's model [20] and NREL [142], high drag coefficients seem to result in larger dam effects, wherefore the drag coefficient can be seen as an indicator for the tower dam influence intensity.

4.4.3 Gust Induced Vibrations

Wind turbulence causes certain characteristic wind frequencies. If these gust frequencies resemble the support structure eigenfrequencies, critical structural vibrations are possible. Therefore, Eurocode 1 DIN EN 1991-1-4 [67] proposes a wind gust frequency model. It provides the gust energy, represented by the dimensionless spectral density $S_L(z, f)$ according to

$$S_{\mathrm{L}}(z,f) = \frac{f\, S_{\mathrm{v}}(z,f)}{\sigma_{\mathrm{v}}^2} = \frac{6.8\, f_{\mathrm{L}}(z,f)}{(1+10.2\, f_{\mathrm{L}}(z,f))^{5/3}}, \tag{4.23}$$

where f is the considered frequency, $S_v(z, f)$ is the auto spectrum of the turbulent wind, σ_v is the standard deviation of the turbulence and f_L equals to the dimensionless frequency

$$f_{\mathrm{L}}(z,f) = \frac{f\, L(z)}{V_{\mathrm{m}}(z)}. \tag{4.24}$$

Equation 4.24 contains the mean wind velocity $V_m(z)$ and the integral length dimension $L(z)$, which is computable through reference length $L_t = 300$ m, reference height $z_t = 200$ m, roughness height z_0 and minimum height z_{min} with

$$\begin{aligned} L(z) &= L_{\mathrm{t}} \left(\frac{z}{z_{\mathrm{t}}}\right)^{\alpha} && \text{for } z \geq z_{\min} \\ L(z) &= L(z_{\min}) && \text{for } z < z_{\min}. \end{aligned} \tag{4.25}$$

The exponent in Equation 4.25 results to $\alpha = 0.67 + 0.05\ \ln(z_0)$. Assuming terrain category II of table 4.1 in DIN EN 1991-1-4 [67] results to $z_0 = 0.05$ m and $z_{min} = 2$ m. Figure 4.6 shows the resulting spectral density of equation 4.23 over frequency f for different heights z. The entered common eigenfrequency region of WT towers reveals that frequent gust excitations are not negligible for the design process. Through non-axisymmetrical profiles, such as introduced by LDD rotatable towers, two different first eigenfrequencies are possible. The eigenfrequency of the weak axis of such towers can move to the left side in Figure 4.6, especially for larger hub heights. Therefore, gust induced vibrations become more relevant for rotatable support structures. The dynamic behaviour of rotatable WT towers is mentioned with more detail in subsection 4.3.3.

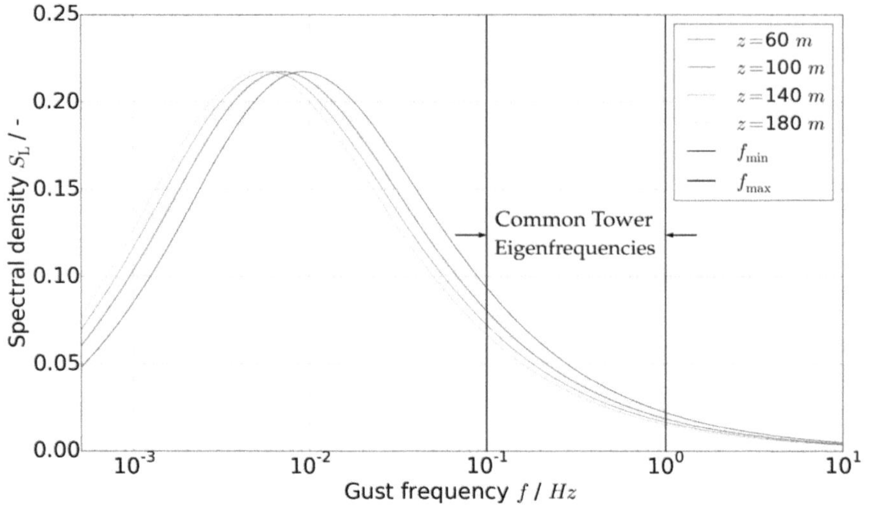

Figure 4.6: Wind gust frequencies in different heights within terrain category II compared to common tower eigenfrequencies

4.4.4 Vortex-Induced Vibrations

In the field of civil engineering, vortex-induction plays the main role regarding building vibrations [170]. These excitations are caused by periodic vortex building behind the obstacle, called Kármán's vortex street and act lateral to the wind direction. If the vortex creation frequency resembles building eigenfrequencies, critical structural vibrations occur. Excitation of bending, torsional, or cross section deformational modes have to be taken into account. Cross section deformation contains effects, such as ovalling of shells without stiffening rings or plate vibrations. In terms of wind energy vortex-induction may be one of several vibration sources and in the most cases it is particularly dangerous during the erection process, when the tower stands without the head mass of the nacelle and the blades [77]. The vortex creation frequency f_v for arbitrary profiles is characterized through the Strouhal number S, the free stream wind velocity V_∞ and the characteristic length l according to

$$f_v = \frac{S\, V_\infty}{l}. \tag{4.26}$$

DIN EN 1991-1-4 [67] provides Strouhal numbers for basic profiles, such as cylinders and rectangles. In contrast to the norm, the Strouhal number is depending on the Reynolds number in the case of cylinders. The Reynolds number dependency for rectangles is very low, but increases with increasing corner radii [175]. Investigations regarding elliptical profiles are documented by Whitbread [193]. Furthermore, in the case of non-pointsymmetric profiles, the flow direction according to the profile does have an influence and can lead to different Strouhal numbers. Analytical solutions for the aero-elastic problem of vortex-induced vibrations are not achievable. Merely the rigid cylinder with small Reynolds numbers, which are outside of relevant civil engineering applications, can be calculated analytically. For arbitrary profiles and aerodynamic conditions, comprehensive experiments in combination with CFD are required to produce Strouhal numbers. In terms of structural dynamics, vortex-induced vibrations are modelled according to a stability border, whereby a second aerodynamic damping term is added to the equation of motion. This second term is able to have a negative sign, wherefore excitations of the system may be the result. Regarding subsection 4.3.3, LDD support structures can have two different first bending eigenfrequencies for each bending axis. The first eigenfrequency of the weak axis tends to be lower than the eigenfrequency of a comparable conventional tower, because of the less necessary stiffness. Assuming vortex-induced excitations, which cause motion in the corresponding eigenmode, this eigenfrequency reduction is an advantage, because it moves away from the higher vortex creation frequency, given by Equation 4.26. Furthermore, vortex-induced excitations along the wind direction are negligible, according to Sockel [175]. Assuming this findings combined with a more aerodynamically formed profile indicate, that LDD towers are less vulnerable with respect to vortex-induction. If however large excitations occur, provisions, such as dampers, Scruton-coils [169], or perforated meshes around the tower may be a solution to reduce the loads.

4.4.5 Flutter

Flutter can be described as self excited structural vibration of at least two DOFs under constant wind flow [184], caused by changing aerodynamic loads attributable to structural deformations. Conventionally, these DOFs are bending and torsional ones, whereby excitations from vortex-induction may be involved [175]. The characteristic flutter motion depends on the bending and torsional eigenfrequencies. If the difference of these eigenfrequencies is large, motions regarding one of the corresponding DOFs may be dominant. DIN EN 1991-1-4 [67] provides three conditions, which have to be fulfilled for a structure to be endangered to flutter. In terms of WT towers, they can be summarized as:

1. The ratio of the main profile dimensions is less than $b/d = 0.25$, where d equals to the dimension in wind direction and b to the dimension orthogonal to it
2. The torsional axis should have a distance of $>= d/4$ downstream from the luv sided edge
3. The lowest eigenfrequency is a torsional one or the torsional eigenfrequency amounts to less than the double of a translational eigenfrequency

As discussed in subsection 4.3.1, the tower profile will have a elongated shape to give an efficient response to the different bending moments for each bending axis. Section 2.5 treats the maximum borders for the tower profiles regarding transportability and condition 1. of the previous enumeration adds another restriction to the profile dimensions in the case that condition 2. and 3. are met. If all three conditions are met, DIN EN 1991-1-4 [67] provides a procedure to calculate a critical divergence wind speed V_{div} according to

$$V_{\text{div}} = \sqrt{\frac{2\,G\,I_{\text{T}}}{\rho_{\text{air}}\,d^2\,\frac{\mathrm{d}c_{\text{M}}}{\mathrm{d}\varphi}}}, \tag{4.27}$$

where $\frac{\mathrm{d}c_{\text{M}}}{\mathrm{d}\varphi}$ equals to the derivative of the aerodynamic moment coefficient with respect to the twist angle of the profile. The divergence wind speed should exceed $2\,V_{\text{m}}$ to avoid flutter. Further investigations for tower shapes with section-wise critical dimensions require comprehensive experiments and aero-elastic simulations to determine the potential of flutter instabilities.

4.5 Methodical Approach for Evaluation of the Tower Concept

The decision for one specific LDD support structure design requires a comparison of each possible solution to come as close to the optimal tower as possible. Attributable to the limited scope of this work and the tremendous amount of parameters to be considered for a comprehensive study, some shortcuts need to be introduced. These shortcuts were made by arguing between different solutions on a conceptual level, rather than developing detailed designs for each possible tower. Designs differ in the underlying tower concept, such as conventional tubular steel, slip joint tubular steel, pre-fabricated concrete, on-site fabricated concrete, tubular hybrid steel concrete, lattice steel, guyed tubular steel, guyed lattice steel, and covered lattice wooden towers. Especially the lattice and guyed concepts can be found in many diverging shapes. The following chapters build on each other, because they represent the journey from one LDD concept to the next. Starting from the conventional circular cross section and walking over to elliptical and aerodynamically shaped cross sections in chapter 5 establishes the overview about shell like structures for LDD WT towers. After these tubular shell concepts, the next step is the discussion of lattice structure concepts in chapter 6. The last conceptual discussion treats the additional possibility of tower inclination and presents some other special concepts, before in the end one concept was chosen to be investigated in more detail.

Preliminary Exclusion of Underlying Concepts

The concepts, discussed in the following chapters, did not account for all underlying concepts listed in section 4.5, because some concepts could already be excluded for different reasons. Consideration of the non-rotatable tower concept comparison in Figure 4.7 and other references, such as Gasch [77], revealed concrete related towers as high weight and high CO_2-emission causing concepts, while only small cost savings were achievable for the hybrid concept compared to the conventional welded tubular steel tower. The values in Figure 4.7 can be questioned for other hub heights and other rated power WTs, but it was not assumed that the previous mentioned points diverge for other configurations in a significant way.

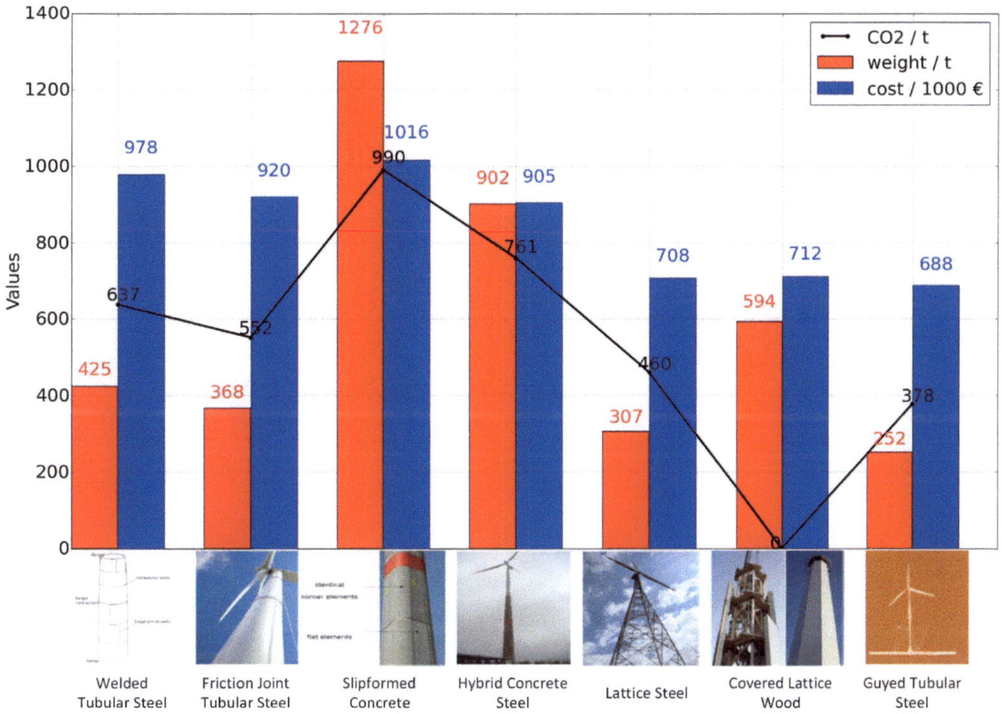

Figure 4.7: Tower concept comparison in terms of weight, costs, and CO_2-emissions for a 3 MW, 100 m rotor diameter, IEC 1B WT at 125 m hub height. Assembled from pictures with permission to use from [68] and from Rudolf's data in [164].

Owing to its nature, wooden towers would lead to very low CO_2-emissions throughout the production chain. On the other hand, because of the lack of long time experience to this material for WTs and its maintenance effort over lifetime, they had to be excluded in this study. Nevertheless, rotatable wooden lattice towers may be interesting for future research projects. Guyed and lattice steel towers are very material efficient concepts already. Guyed towers need a lot of space for their cables and are therefore problematic for agricultural used land where many land-based WTs are installed. Furthermore, Hau [92] indicates that the cable system and its additional foundations tend to cause high acquisition and maintenance costs. The later arise from the requirement of frequent cable pre-tension checks. However, Koppány explains that cable manufacturers can reduce or avoid cable pre-tension check intervals for their cables when they are exposed to defined load cycles during the manufacturing process [124]. Hau [92] and Gasch [77] attest guyed towers to be suitable for smaller hub heights, whereby lattice towers become more interesting for larger hub heights. In summary, tubular and lattice towers remain the favourite concepts and are mentioned in more detail throughout the following two chapters.

5 Rotatable Tubular Concepts

5.1 Comparison of Tubular Cross Sections

Major influencing factors to LDD tubular tower designs are the area stiffness, manufacturing effort, the aerodynamic drag, and local shell/plate stability. The area stiffness (radius of gyration) $i^2 = \frac{I}{A}$ is a measure for material efficiency in terms of bending stiffness. The manufacturing effort of a shell depends on the amount of necessary welding seams, the amount of cross sectional parts, and the complexity of the shape. In this case, complexity is a synonym for change in curvature along the profile, which has a relevance for the rolling process. The aerodynamic drag in FA direction is assumed to have no high influence to the design, because the induced bending moments from this line load along the tower are significantly smaller than the bending moments, induced by rotor thrust. This assumption is supported by findings from section 3.3 and subsection 3.5.3. Section 4.4 discusses the aerodynamics of LDD structures in more detail. In terms of local stability of LDD tubular structures, plate like, ideal shell like, and intermediates are conceivable. Curvature of a shell plays an important role, such as the little discourse in Appendix 14.11 indicates. The table in Figure 5.1 compares different LDD shell concepts on a qualitative level in terms of area stiffness, manufacturing, stability, and aerodynamic drag. Plus signs refer to favoured and minus signs refer to unfavoured properties, respectively.

The sketches on the left side show, how different cross sectional shapes would fit into a rectangular design space, such as introduced in Figure 5.2. Such a transport constraint forces a circular cross section (**a**) to have a small diameter, while a rectangular one (**b**) would fit perfectly into it. The small diameter of concept **a** leads to higher curvature and therefore good local stability, but because of large material distances from the bending axes, concepts **b** and **c** reach better material efficiency in terms of bending stiffness. On the other hand, they have low local stability properties, due to the plate like walls. In terms of manufacturing, concepts **a** and **b** are assessed to be the same, because **a** requires rolling of one steel sheet and one longitudinal welding seam, while **b** requires folding of one steel sheet and likewise one longitudinal welding seam. Concepts **g** and **h** obtain less aerodynamic resistance and intermediate local stability, but on the other hand they lack of a lower area stiffness and show more complex curvatures, wherefore a more expensive manufacturing process is expected. The elliptical cross section (**f**) was found to be a good compromise between the different aspects, wherefore it is considered in more detail throughout the next section.

	Area Stiffness	Manufacturing	Stability	Aerodynamic Drag
a	- -	++	++	-
b	++	++	- -	- -
c	++	-	- -	-
d	+	+	-	-
e	-	+	-	+
f	-	-	+	+
g	-	- -	+	+
h	- -	- -	+	++

Figure 5.1: Shell concept qualitative evaluation matrix

5.2 Elliptical Tower

The popular circular cross section for tubular welded steel towers has been a good choice for WTs with hub heights < 100 m [92]. But as mentioned in sections 2.4 and 2.5, this concept reaches its limits for taller hub heights. The next step is to compress the circular cross section in one direction and elongate it in the other direction towards an elliptical shape. A transport constraint considering elliptical cross section reaches higher FA stiffness as a transport constraint considering circular one. On the other hand, steel sheet curvatures become lower at the ellipses flanks. This reduced curvature increases the danger of local shell buckling failure. Hau [92] states, that stiffness requirements are the most common design drivers, but notes that local shell stability becomes more important for optimized thin walled tubular steel towers. A cheap manufacturing is given by the same rolling procedures as used for conventional circular steel tubs. Note that circumferential welding of circular cross sections are done by the welding machine being placed at a fixed position at the top while the cross section itself will be rotating. Consequently the welding processes of an elliptical cross section has to take into account different welding heights as a function of the rotation angle. Elliptical profiles are not warping free, wherefore additional normal stresses occur based on torsional moments and have to be included in a design process.

A parameter study of a circular and an elliptical cross section can show how much material can be saved if material strength, local shell buckling, and transport constraints are used. Fatigue may also play a role for the chosen tower wall thickness, but it was together with warping caused stresses neglected for simplicity in this approach. For this study, a design space, such as shown in Figure 5.2 with a width $B = 7.5$ m and a height $H = 4.3$ m was defined. These values were chosen according to the land-based transport constraints, discussed in subsection 2.5. The parameter study was carried out for the tower base, because it was assumed to have the most significant material saving potential. The own ASE load simulation in subsection 3.5 provided the applied loads. For this first step the resulting maximum absolute values of each load component at the tower base were assumed. In this preliminary investigation, the structure-load interaction was not considered. Thus changing tower shapes were not assumed to change the loads, which is a simplification, but was assumed to be sufficient to gain a first approximation for the material saving potential of LDD support structures.

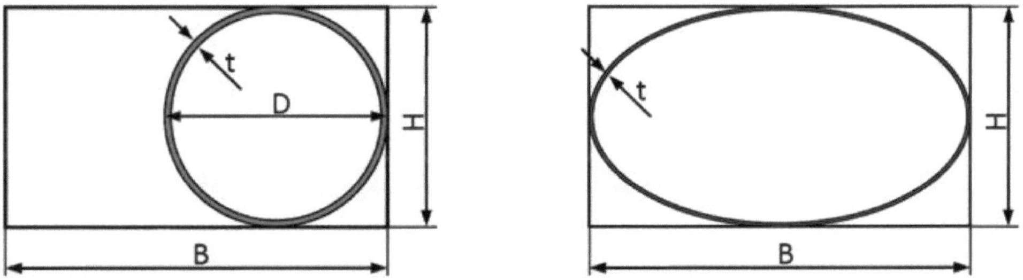

Figure 5.2: Design space for circular and elliptical cross section

Three cases were compared to each other where their wall thicknesses were increased successively until all strength and shell stability requirements were fulfilled for different SS to FA bending moment ratios. The circular NREL tower base cross section with an outer diameter of $D = 6.0$ m should serve as a reference [109], whereby it would not fit into the design space. As second case, a circular cross section with an outer diameter of $D = 4.3$ m and as third case an elliptical cross section with $B = 7.5$ m and $H = 4.3$ m was chosen to fit into the transport design space. Table 5.1 summarizes the assumed cross sectional dimensions.

Name	**Case 1**	**Case 2**	**Case 3**
shape	circular	circular	elliptical
diameter D	6 m	4.3 m	–
width B	–	–	7.5 m
height H	–	–	4.3 m
transport constraint	not valid	valid	valid

Table 5.1: Cross sectional properties, including transportability

5.2.1 Strength Analysis

The cross sectional strength is checked by the equation

$$\gamma_f \, F_k \leq \frac{1}{\gamma_n} \frac{1}{\gamma_m} \, f_y, \tag{5.1}$$

where γ_f, γ_n, and γ_m are the partial safety factors for the kind of load, the consequences of failure of the component, and the material. f_y is the yield strength of the material and F_k is the acting dimensioning equivalent stress, expressed by the interaction equation

$$F_k = \sigma_{eq,Ed} = \sqrt{\sigma_{x,Ed}^2 + \sigma_{\theta,Ed}^2 - \sigma_{x,Ed}\,\sigma_{\theta,Ed} + 3\left(\tau_{x\theta,Ed}^2 + \tau_{xn,Ed}^2 + \tau_{\theta n,Ed}^2\right)}, \tag{5.2}$$

where

$$\sigma_{x,Ed} = \frac{F_z}{A} + \frac{M_x}{I_x} y + \frac{M_y}{I_y} x, \tag{5.3}$$

$$\sigma_{\theta,Ed} = p_n \frac{r}{t}, \tag{5.4}$$

and

$$\tau_{x\theta,Ed} = \frac{M_z}{2\,A_m\,t} \tag{5.5}$$

are the meridian stress, the circumferential stress, and the shear stress. Thereby, F_z, M_x, M_y, and M_z are the acting normal force, bending moments around x and y axes, and the torsional moment. $p_n = \frac{1}{2}\,\rho_{air}\,c_w\,v_w^2$ is the air pressure on the shell surface and A_m is the enclosed area of the cross section wall middle line. For simplicity, F_x, F_y, $\tau_{xn,Ed}$ and $\tau_{\theta n,Ed}$ were assumed to be zero.

5.2.2 Local Shell Buckling

The need of material efficiency and more realistic comparison requires to respect the r/t limits, presented in Equations 14.1 to 14.3, and to carry out buckling proofs. The DNV GL standard, support structures for wind turbines [63], states that shell stability can be checked according to Eurocode 3 DIN EN 1993-1-6 [44]. For the following parameter study, the manual method from Annex D in DIN EN 1993-1-6 [44] was applied within a python script. It considered a cylindrical shell, such as shown in Figure 5.3. In the elliptical case, a curvature equivalent cylindrical shell was assumed for each point around the cross section. Gardner, Chan, and Abela may be reviewed to see the derivation of the theoretical buckling initiation point for an elliptical hollow section under combined compression and uniaxial bending [76]. In this work, the interaction proof of Eurocode 3 DIN EN 1993-1-6 [44] had to be valid for all points around the circumference. It is

$$\left(\frac{\sigma_{x,Ed}}{\sigma_{x,Rd}}\right)^{k_x} - k_i \left(\frac{\sigma_{x,Ed}}{\sigma_{x,Rd}}\right)\left(\frac{\sigma_{\theta,Ed}}{\sigma_{\theta,Rd}}\right) + \left(\frac{\sigma_{\theta,Ed}}{\sigma_{\theta,Rd}}\right)^{k_\theta} + \left(\frac{\tau_{x\theta,Ed}}{\tau_{x\theta,Rd}}\right)^{k_\tau} \leq 1 \tag{5.6}$$

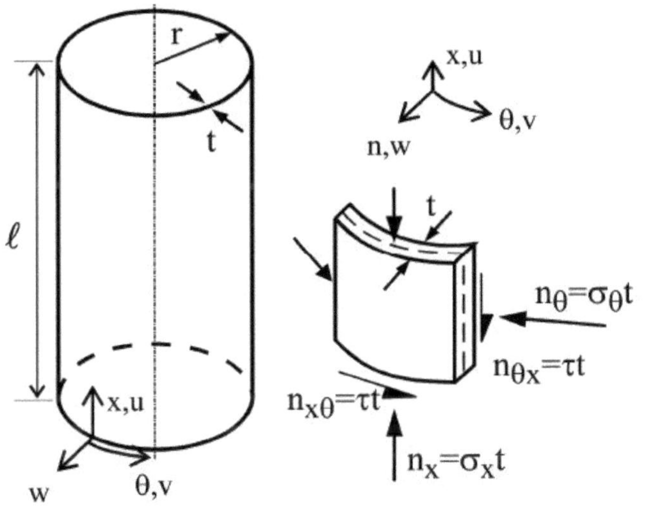

Figure 5.3: Membrane forces of a cylindrical shell section with constant wall thickness from [44]

The dimensioning meridian buckling stress $\sigma_{x,Rd}$, dimensioning circumferential buckling stress $\sigma_{\theta,Rd}$, dimensioning shear buckling stress $\tau_{x\theta,Ed}$, and the buckling interaction parameters k_x, k_θ, and k_τ are given according to Annex D in DIN EN 1993-1-6.

5.2.3 Material Saving Potential of a Cross Section

Parameter	Value	Reference
Steel yield strength	$t <= 40\text{ mm} : f_y = 355\ \frac{\text{N}}{\text{mm}^2}$ $t > 40\text{ mm} : f_y = 335\ \frac{\text{N}}{\text{mm}^2}$	DIN EN 1993-1-1 [42]
Young's modulus	$E = 210{,}000\ \frac{\text{N}}{\text{mm}^2}$	DIN EN 1993-1-1 [42]
Partial safety factor for material	$\gamma_\text{m} = 1.1$	DIN EN 61400-1 [50]
Partial safety factor for consequence of failure	$\gamma_\text{n} = 1.0$	DIN EN 61400-1 [50]
Partial safety factor for type of load (already considered in loads)	$\gamma_\text{f} \rightarrow$ see Tables 2.1 and 2.2	DIN EN 61400-1 [50]
Shell fabrication tolerance quality class	B	DIN EN 1993-1-6 [44]
Shell boundaries	end 1: BC1, end 2: BC2	DIN EN 1993-1-6 [44]
Unsupported shell length	$l_\text{shell} = H_\text{T,NREL}/3 = 29.2\text{ m}$	DIN EN 1993-1-6 [44]
Load case	$M_x = \{0...1\}M_y,\ M_y = 1.64\text{e}8\text{ Nm}$ $F_z = -1.05\text{e}7\text{ N},\ M_z = 3.03\text{e}7\text{ Nm}$	ASE analysis in section 3.5
Wind speed	$v_\text{w} = 20\ \frac{\text{m}}{\text{s}}$	NREL Study [110]
Air density	$\rho_\text{air} = 1.25\ \frac{\text{kg}}{\text{m}^3}$	DIN EN 1991-1-4 [67]
Ellipse drag coefficient	$c_\text{w,ell} = 0.5$	Interpolated from [194, p. 483]
Cylindrical drag coefficient	$c_\text{w,cyl} = 0.3$	[151] and [163]

Table 5.2: Parameter set for material saving comparison between circular and elliptical tower cross sections under strength, local shell stability, and transport boundary conditions

With the assumptions from subsections 5.2.1 and 5.2.2, a material usage estimation can be carried out. Table 5.2 summarizes the assumed parameter set. A common steel for wind turbines with yield strength of 355 $\frac{\mathrm{N}}{\mathrm{mm}^2}$, shell fabrication tolerance quality class B and a tower base section length of $H_{\mathrm{T,NREL}}/3$ = 29.2 m until the next flange stiffening occurs was chosen. Figure 5.4 shows the material savings and necessary wall thicknesses for the different tower shapes and tower base bending moment ratios under material strength, shell stability, and transport constraints. Material saving is expressed with respect to the proposed tower base cross sectional area of the NREL reference tower, denoted by $A_{\mathrm{NREL}} = 0.658\ \mathrm{m}^2$. Note that case 1 assumed the outer diameter of the NREL reference tower, but its wall thickness was recalculated in this study. Thus, the reference tower with the new wall thickness was compared to the one with the proposed wall thickness from Jonkman et al. [109]. All results are displayed in a step like manner, with respect to discrete wall thickness steps of 1 mm. The dashed lines, labelled with SO in the legend represent results without shell stability constraints. They indicate that material strength becomes more important for smaller shell diameters, such as in case 2 and for small $\frac{M_{\mathrm{x}}}{M_{\mathrm{y}}}$ ratios.

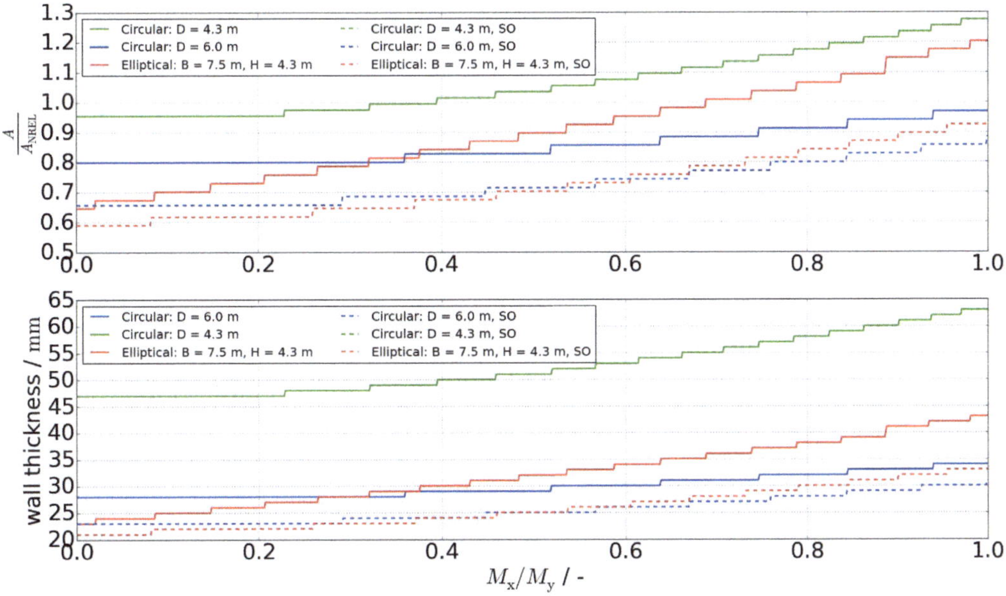

Figure 5.4: Material saving comparison between circular and elliptical tower cross sections under strength, local shell stability, and transport boundary conditions

The results contain the required material usage of the transport unconstrained reference cross section (case 1), the transport constrained elliptical cross section (case 3), and the transport constrained circular cross section (case 2) compared to the NREL reference tower for all bending moment ratios. Assumption of the maximum occurring SS tower base bending moment

$M_x = 9.52e7$ Nm of the own ASE load simulation in section 3.5 revealed the corresponding bending moment ratio of $M_x/M_y = 0.582$ and corresponding material savings of 32.3% and 30.9% in cases 1 and 3 compared to case 2, respectively. The difference is, that in the elliptical case, the transport constraints were valid, but not in case 1. Even for a hypothetical tower base bending moment ratio of $M_x/M_y = 1.0$, material savings of 16.2% and 12.4% of case 1 and 3, compared to case 2 would be achievable. The bottom diagram in Figure 5.4 indicates manufacturing challenging wall thicknesses of $\geq$ 50 mm for case 2 for all bending moment ratios above 0.4, while the transport constraint considering elliptical solution remains in more manageable wall thickness regions. These findings become even more relevant for taller hub heights, because the increasing loads force the wall thickness to be increased likewise. Material savings may be expected also for middle and top tower sections, because the blade tip to tower clearance becomes an additional constraint to the outer dimensions of non-rotatable towers. In case of a LDD tower, blade tip to tower clearance must only be given at one side of the tower, wherefore it will not limit the outer tower dimensions.

5.2.4 Aerodynamic Loads on the Elliptical Tower

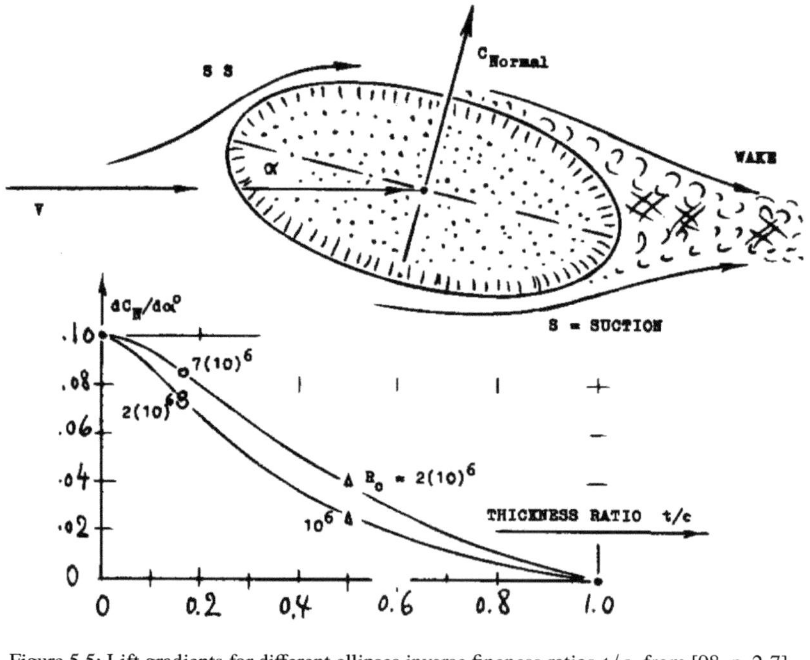

Figure 5.5: Lift gradients for different ellipses inverse fineness ratios t/c, from [98, p. 2-7]

The elliptical tower is a compromise between aerodynamic drag, stiffness, stability and manufacturing effort, such as discussed in subsection 5.1. In the previous preliminary study, the extreme loads from the ASE analysis of a conventional WT configuration with tower drag

Fineness ratio c/t	**Laminar** c_D	**Turbulent** c_D
1	1.2	0.3
2	0.6	0.2
4	0.35	0.15

Table 5.3: Drag coefficients for elliptical profiles with different fineness ratios in a region of $Re > 10,000$

loads were used for a material saving potential analysis. There, the relation between extreme SS and FA tower bending moments played a crucial role in terms of material saving. But the relation itself is not the only relevant factor. The absolute load values themselves are likewise important. One aspect was not accounted for in the previous investigation, namely the profile lift. Especially the profile lift, leads not only to an increased tower base SS to FA bending moment ratio, but also to general increased loads for certain angles of attack on an elliptical profile.

The aerodynamic lift and drag coefficients c_L and c_D are a function of the Reynolds number Re, such as discussed in subsection 4.4.1. The drag coefficient for cylindrical profiles are well evaluated by Prandtl and Tietjens [151] and by Roshko [163]. Furthermore, aerodynamic loads for cylindrical tubes are proposed in the norm DIN EN 1991-1-4 [67] and its Annex [41]. For the tower dimensions of the 5 MW NREL wind turbine and standard atmospheric conditions, the wind velocity dependent Reynolds numbers are between 10^4 and 10^7. Therefore laminar, transient, and turbulent flows occur. White [194, p. 483] provides drag coefficients for elliptical profiles within laminar and turbulent flows in a region $Re > 10^4$, which are listed in Table 5.3.

Depending on the Reynolds number and on the fineness ratio c/t, elliptical profiles within the design space ($c/t = B/H = 1.74$) obtain about 2/3 of the drag coefficient of cylindrical profiles with $c/t = 1$. A linear interpolated drag coefficient of $c_D = 0.226$ for an angle of attack of $\alpha_{AoA} = 0$ deg in turbulent flow from Table 5.3 and $c_D = 0.31$ at $\alpha_{AoA} = 15$ deg, extrapolated from figure 6 in [198], was assumed to get

$$c_D(\alpha_{AoA}) = 0.226 + \frac{0.31 - 0.226}{15 \text{ deg}} \alpha_{AoA} \tag{5.7}$$

A review of figure 12 in [98, p. 2-7] shows, that the lift coefficient gradient $\partial c_L / \partial \alpha_{AoA}$ is constant for angles of attack $\alpha_{AoA} \leq 15$ deg for a wide range of ellipses fineness ratios. With this knowledge one can read the corresponding lift gradient from Figure 5.5, where it is given as function of the inverse fineness ratio. Assuming a $t/c = H/B = 4.3$ m / 7.5 m = 0.57 at the upper curve results in $\partial c_L / \partial \alpha_{AoA} \approx 0.033$. The lift coefficient follows as

$$c_L(\alpha_{AoA}) = \frac{\partial c_L}{\partial \alpha_{AoA}} \alpha_{AoA} \tag{5.8}$$

For a rough assessment of the aerodynamic influence to the SS tower base bending moment, the previous assumed values can be used in Equation 3.6 analogously. In this approach, it was assumed that all outer tower dimensions and the Reynolds number remain constant along the tower height, such as indicated in Equation 5.9.

$$\begin{aligned} M_{\mathrm{W,x}}(0) = & \left[\frac{1}{2}\,\rho_{\mathrm{air}}\,c_{\mathrm{D}}\left(\alpha_{\mathrm{AoA}}\right)\left(\frac{V_{\mathrm{H}}}{z_{\mathrm{H}}^{\alpha}}\right)^{2}\frac{H}{4\,\alpha^{2}+6\,\alpha+2}\left(z_{\mathrm{H}}^{2\,\alpha+2}\right)\right]\sin\left(\alpha_{\mathrm{AoA}}\right) \\ & + \left[\frac{1}{2}\,\rho_{\mathrm{air}}\,c_{\mathrm{L}}\left(\alpha_{\mathrm{AoA}}\right)\left(\frac{V_{\mathrm{H}}}{z_{\mathrm{H}}^{\alpha}}\right)^{2}\frac{H}{4\,\alpha^{2}+6\,\alpha+2}\left(z_{\mathrm{H}}^{2\,\alpha+2}\right)\right]\cos\left(\alpha_{\mathrm{AoA}}\right) \end{aligned} \tag{5.9}$$

By usage of an exponential wind profile with $\alpha = 0.2$, air density $\rho_{\mathrm{air}} = 1.225$ kg/m^3, and the reference hub height $z_{\mathrm{H}} = 90$ m the SS tower base bending moments in Figure 5.6 resulted. Especially for extreme wind speeds $\geq$ 50 m/s, significant additional SS bending moments $\geq$ 9 MNm were induced by the tower aerodynamics for $\alpha_{\mathrm{AoA}} = 15$ deg. A comparison with the ASE results in Figure 3.7 revealed that the lateral lift loads, caused by skewed inflow of $\alpha_{\mathrm{AoA}} = 15$ deg, result in an 11.8 % increased extreme tower base bending moment for extreme wind speeds of 50 m/s at parking situations (DLC 6.2). Considering the SS tower base extreme bending moment driving DLC 2.1 and assuming a simultaneously occurring gust of 30 m/s, a SS moment increase of $\approx$ 3.9 % can be observed. Note that the corresponding PSFs were applied to the previous numbers to achieve comparability with Figure 3.7. These results show that the profile lift has an impact to the tower base extreme loads, but it is not as high as expected and should not be a reason to discard tubular rotatable support structures for WTs in general.

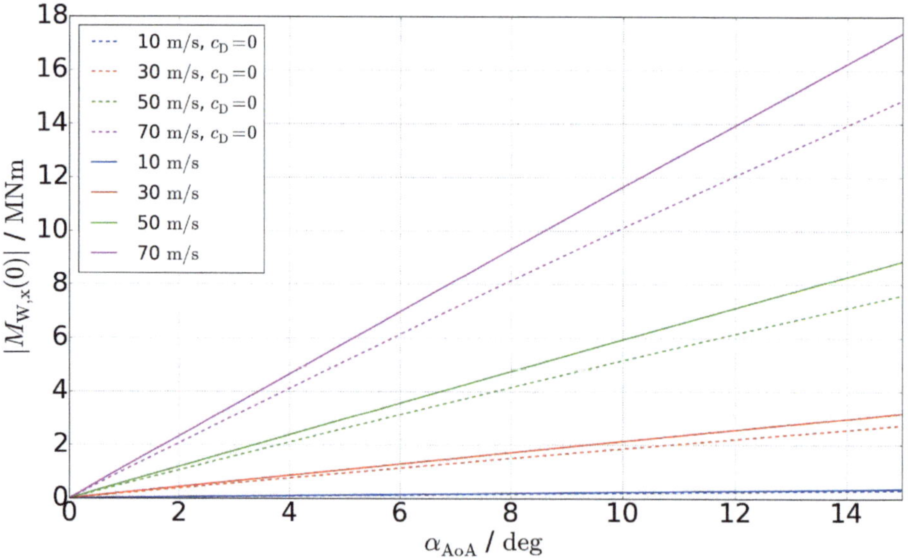

Figure 5.6: SS tower base bending moments for an elliptical tower with a width of 4.3 m and a length of 7.5 m resulting from tower lift and drag for different wind speeds

Furthermore, investigations about the aerodynamic characteristics of an elliptical radar antenna [88] show that the size and momentum deficit of the wakes behind an elliptical profile are reduced compared to a cylindrical one, where lower wake interactions for corresponding towers can be expected.

5.2.5 Material Saving Potential for Towers

In this subsection, material saving estimations for different shell tower configurations with changing hub heights are considered. A review of the previous two subsections indicates, that an elliptical tower shape provides material saving potential compared to a transport constrained circular one. Note that no fatigue considerations were implied in the previous subsection 5.2.3. In terms of the preliminary nature of this work, no additional ASE simulations were carried out to investigate the proposed tubular towers in more detail. Instead, a simplified approach should serve to estimate tubular tower masses. This required a well documented and investigated WT and tower as reference. Again, the 5 MW baseline WT was appropriate with respect to these requirements and its properties may be reviewed in section 3.2.

As reasonable simplification for this study, the towers were designed for extreme loads, such as suggested by the previous subsection 5.2.3. This assumption is not self-evident, because the design report of the NREL reference WT states that the tower wall thickness had to be scaled up about 30 % to avoid resonance within all operational conditions. Thus, it was designed with respect to its eigenfrequency and not with respect to ultimate loads. A review of the WTs Campbell diagram in Figure 4.4 shows, that this reason for increased wall thickness is not really comprehensible. In fact, the eigenfrequency of the reference tower is very close to the 3p excitation frequency at cut-in rotational speed. A discussion with the developers of the NREL reference WT reveals, that earlier versions of the baseline WT had a higher cut-in rotational speed and that the tower wall thickness increase is an artefact from that time to address earlier resonance problems. However, this adjustment was never been corrected with respect to the new operational rotational speed range. The mentioned discussion may be reviewed in Appendix 14.12. Regardless of the real driving factor, material strength and buckling were assumed to govern the wall thickness of the tower designs for this comparison study with transport constraints to the outer tower dimensions, such as indicated by Figure 5.2 and listed in Table 5.4. Case 1 corresponded to the outer dimensions of the reference WT and case 2 used the transport constraint at the tower top and bottom, while the tower top diameter remained the same as for case 1. This decision was made attributing to the assumption of standardized yaw bearing diameters. In case 3, an elliptical cross section with transport considering outer dimensions at the tower base and tower top was assumed.

Name	Case 1	Case 2	Case 3
shape	circular	circular	elliptical
top diameter D_T	3.87 m	3.87 m	–
bottom diameter D_B	6.0 m	4.3 m	–
width B	–	–	7.5 m
height H	–	–	4.3 m
transport constraint	not valid	valid	valid

Table 5.4: Top and bottom tower cross sectional properties, with the additional aspect of transportability

For each of these cases a minimum wall thickness was iteratively calculated at the tower top and bottom to withstand the respective maximum loads from the ASE load simulation in subsection 3.5.3. In terms of this preliminary study, all outer dimensions and wall thickness were linear interpolated between the top and bottom of the tower, whereby the designing loads are summarized in Table 5.5. Note that shear forces were neglected, because of their low shear stress contribution in terms of the cross sectional utilization.

Tower Section	$F_{z,ref}$ / N	$M_{x,ref}$ / Nm	$M_{y,ref}$ / Nm	$M_{z,ref}$ / Nm
Top	−5.40e6	−2.58e7	−2.98e7	−3.03e7
Bottom	−7.03e6	−9.52e7	1.64e8	−3.03e7

Table 5.5: Tower top and bottom extreme loads from the reference ASE load simulation in subsection 3.5.3

The SS and FA bending moments at the tower base were assumed to increase linear with the tower hight according to Equation 5.10.

$$\begin{aligned} M_x &= M_{x,ref} \frac{H_T}{H_{T,NREL}} \\ M_y &= M_{y,ref} \frac{H_T}{H_{T,NREL}} \end{aligned} \tag{5.10}$$

This assumption is not accurate, because of the moments from wind drag on the tower, but should serve as simplified approach for this study. Note that tower drag was already included in the loads of Table 5.5 and had no significant influence to the extreme loads, such as discussed in subsection 3.5.3. The lift induced additional SS bending moment at the tower base was included for another study case of the ellipse (C3,wl) to quantify its relevance for the whole tower design. Therefore, Equation 5.9 was used with the same parameters as in subsection 5.2.4 to be added to the calculated SS tower base bending moment from Equation 5.10. Since the SS tower base bending moment was driven by a production load case (2.1), a conservative high wind speed of 30 m/s with an angle of attack of 15 deg was assumed to calculate the additional SS base moment from lift for the different tower heights. Except for the loads, the shell boundary conditions, and the unsupported shell lengths, which should not exceed 30 m according to Equation 5.11, all other parameters remained the same as in Table 5.2. Although tower top and bottom cross sections were optimized, shell boundary conditions of BC2 were assumed for both ends of shells to account for the majority of shell segments in the tower, which were seen as long ring stiffened shells according to Figure 8.1f in DIN EN 1993-1-6 [44, p. 39]. In terms of tubular WTs, tower flanges are these ring stiffeners.

$$l_{\text{shell}} = \frac{H_{\text{T}}}{\lceil H_{\text{T}}/30\text{ m}\rceil} \tag{5.11}$$

The tower's eigenfrequencies were calculated by means of the Lagrangian equations of motion, such as explained in Appendix 14.13. An iterative wall thickness optimization for each tower case at each tower hight under the previous discussed assumptions and constraints led to the results in Figure 5.7. It shows the tower mass in the first diagram, the tower top and bottom wall thickness in the second diagram, the specific tower mass in the third diagram, and the tower's eigenfrequencies in the last diagram for tower heights from 80 m to 180 m. In this design study, the mass of the tower with original outer dimensions (C1) resulted to 229.121 t for the reference tower height of $H_{\text{T,NREL}}$ = 87.6 m and was therefore significantly lower than the documented mass of 347.460 t for the reference tower [109] with the same outer dimensions. The reference tower mass would decrease to 267.861 t if the 30 % wall thickness increase attributable to (unjustified) eigenfrequency requirements would have been neglected. A remaining difference of 38.740 t in both designs may be explained through the adjusted control settings to reduce the extreme SS tower base bending moment and the tower torsional moment, such as discussed in subsection 3.5.3.

Tower masses of all cases increased almost quadratic with a very small cubic part, because of the assumption that rotor thrust forces would not change for larger hub heights. An increased hub height means that the rated wind speed is reached earlier. This has not necessarily implications to the extreme rotor thrust, because rotor blades will be pitched out of the wind to control the rotor power to be at rated level and thereby the rotor thrust is held down. One extreme rotor thrust force changing aspect is the different turbulence intensity at different hub heights for the same ground roughness. Such effects were neglected in this preliminary study for tubular tower concepts. Note that the coefficients for a fitted cubic polynomial to the results of the first diagram are documented in Appendix 14.14. The overall tower mass of the tower with original outer tower dimensions (C1) required 27.1 % less material at H_{T} = 80 m and 31.38 % less material at H_{T} = 180 m, compared to the tower with transport constrained outer dimensions (C2). An elliptical tower (C3) required 28.94 % and 26.69 % less material for both tower heights, compared to C2. This shows, that an transport constraint considering elliptical tower concept (C3) is capable to reduce the required tower mass to almost the same as the not transport constraint considering tower with original outer dimensions (C1), compared to the constrained circular concept (C2). Inclusion of the conservatively assumed lift component at the elliptical tower (C3,wl) did not significantly change the results, such as indicated by the black line in the first diagram. Note that the discontinuity at tower heights between 130 m and 140 m was caused by the thickness depending material strength, such as listed in the first row of Table 5.2.

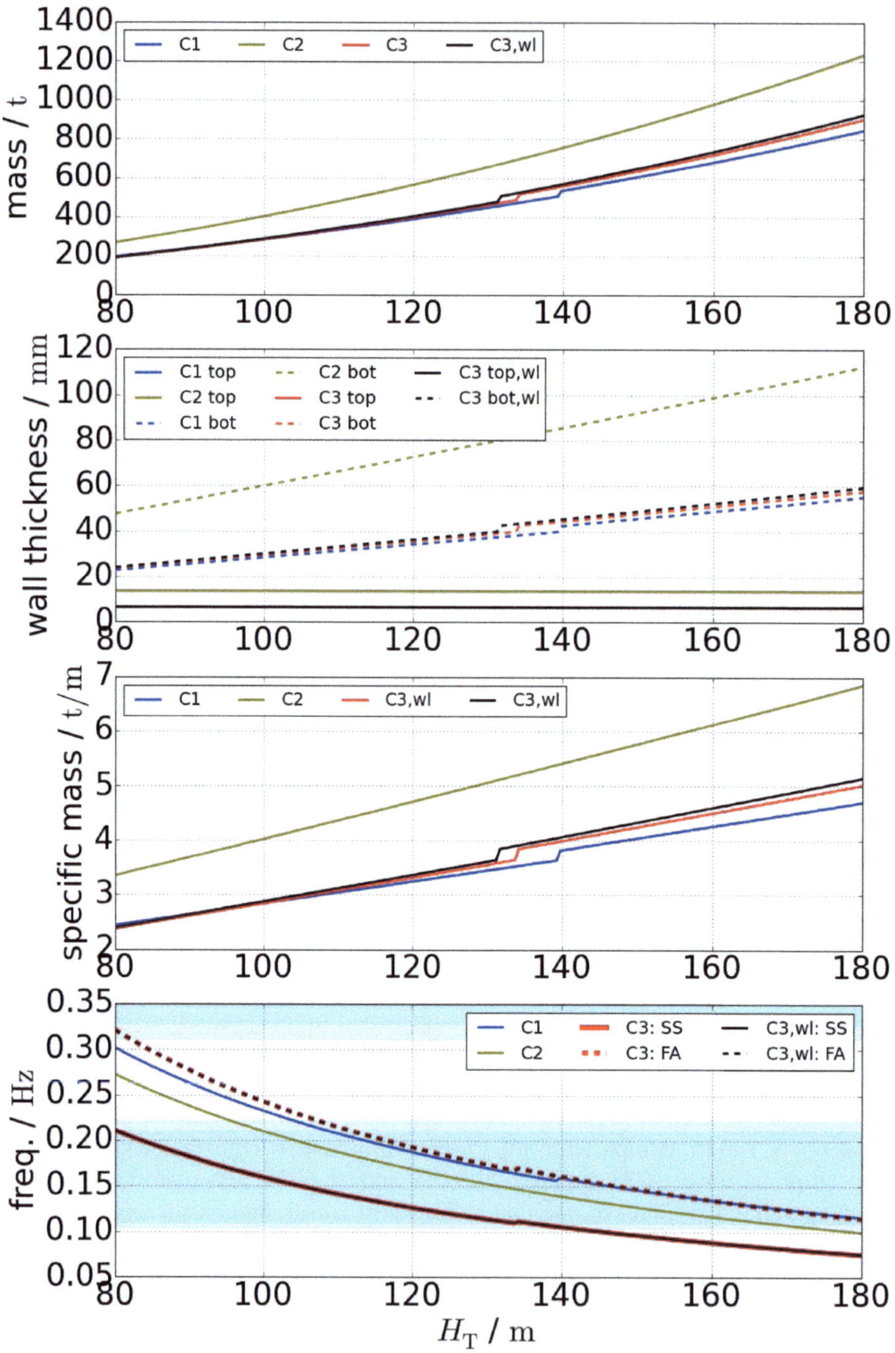

Figure 5.7: Tower mass, wall thickness, specific mass, and eigenfrequencies over the height of different tubular tower concepts

A good manufacturable wall thickness of < 50 mm at the tower bottom was achieved for tower heights of < 164.9 m, < 84.0 m, < 156.9 m, and < 153.4 m for cases C1, C2, C3, and C3,wl. This shows, that the elliptical tower concept is feasible and a material saving solution for transport constrained towers with large hub heights. The specific tower mass increased linear with the tower height as it is the derivative of the mass with respect to the tower height. Thereby, different slopes corresponded to different cases. The last diagram in Figure 5.7 shows the first bending eigenfrequency of each tower. It reveals, that the most configurations had tower eigenfrequencies within the 1p excitation frequency of the operational rotational speed range, indicated by the bottom light blue rectangular patch. In cases C1 and C2, the +5 % frequency safety range is reached at tower heights of 103.9 m and 95.5 m. Since the elliptical tower had different bending eigenfrequencies in the FA and SS direction, no tower of cases C3 or C3,wl exists, which would not hit either the 1p or 3p excitation frequency within the operational range. Note that the 3p excitation frequency range is indicated by the light blue rectangular patch at the top of the fourth diagram. In terms of eigenfrequencies, none of the present cases was appropriate for large hub heights if no further improvements of the dynamic structural properties would be incorporated. If the outer tower dimensions should remain the same, following approaches may be applied to reduce or to avoid resonance of such tower configurations:

- increase the wall thickness, especially at the tower bottom to increase the tower's eigenfrequencies
- increase the tower head mass to decrease the tower eigenfrequencies
- use a damping system to convert energy into heat, whereby resonance amplitudes can be reduced
- use a monitoring system on the WT to detect tower resonances and drive through them as quick as possible

Each of these options is either connected to increased costs or decreased WT efficiency and is not further evaluated in terms of this preliminary study.

5.3 Other Tubular Tower Concepts

This section provides a brief discussion about other possible tubular tower shapes. Thereby, Kamm-back and outer skin modified towers are treated.

5.3.1 Kamm-back Tower

Kamm-back profiles are aerodynamically shaped, but have no real trailing edge, such as shown in Figure 5.8. Their invention was done by Kamm, W. [100] with the intention to realize a low aerodynamic drag for cars with the compromise of a practical shape. Therefore, he proposed to cut off the trailing edge of a streamlined car at a point of 50 % of the maximum cross sectional thickness. This sharp straight edge in the back induces a triangular vortex area behind the car and acts as an extended aerodynamic profile. Car velocity depending Reynolds numbers are between 10^4 and 10^7 and have drag coefficients of $c_D \approx 0.3$ if a Kamm-back is applied [100].

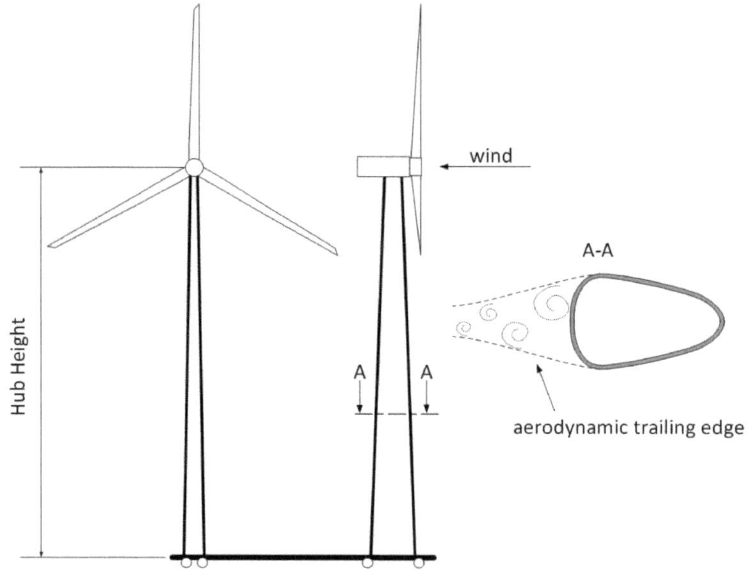

Figure 5.8: Concept - Kamm tail tower shape and yaw bearing at tower base

The cross section in Figure 5.8 has a curved shape instead of a sharp edge at its back due to manufacturing constraints. The physical behaviour of flows tending to stay attached at curved surfaces is known as the Coanda effect [173]. Nevertheless flows detach from curved surfaces, known as boundary layer separation, if the curvature or the Reynolds number increases [151, p. 69-70]. Consequently, forming of the turbulence field which serves as aerodynamic trailing edge behind the profile is therefore more dependent on the Reynolds number compared to a sharp edged design. However, the shape has an advantage with respect to stiffness compared to other mentioned shapes. Material distributions resemble a rectangular box, wherefore less material is necessary to realize the same stiffness as for an elliptical shape. Plate-like areas have low buckling resistance and have to be locally stiffened, depending on the exact design and loads. The rolling process can be carried out similar to conventional circular profiles with a difference regarding rotation angle depending welding heights for circumferential welding seams. No limitations to the exact shape achieves flexibility for a good compromised design between aerodynamics, stiffness, stability and manufacturing. Kamm-back profiles are not warping free, wherefore additional normal stresses occur based on torsional moments and have to be considered during a design process.

5.3.2 Outer Skin Modification

This concept describes an aerodynamically modified outer skin for a conventional bottom fixed tower, such as shown in Figure 5.9 and is based on patents such as [133], [185] and [185]. The overarching goal of the outer skins is wind drag reduction through aerodynamically formed shapes. Airfoil profiles and Kamm-back profiles are both imaginable in this context. The skins may be connected to the rotatable nacelle to have the same alignment to the wind as the rotor. Coverings can end at a certain height of the tower or otherwise be designed along the whole tower with a bearing at the bottom of the tower for the skin. This latter concept achieves rotor independent skin alignments and thereby faster reactions with respect to wind direction changes.

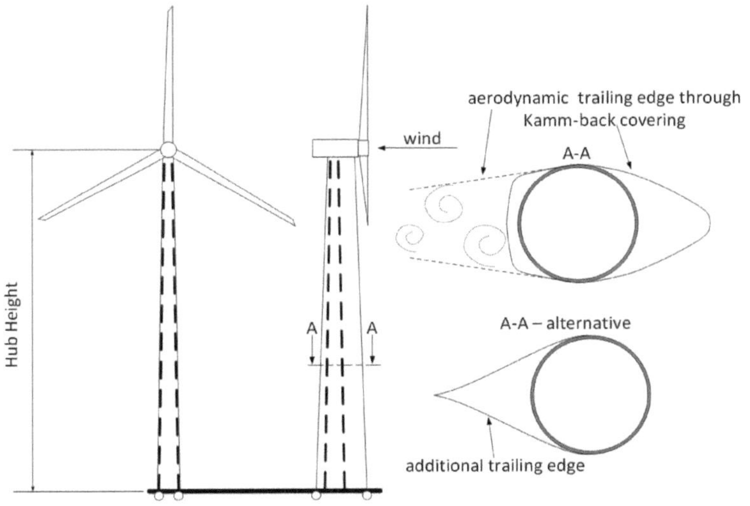

Figure 5.9: Concept - Conventional tower with outer skin modification and yaw bearing at tower base

Corresponding drag reductions achieve smaller wall thicknesses or diameters of the tower. Such smaller tower diameters help regarding transport problems up to a certain point. Furthermore, no load direction oriented stiffness design is applied to the tower and the material saving potential is assumed to be low, because tower drag is not a huge design driving factor, such as indicated by the ASE analysis in subsection 3.5.3.

Furthermore, luv WT approaches have to take into account the smaller blade clearance, because of the protruding skin, which leads to shorter or more expensive blades. Another option is to apply only a trailing edge, such as the alternative cross section in Figure 5.9 or to use a downwind rotor concept instead. In summary, this approach reveals maybe a little material saving potential for WT towers, but does not give satisfactory answers according to transport problems of larger hub height dimensions. Circular profiles are warping free, wherefore no additional normal stresses with respect to torsion occur.

6 Rotatable Inclined Lattice Tower Concepts

6.1 Tower Cross Sectional Concepts

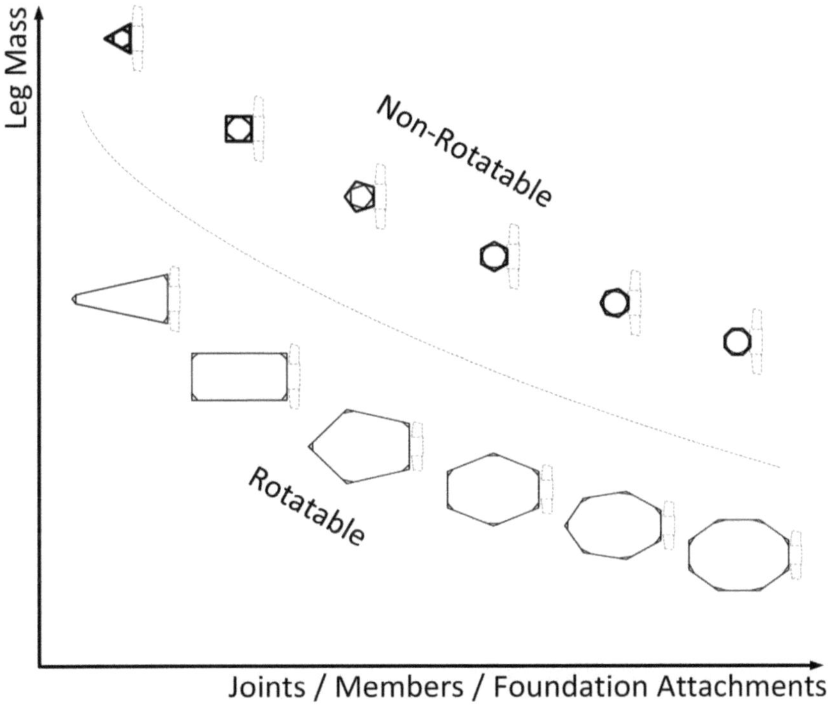

Figure 6.1: Conceptual comparison between different non-rotatable and rotatable lattice towers

Rotatable lattice towers have a significant aerodynamic advantage, because they work not as a wing in contrast to aerodynamically shaped tubular concepts. This aspect becomes even more important for extreme wind speed parking situations with yaw system fault, where inclined airflow may cause tremendous lift forces, which would act along the weak direction of a rotatable tower. As suggested in subsection 2.7, many different types and shapes of lattice towers are conceivable. Figure 6.1 shows some different non-rotatable and rotatable lattice tower cross sectional concepts. A warping free three legged design would thereby reduce the amount of members and joints within the structure compared to a polygonal tower cross section with more legs. On the other hand, more legs lead to lower axial loading per leg and therefore lower leg masses. In general, the advantage of rotatable lattice towers with respect to non-rotatable ones lies in the increased space for the outer tower dimensions. Non-rotatable lattice towers must have small outer dimensions at least for the tower sections where the blades pass, such as indicated by Figure 2.12. A review of Equation 2.50 reveals that in case of a rectangular tower cross section, leg forces reduce linear with their distance to each other. For all rotatable concepts, the tower alignment with respect to the RNA is such that the straight front edge reveals good blade clearance. How the tower looks like behind the front edge does not matter for the blade clearance, because it is fixed with respect to the rotor alignment.

This freedom in the outer shape of the lattice tower cross section means that it can be designed with respect to the main load direction, such as explained in section 4.1. Therefore, a square tower cross section may become a stretched rectangular one or a polygonal tower cross section may be stretched to be adjusted to the higher FA loadings, such as suggested by Figure 6.1. However, rotatable lattice towers reveal material saving potential even for loads, which are equal in SS and FA direction because of the possibility of increased outer dimensions, compared to non-rotatable ones. In the end, the amount of legs is an optimization problem with costs and CO_2-emissions being the values to be minimized. Counting the bolts of the tower in Figure 2.12 reveals that each x-bracing or k-bracing, which consists out of four members, requires about 48 bolts for its five joints. Furthermore, the connection between each of the 8 leg segments per leg requires about 52 bolts. Summing the bolts for the leg connections and the 21 bracings per leg up and assuming the lockbolt prices from the offer in Appendix 14.15, results in Figure 6.2 for different amounts of lattice tower legs. Note that the Butzkies tower in Figure 2.12 contains also bolts to merge the two cross sectional parts to one member along its lengths. These bolts were neglected in Figure 6.2 where only joint bolts were considered. Furthermore, the 2.5 MW WT on the Butzkies tower leads probably to lower tower head loads than the 5 MW NREL reference WT, but the first tower is 12.4 m taller than the reference one. Therefore, the results were not completely representative for an analogous lattice tower for the reference WT.

However, assuming the larger 25.4 mm Bobtail lockbolts for the whole tower, revealed costs between $25,416$ € and $67,775$ € for towers with 3 and 8 legs, respectively. This cost difference of $42,359$ € appears relatively small, but note that only purchase costs are considered. The assembly of all the joints and their observation of the pre-load in the bolts lead to additional costs over the lifetime of the WT. Since the lockbolts advantage are the low maintenance costs, these costs were not evaluated for the preliminary nature of this work.

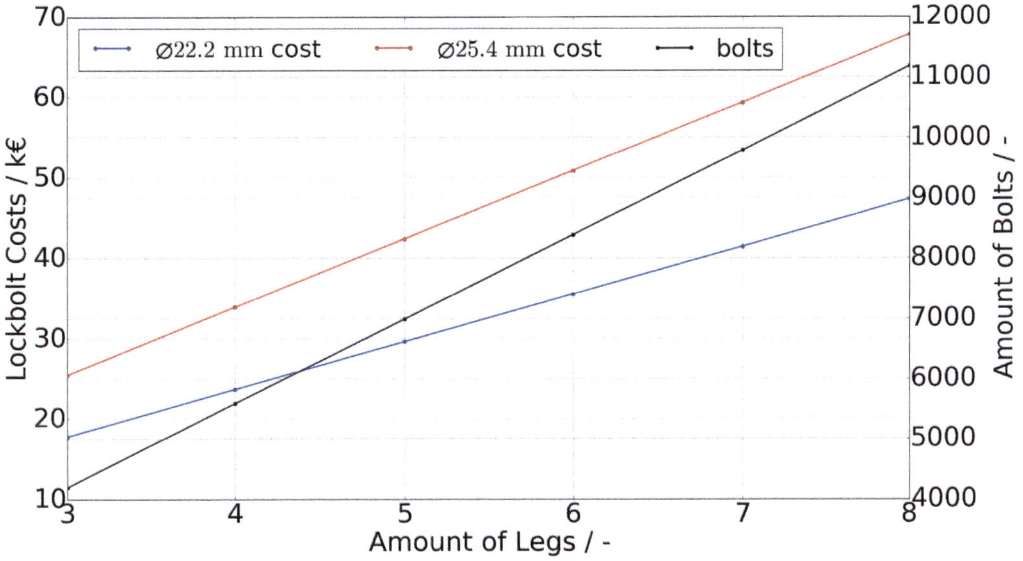

Figure 6.2: Bobtail lockbolt costs for different bolt diameters and different amounts of lattice tower legs based on the Butzkies tower in Figure 2.12 and an lockbolt offer from KVT-Fastening GmbH

6.2 Tower Inclination

Apart from the tower's ability to rotate, another innovative aspect is a tower inclination. It can be applied to tubular and lattice concepts, such as shown in Figure 6.3. The tower inclination leads to additional bending moments within the tower. These bending moments are induced through the lever arms from the center of mass from the RNA and the tower masses to the original vertical tower base centerline. Such bending moments act within the tower during low wind speeds and will decrease with higher thrusts and wind drag on the tower. At a certain point, the bending moment changes its sign and increase towards the opposite direction. This preloaded tower can be used to design the tower cross section according to a tensile and pressure side, such as mobile crane boom cross section concepts present as an example, shown in Figure 2.38c.

Extreme bending moment around the rotor parallel axis	Value / kNm
$M_{y,max,tb}$	$1.53 \cdot 10^5$
$M_{y,min,tb}$	$-3.53 \cdot 10^4$
$M_{y,max,t50}$	$7.89 \cdot 10^4$
$M_{y,min,t50}$	$-1.83 \cdot 10^4$

Table 6.1: Extreme tower bending moments at tower base and 50 % tower height stages around the rotor parallel axis y of the land-based 5 MW NREL reference wind turbine [110, p. 204]

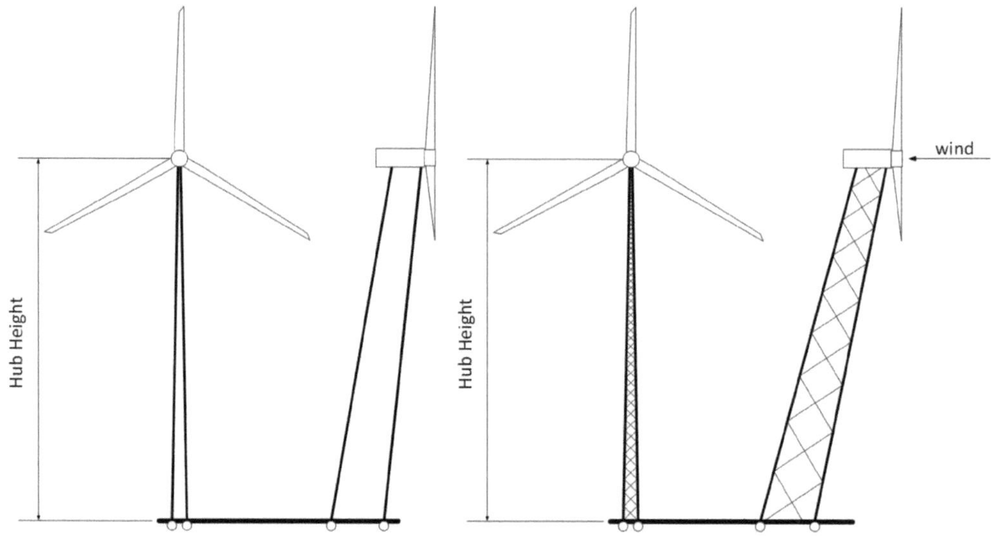

Figure 6.3: General inclined wind turbine tower concepts

6.2.1 Inclination Consequences to Extreme Tower Loads

Another option is to adjust the inclination angle in such a way that an equilibrium of the unequal extreme bending moments from the turbine operation in each direction with and against the wind stream is achieved. This reduces the overall extreme bending moments and facilitates further material savings. For the purpose of demonstration, extreme bending moments from literature ASE load simulations of the land-based 5 MW NREL reference WT at the tower base and 50 % tower height are presented in Table 6.1 [110, p. 204]. All listed bending moments act around the rotor parallel axis y. Assuming a full balanced design according to extreme loads and $M_{y,max} - M_{y,min} = \Delta M_y = \text{const.}$, load reductions of

$$\begin{aligned}\text{Reduct.}_{y,tb} &= 100\,\% - \frac{100\,\%}{\max\left(|M_{y,max,tb}|, |M_{y,min,tb}|\right)} \frac{|M_{y,max,tb}| + |M_{y,min,tb}|}{2} \\ &= 38\,\%\end{aligned} \tag{6.1}$$

for the tower base and

$$\begin{aligned}\text{Reduct.}_{y,t50} &= 100\,\% - \frac{100\,\%}{\max\left(|M_{y,max,t50}|,|M_{y,min,t50}|\right)}\,\frac{|M_{y,max,t50}|+|M_{y,min,t50}|}{2} \\ &= 38\,\%\end{aligned} \tag{6.2}$$

for the 50 % tower height stage resulted. A tower inclination parallel to the rotor plane revealed reductions of extreme loads up to $\text{Reduct.}_{x,tb} \approx 16\,\%$ and $\text{Reduct.}_{x,t50} \approx 27\,\%$, respectively. Note that these load reductions were from theoretical nature and have to be validated in an ASE load simulation. Furthermore, loads found in literature only accounted for DLCs 1.x in their extreme load evaluation. Other ratios may result from the inclusion of more DLCs. However, the proposed load reductions would come additionally to load reductions from lower aerodynamic tower drag in case of a lattice tower and lower periodic excitations, caused by the smaller tower dam influence from a larger blade clearance.

6.2.2 Inclination Consequences to Tower Fatigue

Fatigue loads are also influenced by tower inclination if mean stress levels are considered. According to Eurocode 3 DIN EN 1993-1-9 [47], mean stresses should not be mentioned for welded joints, except stress-relief annealing is applied. DNV GL [63] proposes a reduction factor f_m for hot spot stress ranges $\Delta\sigma$ for welded structural details under the following conditions: (1) They have to be subject to post weld heat treatment or (2) corresponding low residual stresses must be documented. The approximation of the tower base bending moment as an ideal sinusoidal function around a constant mean value $M_{mean} \neq 0$ would lead to qualitative normal stresses, such as indicated in Figure 6.4.

The black dashed line in Figure 6.4 represents normal stresses, caused by the structural weight. Red and blue lines are the mean and periodic cycling stresses on the tensile and on the compression side of an arbitrary tower. In many fatigue assessment cases, the highest evaluated lifetime damage at a certain cross sectional point drives the necessary wall thickness of a whole tower section. Thereby, a decreased mean stress level decreases the reduction factor f_m, which leads to material savings. The factor is determined by Equation 6.3.

$$f_m = \begin{cases} 0.8 & \text{for} \quad \sigma_m \leq -\Delta\sigma/2 \\ 0.9 + 0.2\dfrac{\sigma_m}{\Delta\sigma} & \text{for} \quad -\Delta\sigma/2 < \sigma_m < \Delta\sigma/2 \\ 1.0 & \text{for} \quad \sigma_m \geq \Delta\sigma/2 \end{cases} \tag{6.3}$$

With increasing tower inclination against the wind direction, mean normal stresses on the compression and tensile side will come closer to the normal stress caused by the structural weight $\sigma_{m,W}$, such as implied in Figure 6.5.

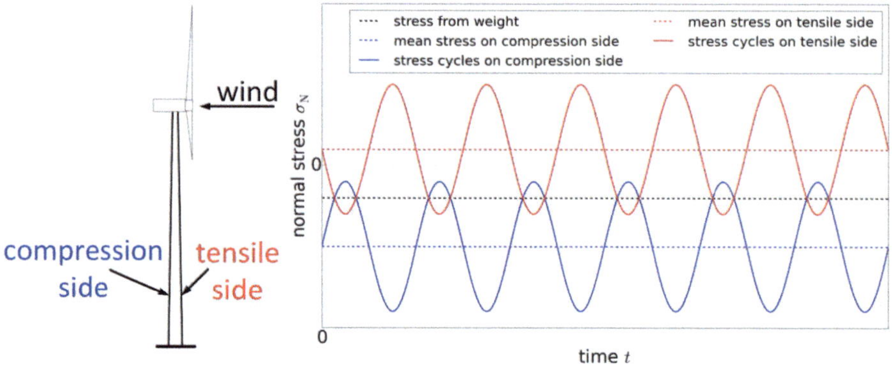

Figure 6.4: Idealized stress cycles of the tensile and compressive side of a wind turbine tower

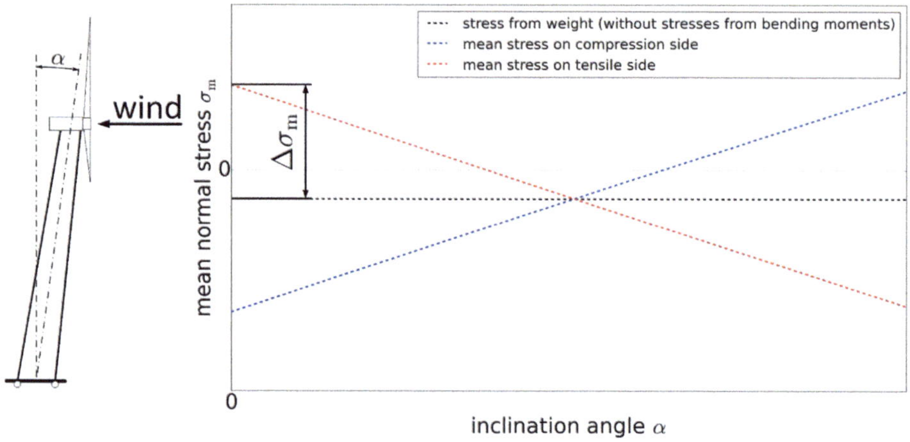

Figure 6.5: Idealized mean normal stress as function of tower inclination

Thereby, mean stress reductions of

$$\Delta\sigma_{\mathrm{m}}(\alpha_{\mathrm{TI}}) = \sigma_{\mathrm{m,T}}(\alpha_{\mathrm{TI}}) - \sigma_{\mathrm{m,W}} \tag{6.4}$$

with $\sigma_{\mathrm{m,T}}$ as mean normal stress on the tensile side as function of the inclination angle α_{TI}, are achievable. From Equation 6.3 one can see that fatigue damage reductions of up to 20 % are achievable under the previous described conditions.

6.2.3 Blade to Tower Clearance

Another advantage of tower inclination is the increased blade to tower clearance. It enables larger or cheaper blades, compared to conventional vertical centerline tower shapes, because modern WT blades are commonly designed with respect to their stiffness. Thereby, blade to tower clearance is an important design criterion to the blade structural properties and is checked within the SLS, such as explained in subsection 2.2.8. The allowed rotating blade to tower clearance amounts to 30 % to 20 % with respect to the unloaded state, dependent on the measured stiffness reliability of the real blades. For parked situations, the blade to tower clearance should not be less than 5 % of the unloaded state [61, p. 32]. Figure 6.6 shows the difference in the blade to tower clearance between a conventional and an inclined tower. Thereby, α_{TI} is the tower inclination angle, α_{ST} is the rotor shaft tilt, α_{PC} is the blade precone angle, H_{T} is the tower height, l_{OH} is the overhang length from the tower top center at height $H_{\mathrm{T}} + l_{\mathrm{TS}}$ to the blade root center in the hub, and $d_{\mathrm{T}} = f(z)$ is the horizontal distance from the tower center line to its outer surface at the height of the blade tip z_{Tip}. In this consideration, blade pre-bends are neglected.

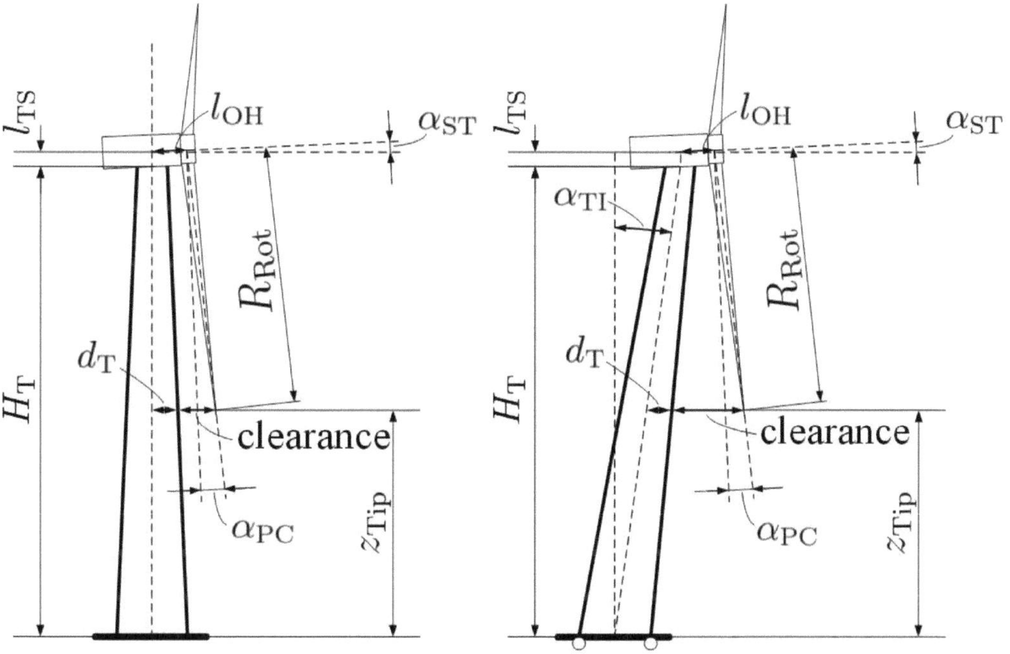

Figure 6.6: Comparison between conventional and rotatable inclined wind turbine tower with respect to the blade to tower clearance

From Figure 6.6, the blade to tower clearance in unloaded condition calculates as

$$\text{clear.} = \cos(\alpha_{\mathrm{ST}})\, l_{\mathrm{OH}} + \tan(\alpha_{\mathrm{TI}})\,(H_{\mathrm{T}} - z_{\mathrm{Tip}}) + \sin(\alpha_{\mathrm{ST}} + \alpha_{\mathrm{PC}})\, R_{\mathrm{rot}} - d_{\mathrm{T}}(z_{\mathrm{Tip}}), \tag{6.5}$$

where the blade tip height is

$$z_{\mathrm{Tip}} = H_{\mathrm{T}} + l_{\mathrm{TS}} + \sin(\alpha_{\mathrm{ST}})\, l_{\mathrm{OH}} - \cos(\alpha_{\mathrm{ST}} + \alpha_{\mathrm{PC}})\, R_{\mathrm{rot}} \tag{6.6}$$

and the outer tower radius at the height of the blade tip becomes

$$d_{\mathrm{T}}(z_{\mathrm{Tip}}) = \frac{D_{\mathrm{B}} + \frac{(D_{\mathrm{T}} - D_{\mathrm{B}})}{H_{\mathrm{T}}} z_{\mathrm{Tip}}}{2}, \tag{6.7}$$

where D_{B} and D_{T} are the tower base and top diameters, respectively. Application of the 5 MW NREL reference WT parameters, documented in [109], resulted in the blade tip to tower clearances in Figure 6.7. In this example, the conventional tubular steel tower of the reference WT was used. The black dashed line marks the original configuration of the reference WT and the red curve shows, how the blade to tower clearance increased with increasing tower inclination angle α_{TI}. The blue, green, and yellow curves indicate the clearances for different RNA configurations, where the shaft tilt angle α_{ST} or/and the blade precone angle α_{PC} were changed. Thereby, setting $\alpha_{\mathrm{PC}} = 0$ deg increased the rotor area and therefore the wind energy harvest about 1 %. A corresponding tower inclination angle adjustment of $\alpha_{\mathrm{TI}} = 2.61$ deg achieves the same blade to tower clearance as the reference configuration. Analogously, a setting with original precone angle $\alpha_{\mathrm{PC}} = 2.5$ deg, but with changed shaft tilt angle $\alpha_{\mathrm{ST}} = 0$ deg increased the rotor area about 1.5 %. In this situation, compensation of the reduced blade to tower clearance, required a tower inclination of $\alpha_{\mathrm{TI}} = 5.13$ deg. The largest rotor area was achieved by setting the shaft tilt angle and the blade precone angle to $\alpha_{\mathrm{ST}} = 0$ deg and $\alpha_{\mathrm{PC}} = 0$ deg. With this adjustment, the rotor area and energy yield efficiency increased about 1.7 % with respect to the reference configuration and a tower inclination of $\alpha_{\mathrm{TI}} = 7.67$ deg was required for blade to tower clearance compensation.

The energy harvest efficiency increase of 1.7 % through tower inclination is an additional way to improve the economics of rotatable WT towers. A more detailed consideration on how much the tower inclination affects the economics of rotatable WTs is given for the final design in subsection 10.2. Another way is to remain the rotor shaft tilt and the blade precone angles at the original values and to reduce the stiffness of the blades by using less material. In this case, the tower inclination angle must be adjusted with respect to the decreased rotor blade stiffness. This could likewise lead to improved economics, because of the saved blade material but is out of scope for this work.

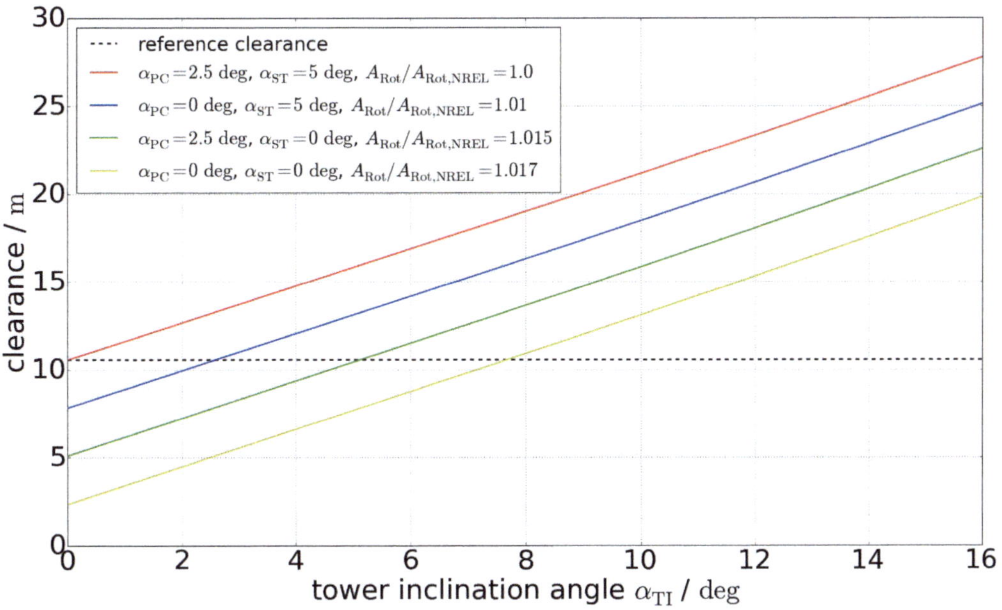

Figure 6.7: Blade to tower clearance for different tower inclination angles α_{TI} and different RNA configurations with respect to the 5 MW NREL reference wind turbine

Tower Dam

An additional option is to use the tower inclination to increase the blades distance to the tower and therefore to the tower dam area. Note that the tower dam effect is introduced in subsection 4.4.2. The larger blade distance to the tower dam area decreases the 1p periodic excitation load on each blade and the 3p periodic excitation load on the tower, wherefore fatigue in the components is reduced. Results of recent research activities, which are presented in Figure 6.8, show that for a certain configuration, the tower dam effect of conventional tubular steel towers is responsible for 34 % de-loading of the tower drag, 11.6 % de-loading of the blade shear force and for a reduction of 25 % of the rotor torque during blades passage [171]. In that study, detailed experimental and CFD analyses were used to investigate the blade to conventional tubular tower dam interaction [171].

Although these values may be different in other WT configurations, they indicate that increased blade to tower clearance through tower inclination leads to dynamical load reductions on the blades, the tower and the rotor shaft and to increased energy harvest efficiency. Up to now, no results for lattice structures are published and are still a task for future research.

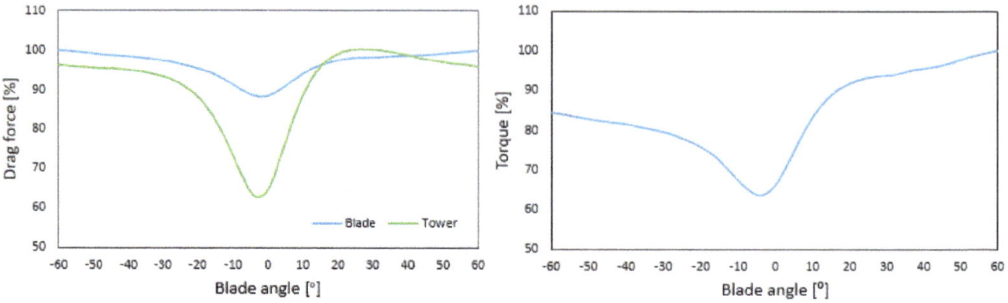

Figure 6.8: Conventional tubular tower dam effect to tower, blade and rotor shaft loads, taken from Shkara et al. [171]

6.2.4 Loads from Gyroscopic Motion

Inclination of a WT tower leads to a higher distance between the rotor plane and the yaw axis, wherefore this subsection treats the influence of the inclination with respect to gyroscopic loads. The following text distinguishes between so called gyroscopic moments and torsional moments, where the first designation points to moments caused by the gyroscopic effect. In this context, torsional moments are specified as moments caused by a angular acceleration of mass inertia. The section is focused on the effects of rigid body motion of the blades and the resulting moments regarding a stationary coordinate system in space. Hence, the first part shows the derivation of an appropriate physical model and the second subsection presents the results, extracted from the model.

Physical Model of Motion
A physical model, such as shown in Figure 6.9 is used to apply Euler's second law. The law can be used to compute the acting moments around the axis of an inertial reference frame $\mathcal{F}_0$. Bauchau [8, p. 99] cites Euler's second law in English as

> *The time derivative of the angular momentum vector of a system of particles equals the sum of all moments externally applied to the system, when these quantities are evaluated with respect to a common inertial point.* Bauchau [ibid.] gives the references [69], [70] for his translation.

From there the description of the position vector $\underline{r}_j^I$ of one mass point m_j resolved in the inertial reference frame $\mathcal{F}_0 = \left[\mathbf{O}, \mathcal{I} = \left(\bar{i}_1, \bar{i}_2, \bar{i}_3\right)\right]$ is required to evaluate the angular momentum and its derivative.

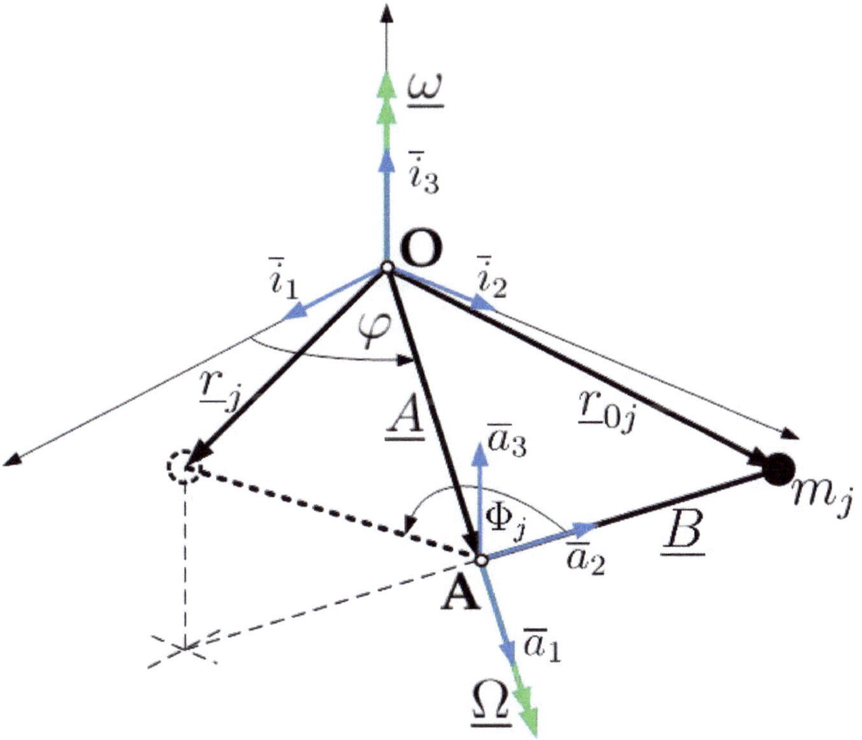

Figure 6.9: Gyroscopic physical model for one blade. Drawing: A. Struve

The inertial reference frame $\mathcal{F}_0$ is defined according to its orthonormal basis $\mathcal{I}$, which has the origin point **O**. Unit vectors, such as $\bar{i}_1$ are declared with a bar on the top, vectors, such as $\underline{r}_j^I$ are denoted with a bar on the bottom and matrices are marked by two bars on the bottom. In this model one mass point m_j represents the mass of one WT blade. After the mathematical description of the system is accomplished, it is no problem to add more masses to consider more blades, such as $n_{\text{bld}} = 3$ blades for conventional WTs. The position vector $\underline{r}_j^I$ can be derived from the description of a second reference frame $\mathcal{F}_1 = [\mathbf{A}, \mathcal{A} = (\bar{a}_1, \bar{a}_2, \bar{a}_3)]$, which lays in the rotor hub center **A**. $\mathcal{F}_1$ rotates around yaw axis $\bar{i}_3$ and its unit vectors $\bar{a}_1$ and $\bar{a}_2$ will always lay within plane $\mathcal{P}_{\text{yaw}} = \left(\mathbf{O}, \bar{i}_3\right)$, where **O** describes the location of the plane and $\bar{i}_3$ is the normal vector of this plane, which further implies $\bar{a}_3 || \bar{i}_3$. Vector $\underline{A}$ describes the position of the hub center and its norm $||\underline{A}||$ is equal to the orthogonal distance of the hub with respect to the yaw axis $\bar{i}_3$. Vector $\underline{B}$ represents the position of the blade mass regarding orthonormal basis $\mathcal{A}$. Yaw motion is represented through the yaw angular speed $\omega = \underline{\omega}^T\, \bar{i}_3$ and the resulting angle

$$\varphi = \omega\, t + \varphi_0 \tag{6.8}$$

between $\bar{i}_1$ and $\underline{A}$, where scalar t represents time. Incidentally all position vectors, unit vectors $\bar{a}_1$ and $\bar{a}_2$, and angular speed $\underline{\omega}$ are time dependent, but denotation (t) was neglected to make the equations more readable. The position $\underline{r}_j^{\mathcal{I}}$ of m_j resolved in $\mathcal{F}_0$ will be described through a rotation of initial position vector $\underline{r}_{0j}^{\mathcal{I}}$ around axis $\overline{a}_1$. The corresponding rotation angle is called Φ_j and describes the rotation of blade j around the rotor axis resolved in reference frame $\mathcal{F}_1$. It is calculated through the rotor angular velocity $\Omega = \underline{\Omega}^T\, \overline{a}_1^{\mathcal{A}}$

$$\Phi_j = \Omega\, t + \Phi_{0j} \tag{6.9}$$

where Φ_{0j} is equal to the initial angle of each blade j. Unit vectors $\overline{a}_1^{\mathcal{I}}, \overline{a}_2^{\mathcal{I}}$ and $\overline{a}_3^{\mathcal{I}}$ are described through a rotation tensor $\underline{\underline{R}}_{\varphi}^{\mathcal{I}}$, which brings orthonormal basis $\mathcal{I}$ to $\mathcal{A}$. The corresponding rotation tensor $\underline{\underline{R}}_{\varphi}^{\mathcal{I}}$ is defined as

$$\begin{aligned}
\underline{\underline{R}}_{\varphi}^{\mathcal{I}} &= \underline{\underline{I}} + \sin(\varphi)\widetilde{\overline{i}_3} + (1-\cos(\varphi))\,\widetilde{\overline{i}_3}\,\widetilde{\overline{i}_3} \\
&= \begin{bmatrix} 1 & 0 & 0 \\ 0 & 1 & 0 \\ 0 & 0 & 1 \end{bmatrix} + \begin{bmatrix} 0 & -S_\varphi & 0 \\ S_\varphi & 0 & 0 \\ 0 & 0 & 0 \end{bmatrix} + \begin{bmatrix} -1 & 0 & 0 \\ 0 & -1 & 0 \\ 0 & 0 & 0 \end{bmatrix} + \begin{bmatrix} C_\varphi & 0 & 0 \\ 0 & C_\varphi & 0 \\ 0 & 0 & 0 \end{bmatrix} \\
&= \begin{bmatrix} C_\varphi & -S_\varphi & 0 \\ S_\varphi & C_\varphi & 0 \\ 0 & 0 & 1 \end{bmatrix},
\end{aligned} \tag{6.10}$$

where S_φ and C_φ are abbreviations for $\sin(\varphi)$ and $\cos(\varphi)$, respectively. The used notation $(\widetilde{\bullet})$ denotes the skew-symmetric matrix of a vector and can be used as alternative representation of a cross product through a matrix multiplication. According to a rotation around axis $\overline{i}_3$, unit vectors of $\mathcal{A}$ resolved in $\mathcal{I}$ become

$$\overline{a}_1^{\mathcal{I}} = \underline{\underline{R}}_{\varphi}^{\mathcal{I}}\, \overline{i}_1 = \begin{pmatrix} C_\varphi \\ S_\varphi \\ 0 \end{pmatrix} \tag{6.11}$$

$$\overline{a}_2^{\mathcal{I}} = \underline{\underline{R}}_{\varphi}^{\mathcal{I}}\, \overline{i}_2 = \begin{pmatrix} -S_\varphi \\ C_\varphi \\ 0 \end{pmatrix} \tag{6.12}$$

$$\overline{a}_3^{\mathcal{I}} = \underline{\underline{R}}_{\varphi}^{\mathcal{I}}\, \overline{i}_3 = \begin{pmatrix} 0 \\ 0 \\ 1 \end{pmatrix} \tag{6.13}$$

The second rotation about axis $\overline{a}_1^I$, which describes the blade motion around the hub center takes place through rotation tensor $\underline{\underline{R}}_j^I$:

$$\begin{aligned}\underline{\underline{R}}_j^I &= \underline{\underline{I}} + \sin(\Phi_j)\widetilde{a}_1^I + (1-\cos(\Phi_j))\,\widetilde{a}_1^I\,\widetilde{a}_1^I \\ &= \begin{bmatrix} 1+S_\varphi^2(1-C_\Phi) & S_\varphi\,C_\varphi\,(1-C_\Phi) & S_\varphi\,S_\Phi \\ C_\varphi\,S_\varphi\,(1-C_\Phi) & 1+C_\varphi^2\,(C_\Phi-1) & -C_\varphi\,S_\Phi \\ -S_\varphi\,S_\Phi & C_\varphi\,S_\Phi & 1-S_\varphi^2 C_\varphi^2+\left(S_\varphi^2+C_\varphi^2\right)C_\Phi \end{bmatrix},\end{aligned} \tag{6.14}$$

where S_Φ and C_Φ are abbreviations for $\sin(\Phi_j)$ and $\cos(\Phi_j)$, respectively. With initial position vector $\underline{r}_{0j}^I$

$$\underline{r}_{0j}^I = \overline{a}_1^I\;||\underline{A}|| + \overline{a}_2^I\;||\underline{B}||\;\cos(\Phi_0) + \overline{a}_3^I\;||\underline{B}||\;\sin(\Phi_0) \tag{6.15}$$

the complete motion of m_j is described by

$$\underline{r}_j^I = \underline{\underline{R}}_j^I\;\underline{r}_{0j}^I. \tag{6.16}$$

The angular momentum $\underline{L}^I$ with respect to the inertial reference frame $\mathcal{F}_0$ is the sum of all linear moments $\underline{p}_j^I$ with respect to their position $\underline{r}_j^I$. Following equations are written without subscript ^{I}, because all corresponding variables are described in terms of the inertial reference frame.

$$\underline{L} = \sum_{j=1}^{n_{\mathrm{bld}}} \underline{r}_j \times \underline{p}_j = \sum_{j=1}^{n_{\mathrm{bld}}} m_j\;\underline{r}_j \times \underline{\dot{r}}_j. \tag{6.17}$$

According to Euler's second law a derivative with respect to time, denoted as $\frac{\mathrm{d}}{\mathrm{d}t} = (\dot{\bullet})$ leads to the acting moments around the reference frame axes:

$$\begin{aligned}\underline{M} = \underline{\dot{L}} &= \sum_{j=1}^{n_{\mathrm{bld}}} m_j\;\underline{\dot{r}}_j \times \underline{\dot{r}}_j + \sum_{j=1}^{n_{\mathrm{bld}}} m_j\;\underline{r}_j \times \underline{\ddot{r}}_j = \sum_{j=1}^{n_{\mathrm{bld}}} m_j\;\underline{r}_j \times \underline{\ddot{r}}_j \\ &= \sum_{j=1}^{n_{\mathrm{bld}}} m_j\;\widetilde{r}_j\;\underline{\ddot{r}}_j = \sum_{j=1}^{n_{\mathrm{bld}}} m_j\;\widetilde{\left(\underline{\underline{R}}_j\;\underline{r}_{0j}\right)}\;\left(\underline{\underline{\ddot{R}}}_j\;\underline{r}_{0j} + 2\,\underline{\underline{\dot{R}}}_j\;\underline{\dot{r}}_{0j} + \underline{\underline{R}}_j\;\underline{\ddot{r}}_{0j}\right)\end{aligned} \tag{6.18}$$

The required derivatives of the initial position vector $\underline{r}_{0j}$ are expressed as

$$\underline{\dot{r}}_{0j} = \dot{\overline{a}}_1\;||\underline{A}|| + \dot{\overline{a}}_2\;||\underline{B}||\;\cos(\Phi_0) \tag{6.19}$$

$$\underline{\ddot{r}}_{0j} = \ddot{\overline{a}}_1\;||\underline{A}|| + \ddot{\overline{a}}_2\;||\underline{B}||\;\cos(\Phi_0), \tag{6.20}$$

where the derivatives of the corresponding unit vectors, which are given by Equations 6.11, 6.12, and 6.13 are

$$\dot{\vec{a}}_1 = \begin{pmatrix} -S_\varphi \ (\dot{\omega}\, t + \omega) \\ C_\varphi \ (\dot{\omega}\, t + \omega) \\ 0 \end{pmatrix} \tag{6.21}$$

$$\ddot{\vec{a}}_1 = \begin{pmatrix} -C_\varphi \ (\dot{\omega}\, t + \omega)^2 - S_\varphi \ (\ddot{\omega}\, t + 2\,\dot{\omega}) \\ -S_\varphi \ (\dot{\omega}\, t + \omega)^2 + C_\varphi \ (\ddot{\omega}\, t + 2\,\dot{\omega}) \\ 0 \end{pmatrix} \tag{6.22}$$

$$\dot{\vec{a}}_2 = \begin{pmatrix} -C_\varphi \ (\dot{\omega}\, t + \omega) \\ -S_\varphi \ (\dot{\omega}\, t + \omega) \\ 0 \end{pmatrix} \tag{6.23}$$

$$\ddot{\vec{a}}_2 = \begin{pmatrix} S_\varphi \ (\dot{\omega}\, t + \omega)^2 - C_\varphi \ (\ddot{\omega}\, t + 2\,\dot{\omega}) \\ -C_\varphi \ (\dot{\omega}\, t + \omega)^2 - S_\varphi \ (\ddot{\omega}\, t + 2\,\dot{\omega}) \\ 0 \end{pmatrix} \tag{6.24}$$

$$\dot{\vec{a}}_3 = \ddot{\vec{a}}_3 = \begin{pmatrix} 0 \\ 0 \\ 0 \end{pmatrix} \tag{6.25}$$

Furthermore, derivatives of the rotation tensor $\underline{\underline{R}}_j$ are required to compute Equation 6.18. They may be written as

$$\begin{aligned} \dot{\underline{\underline{R}}}_j =& \frac{d}{dt}\left(\underline{\underline{I}} + \sin(\Phi_j)\widetilde{a}_1 + (1 - \cos(\Phi_j))\, \widetilde{a}_1\, \widetilde{a}_1\right) \\ =& C_\Phi\ \left(\dot{\Omega}\, t + \Omega\right)\, \widetilde{a}_1 + S_\Phi\, \dot{\widetilde{a}}_1 + \dot{\widetilde{a}}_1\, \widetilde{a}_1 + \widetilde{a}_1\, \dot{\widetilde{a}}_1 + S_\Phi\ \left(\dot{\Omega}\, t + \Omega\right)\, \widetilde{a}_1\, \widetilde{a}_1 \\ &- C_\Phi\, \dot{\widetilde{a}}_1\, \widetilde{a}_1 - C_\Phi\, \widetilde{a}_1\, \dot{\widetilde{a}}_1 \end{aligned} \tag{6.26}$$

$$\begin{aligned} \ddot{\underline{\underline{R}}}_j =& \frac{d^2}{dt^2}\left(\underline{\underline{I}} + \sin(\Phi_j)\widetilde{a}_1 + (1 - \cos(\Phi_j))\, \widetilde{a}_1\, \widetilde{a}_1\right) \\ =& - S_\Phi\ \left(\dot{\Omega}\, t + \Omega\right)^2\, \widetilde{a}_1 + C_\Phi\ \left(\ddot{\Omega}\, t + 2\,\dot{\Omega}\right)\, \widetilde{a}_1 + 2\, C_\Phi\ \left(\dot{\Omega}\, t + \Omega\right)\, \dot{\widetilde{a}}_1 + S_\Phi\, \ddot{\widetilde{a}}_1 \\ &+ \ddot{\widetilde{a}}_1\, \widetilde{a}_1 + 2\, \dot{\widetilde{a}}_1\, \dot{\widetilde{a}}_1 + \widetilde{a}_1\, \ddot{\widetilde{a}}_1 + C_\Phi\ \left(\dot{\Omega}\, t + \Omega\right)^2\, \widetilde{a}_1\, \widetilde{a}_1 + S_\Phi\ \left(\ddot{\Omega}\, t + 2\,\dot{\Omega}\right)\, \widetilde{a}_1\, \widetilde{a}_1 \\ &+ 2\, S_\Phi\ \left(\dot{\Omega}\, t + \Omega\right)\, \dot{\widetilde{a}}_1\, \widetilde{a}_1 + 2\, S_\Phi\ \left(\dot{\Omega}\, t + \Omega\right)\, \widetilde{a}_1\, \dot{\widetilde{a}}_1 - C_\Phi\, \ddot{\widetilde{a}}_1\, \widetilde{a}_1 - 2\, C_\Phi\, \dot{\widetilde{a}}_1\, \dot{\widetilde{a}}_1 \\ &- C_\Phi\, \widetilde{a}_1\, \ddot{\widetilde{a}}_1 \end{aligned} \tag{6.27}$$

Finally the derivatives of the skew symmetric matrix of the unit vector $\overline{a}_1$ are given through

$$\begin{aligned}\dot{\tilde{\overline{a}}}_1 &= \frac{d}{dt}\left(\begin{bmatrix} 0 & 0 & S_\varphi \\ 0 & 0 & -C_\varphi \\ -S_\varphi & C_\varphi & 0 \end{bmatrix}\right) \\ &= \begin{bmatrix} 0 & 0 & C_\varphi\ (\dot{\omega}\, t + \omega) \\ 0 & 0 & S_\varphi\ (\dot{\omega}\, t + \omega) \\ -C_\varphi\ (\dot{\omega}\, t + \omega) & -S_\varphi\ (\dot{\omega}\, t + \omega) & 0 \end{bmatrix}\end{aligned} \tag{6.28}$$

$$\begin{aligned}\ddot{\tilde{\overline{a}}}_1 &= \frac{d^2}{dt^2}\left(\begin{bmatrix} 0 & 0 & S_\varphi \\ 0 & 0 & -C_\varphi \\ -S_\varphi & C_\varphi & 0 \end{bmatrix}\right) \\ &= \begin{bmatrix} 0 & 0 & -S_\varphi\ (\dot{\omega}\ t+\omega)^2 + C_\varphi\ (\ddot{\omega}\ t+2\ \dot{\omega}) \\ 0 & 0 & C_\varphi\ (\dot{\omega}\ t+\omega)^2 + S_\varphi\ (\ddot{\omega}\ t+2\ \dot{\omega}) \\ S_\varphi\ (\dot{\omega}\ t+\omega)^2 - C_\varphi\ (\ddot{\omega}\ t+2\ \dot{\omega}) & -C_\varphi\ (\dot{\omega}\ t+\omega)^2 - S_\varphi\ (\ddot{\omega}\ t+2\ \dot{\omega}) & 0 \end{bmatrix}\end{aligned} \tag{6.29}$$

Model Application and Results

For the computation of the acting moments, parameters of the NREL 5 MW definition WT [109] were applied to the previous established model, depicted in Figure 6.9. Thereby, one of the $n_{\text{bld}} = 3$ blade masses is equal to $m_j = 17{,}740$ kg, the radius to the blade mass with respect to the second mass moment of inertia $J^{\mathcal{A}}_{\text{rot},\overline{a}_1}$ of one blade becomes

$$||\underline{B}|| = \sqrt{\frac{J_{\text{rot},\overline{a}_1^{\mathcal{A}}}}{m_j}} = \sqrt{\frac{11{,}776{,}047\ \text{kg m}^2}{17{,}740\ \text{kg}}} = 25.76\ \text{m} \tag{6.30}$$

and the orthogonal hub distance from the yaw axis equals to $||\underline{A}||_{\text{NREL}} = 5$ m. The rotor rotational speed was chosen to be constant and equal to the rated rotational speed

$$\Omega = 12.1 \frac{1}{\text{min}} \frac{\text{min}}{60\ \text{s}}\ 2\,\pi = 1.27 \frac{\text{rad}}{\text{s}} \tag{6.31}$$

$$\dot{\Omega} = \ddot{\Omega} = 0. \tag{6.32}$$

Yaw motion is modelled by a certain acceleration from $\omega(t = 0\ \text{s}) = 0\frac{1}{\text{s}}$ to $\omega(t = \infty) = \omega_\infty = 0.3\frac{^\circ}{\text{s}}\frac{\pi}{180^\circ}$ through the following relation

$$\omega(t) = \omega_\infty\ \left(1 - e^{-\alpha_\omega\ t}\right) \tag{6.33}$$

$$\dot{\omega}(t) = \omega_\infty\ \alpha_\omega\ e^{-\alpha_\omega\ t} \tag{6.34}$$

$$\ddot{\omega}(t) = -\omega_\infty\ \alpha_\omega^2\ e^{-\alpha_\omega\ t}, \tag{6.35}$$

where α_ω adjusts the steepness of the yaw acceleration curve. Figure 6.10 shows the results with respect to an arbitrary chosen $\alpha_\omega = 2$ for the first 5 seconds of rotor and yaw motion.

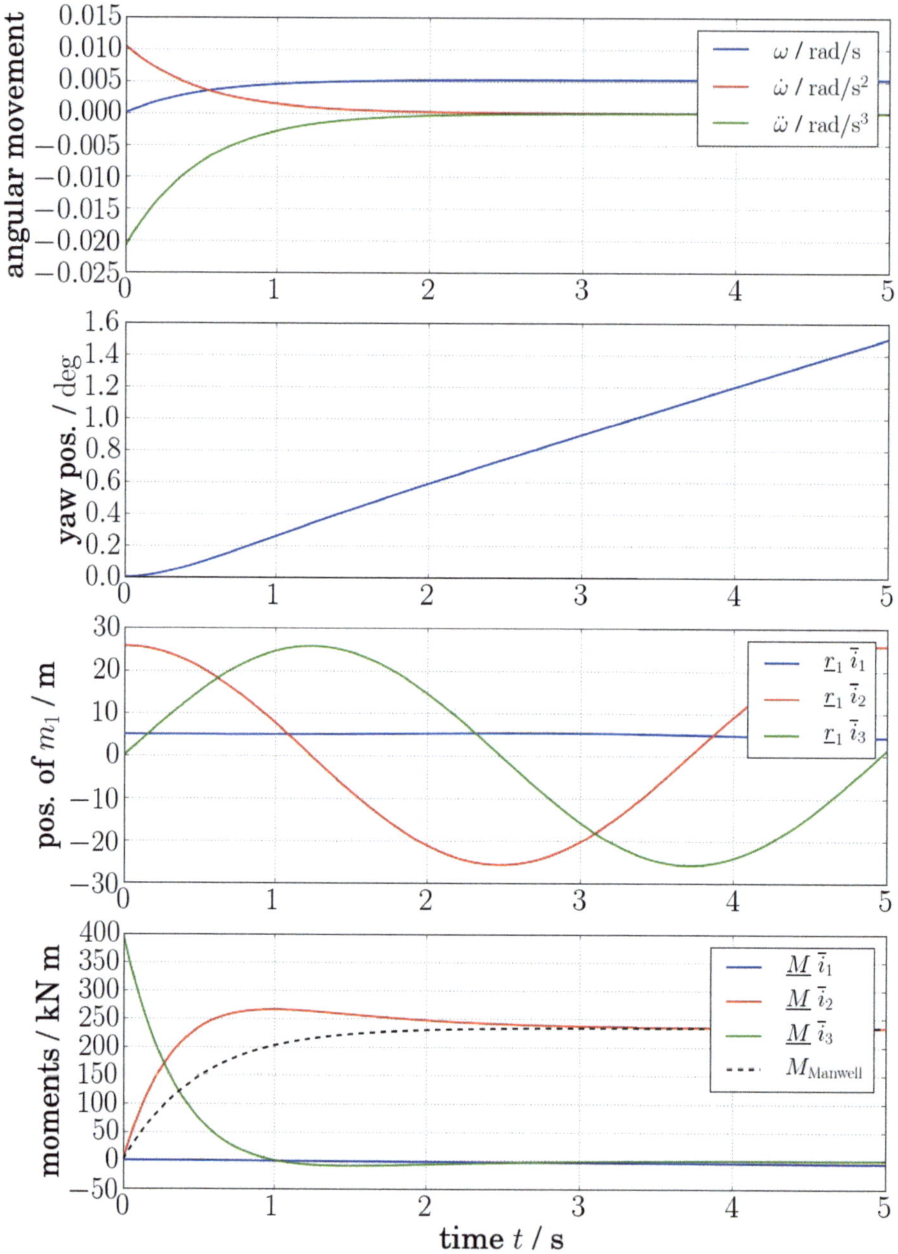

Figure 6.10: Acting moments according to rigid body motion of the rotating NREL 5 MW rotor during yaw motion

The first two diagrams in Figure 6.10 show the angular speed and its derivatives with respect to time and the corresponding yaw position of the nacelle, which is equal to φ converted to degrees instead of radians. Moreover, diagram three shows the position of m_1 with respect to the inertial orthonormal basis $\mathcal{I}$. From this diagram it is clear, that the rotor plane $\mathcal{P}_{\text{rot}} = (\mathbf{A}, \bar{a}_1)$ is parallel to $\mathcal{P} = \left(\mathbf{O}, \bar{i}_1\right)$ in the beginning of the simulation ($t = 0$ s). Due to small yaw angles, no significant changes in the $\bar{i}_1$ direction were observed. The fourth diagram in Figure 6.10 shows the resulting moments within the inertial orthonormal basis $\mathcal{I}$. Load representation in basis $\mathcal{I}$ is a good choice, because all blade force vectors, which affect gyroscopic moments lie in a plane parallel to $\mathcal{P}_{\text{yaw}}$. This means that the superimposed gyroscopic moments will be equal regardless of the mentioned orthonormal basis. Only the proportion according to each gyroscopic axis will differ because of the rotation of $\mathcal{A}$ in $\mathcal{I}$. The blue line in Figure 6.10 shows that there is no moment around the initial rotational axis of the rotor. This is expectable, because the angular speed Ω of the rotor around its axis is constant in time and a torsional moment would follow the equation

$$M_{\text{T}} = \dot{\Omega}\, J_{\text{rot},\bar{i}_3^{\mathcal{I}}}\,. \tag{6.36}$$

The more interesting red curve represents the gyroscopic moment around axis $\bar{i}_2$. Furthermore, a simplified equation for the gyroscopic moment by Manwell et al. [137, p. 164] is provided by

$$M_{\text{Manwell}} = J_{\text{rot},\bar{a}_1^{\mathcal{A}}}\, \Omega\, \omega. \tag{6.37}$$

and depicted in black. Its curve differs from the result of the own physical model, especially during the first two seconds. Equation 6.37 does not take into account yaw accelerations in contrast to Equation 6.18, wherefore low yaw accelerations, such as observed in the later course of time lead to coincident curves. This means that high yaw accelerations can hive the maximum gyroscopic load above the simplified assumption from Equation 6.37. The green curve represents the yaw moment and decreases linear dependent on the decreasing yaw acceleration down to $\underline{M}\, \bar{i}_3 \approx 0$, which is likewise plausible due to Equation 6.36. Incidentally, lower α_ω will decrease the yaw acceleration and therefore its influence to the gyroscopic load.

With this model it was possible to calculate the influence of an inclined tower by changing the length of $\underline{A}$, because it represents the rotor distance with respect to the yaw axis. Assuming the same WT configurations as before, Figure 6.11 shows the acting moment around the yaw axis $\bar{i}_3$ of orthonormal basis $\mathcal{I}$ as function of distance $||\underline{A}||$. In this plot, $||\underline{A}||$ was normalized by the value from the initial NREL setting and the yaw moment is normalized by the value that occurs for the reference setting from the literature [110] with $\dfrac{||\underline{A}||}{||\underline{A}||_{\text{NREL}}} = 1.0$.

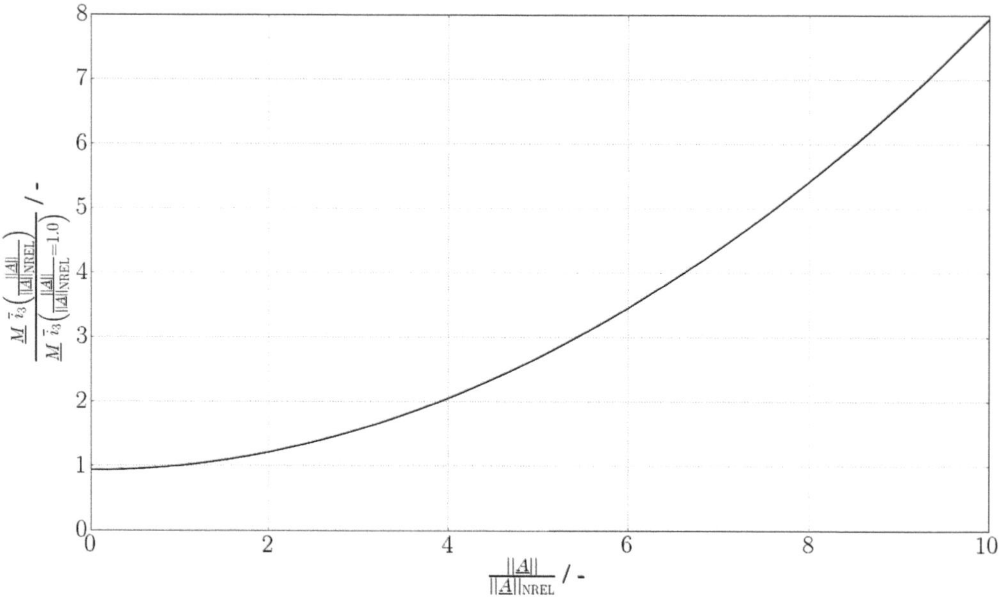

Figure 6.11: Relative yaw moment, related to the yaw moment in the reference setting as function of the normalized distance between the rotor and the yaw axis

Torsional moments caused by yaw motion of an inclined tower will significantly change with respect to the original configuration where $\frac{||\underline{A}||}{||\underline{A}||_{\text{NREL}}} = 1.0$, because the high rotor mass will be placed in another distance from the yaw axis. For example, assuming the NREL reference tower with an inclination angle of $\alpha_{\text{TI}} = 8$ deg, according to Figure 6.7, a distance of $\frac{||\underline{A}||}{||\underline{A}||_{\text{NREL}}} = 2.46$ result. By usage of this distance in Figure 6.11, one can read that the yaw moment is increased about 35.5 % compared to the reference configuration.

The next plot in Figure 6.12 shows the absolute yaw moments, which were normalized by the extreme torsional moment at the tower base $M_{\text{T,max,NREL}} = \max\left(|M_{\text{T,max}}|, |M_{\text{T,min}}|\right) = 1.23 \times 10^4$ kN m from Jonkman's analysis in [110]. Here, different coloured curves represent the yaw moments with respect to varying yaw angular accelerations. As expected, the yaw moments increase with increasing yaw accelerations, but they can be hold on a low level compared to the maximum occurring torsional moment if appropriate yaw accelerations are obtained.

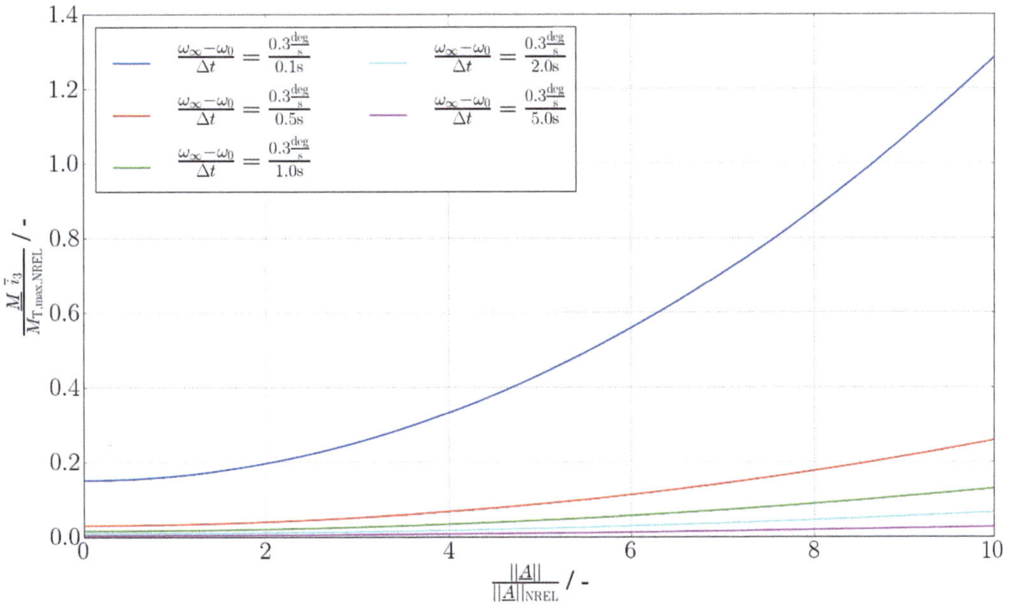

Figure 6.12: Moments due to gyroscopic motion and tower inclination at $t = 1$ s

As another noticeable and more important result of this analysis, the gyroscopic moments around axis $\bar{i}_1$ and $\bar{i}_2$ are independent from the distance $||\underline{A}||$. Therefore, these loads are unaffected by tower inclination and are only sensitive to the angular velocity and acceleration. This result is supported by the simplified assumption from Manwell et al. [137, p. 164] in Equation 6.37 where no rotor to yaw axis distance is necessary to calculate the gyroscopic moment. This means, that for the design of a rotatable inclined tower only the yaw acceleration has to be considered carefully, but the yaw speed can be the same as for conventional designs. In practice, a good control capability of the yaw accelerations at start and stop yaw maneuvers is recommended. Furthermore, skewed wind inflow can lead to additional moments of the yaw axis and tend to act against the aspired yaw alignment. To avoid large counter acting yaw moments, towers with low aerodynamic lateral drag, such as lattice towers should be applied. Another way is given through an eccentric tower position on the yaw bearing at the tower base. This can be applied to use the tower drag to support the yaw alignment and to reduce the yaw moment at the same yaw acceleration through the respective lower yaw inertia.

6.3 Other Concepts

6.3.1 Divided Hybrid Profile

The divided hybrid profile was invented by the Steel Pro Maschinenbau GmbH company and introduced in subsection 2.9.1. Its structure consists of half cylindrical tower shells, which are connected by lattice members. An advantageous feature compared to a closed elliptical or aerodynamic tower shape are the low lift forces for skewed wind inflow at the tower. On the other hand an increased tower drag results at the back half cylindrical shell, because the wind stream attacks its concave side. Rigorous CFD simulations help to evaluate the aerodynamic loads on the tower and reveal important frequencies of the vortexes behind the half cylindrical shells.

Another two advantages are obtained by its structural properties. Since, lattice members are used at the tower flanks, material intense plates can be avoided. Such plates would require a lot of material or additional stiffenings to withstand local plate buckling. Furthermore, the lattice member connection achieves large distances between the half cylindrical shells, whereby the tower's FA bending stiffness and its corresponding eigenfrequency can be adjusted. Attention should be given to the lattice member eigenfrequencies and to the cost rising through complicated joint concepts. Moreover, shell buckling has to be analysed for the half cylindrical shells, which have a reduced buckling resistance compared to closed cylinders.

Although divided hybrid profiles are interesting alternatives for rotatable WT towers, they are not investigated in more detail in this work, because of the combination of disadvantages from lattice and tubular shell towers. The disadvantage of lattice towers is the increased number of joints compared to shell towers and the disadvantage of tubular shell towers is their higher required amount of material compared to lattice towers.

6.3.2 Inclined Guyed Tower

Guying from the fixed ground to the inclined tower, such as shown in Figure 6.13, prohibits the tower to rotate and thereby saves the complete costs of a yaw system. This fixed tower is also proposed to be inclined to enable greater blade clearances, but has to be applied for downwind WTs because of the tower head guying. Tower head guying reduces the bending moments within the tower and thereby leads to further material savings. Such a material and cost saving concept decreases the energy harvest with respect to changing wind directions over lifetime.

Figure 6.14 shows some exemplary wind roses. The one on the top left depicts the occurrence frequency of each wind direction, the top right presents the mean wind speed from each direction, and the bottom one indicates the resulting energy density per year from each wind direction. Such site dependent roses of energy can be used to examine the economic feasibility of inclined guyed towers. In the exemplary case, a WNW or NNW alignment of a WT with guyed inclined tower would bring the most energy yield. This concept may be interesting for certain sites with very dominant main wind direction, but a more detailed investigation is out of scope for this work and should be carried out in future research projects.

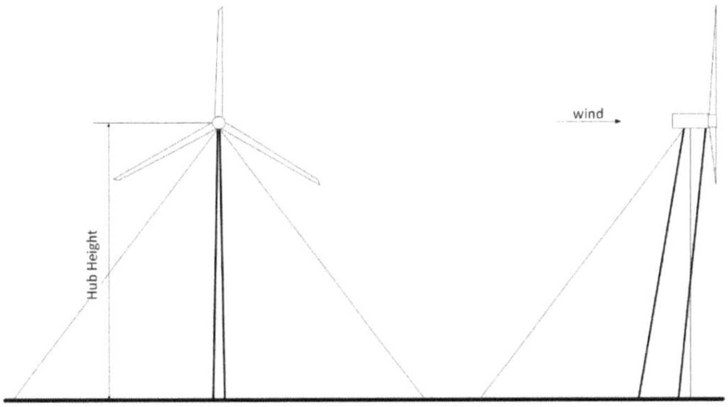

Figure 6.13: Concept - Tilted guyed tower without yaw bearing

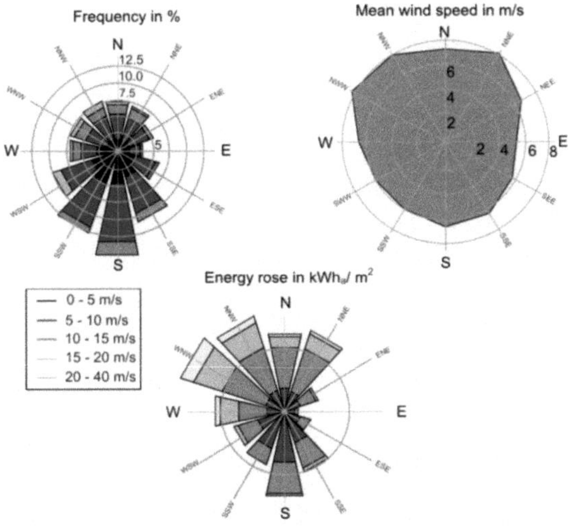

Figure 6.14: Wind roses: frequency rose, rose of mean wind speed and energy rose [77]

7 Rotatable Inclined Lattice Tower Design

7.1 Conceptual Specifications

This section describes the general concept and its parameters to be investigated in more detail throughout the following chapters. Thereby, the first subsection 7.1.1 treats the shape and parameters of the chosen rotatable inclined lattice tower concept, the following subsection 7.1.3 describes the chosen member profiles in more detail, and the last two subsections 7.1.4 and 7.1.2 treat the conceptual joint details and the nacelle attachment.

7.1.1 Shape and Parameters of the Rotatable Inclined Lattice Tower

The decision for the amount of lattice tower legs is an optimization problem with many parameters. In this work, a four legged rectangular lattice tower cross section was investigated for the following reasons:

- Relatively low amount of joints, which means less assembling costs, less bolts and therefore reduced purchase costs, and less maintenance with respect to tower cross sections with more legs
- Relatively torsional stiff compared to triangular tower cross sections
- Lower bracing member buckling lengths compared to triangular cross sections
- The tower top shape resembles the rectangular shape of common machine foundations and provides an appropriate space for the oblong arranged nacelle components

Other tower cross sections may be more material efficient, but have to be investigated in future research projects. Figure 7.1 shows the intended rotatable inclined lattice tower concept with its dimensions. Thereby, a linear tower inclination and tower taper was chosen with respect to the almost linear bending moment distribution, which is mainly driven by the rotor thrust. The tower drag induced nonlinear part of the tower bending moment was assumed to be smaller compared to a tubular shell concept. Furthermore, a linear tower inclination and taper requires less design parameters and is therefore more appropriate in terms of this preliminary work. X-bracings were used along the whole height of the tower in contrast to the Butzkies lattice tower concept in Figure 2.12, where four k-bracings were used for the bottom part and x-bracings were used for the upper tower part. Moreover, no secondary members should be used to reduce any buckling lengths, which saves members and joints to the cost of more severe cross sections. The resulting simple tower geometry may be completely described by 7 parameters, namely the tower height H_{T}, the tower inclination angle α_{TI}, the tower top length a_{T} and width b_{T}, the tower bottom length a_{B} and width b_{B}, and the number of bracing segments N_{seg}. The bracing angle α_{B} should be constant for all N_{seg} bracing segments. Considering the back view parallel to the $y - z$ plane in Figure 7.1, a system of equations is necessary to solve for the vertical joint distances h_i. In this view no tower inclination is visible, but only the tower taper. One can find the slope of the tapered shape by

$$k = \frac{b_{\mathrm{T}} - b_{\mathrm{B}}}{H_{\mathrm{T}}} \tag{7.1}$$

and the tangent of the tower taper angle by

$$\tan\left(\alpha_{\mathrm{TT}}\right) = \frac{b_{\mathrm{B}} - b_{\mathrm{T}}}{2\,H_{\mathrm{T}}} \tag{7.2}$$

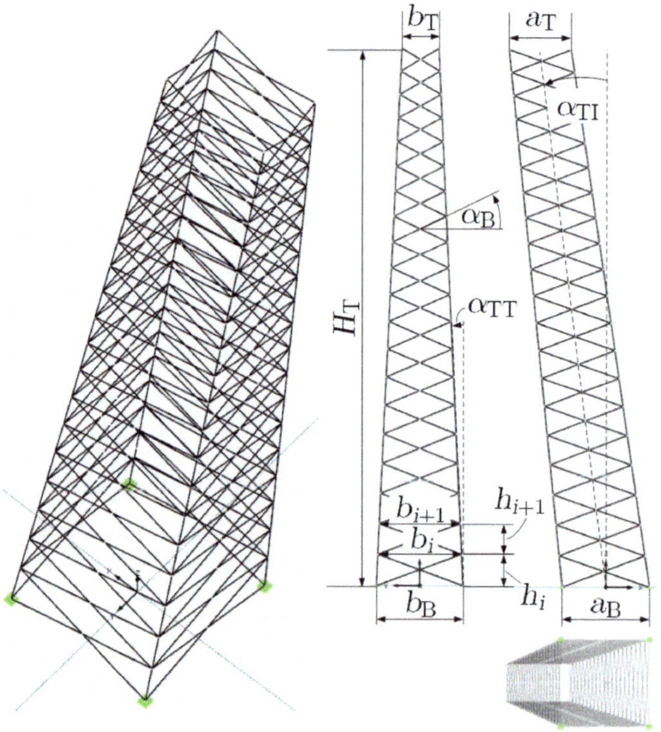

Figure 7.1: Rotatable inclined lattice concept with rectangular tower cross section and dimensions

Next, following equations describe the relations between the tower width b_i and the vertical joint distance h_i as

$$\begin{aligned} b_1 &= b_\mathrm{B} + k\ h_1 \\ b_2 &= b_1 + k\ h_2 \\ &\vdots \\ b_i &= b_{i-1} + k\ h_i \\ &\vdots \\ b_\mathrm{T} &= b_{\mathrm{N_{seg}}-1} + k\ h_{\mathrm{N_{seg}}} \end{aligned} \tag{7.3}$$

From Equation system 7.3 one gets N_seg equations and $2\ N_\mathrm{seg} - 1$ variables. Another relationship is extracted from the line intersection function between a bracing member and a leg with

$$\begin{aligned} -\frac{b_\mathrm{B}}{2} + \frac{z}{\tan(\alpha_\mathrm{B})} &= \frac{b_\mathrm{B}}{2} - \tan(\alpha_\mathrm{TT})\, z \\ \rightarrow z &= b_\mathrm{B} \underbrace{\left[\frac{1}{\tan(\alpha_\mathrm{B})} + \tan(\alpha_\mathrm{TT})\right]^{-1}}_{C} \end{aligned} \tag{7.4}$$

Note that the coordinate system is placed in the middle of the tower base and that Equation 7.4 contains the constant bracing and tower taper angles α_B and α_TT within constant C. With Equation 7.4 and the Equation system 7.3 one can derive another set of relationships between the tower width b_i and the vertical joint distance h_i as

$$\begin{aligned} h_1 &= b_\mathrm{B}\ C \\ h_2 &= b_1\ C = (b_\mathrm{B} + k\ h_1)\ C \\ &\vdots \\ h_\mathrm{i} &= b_{i-1}\ C = \left(b_\mathrm{B} + k \sum_{j=1}^{i-1} h_j\right) C \\ &\vdots \\ h_{\mathrm{N_{seg}}} &= b_{\mathrm{N_{seg}}-1}\ C = \left(b_\mathrm{B} + k \sum_{j=1}^{N_\mathrm{seg}-1} h_j\right) C \end{aligned} \tag{7.5}$$

The Equation system 7.5 contains now N_seg equations and $N_\mathrm{seg} + 1$ variables to be solved. Therefore, a last Equation is necessary where all heights are summed up to the tower height H_T according to

$$H_\mathrm{T} = \sum_{j=1}^{N_\mathrm{seg}} h_j \tag{7.6}$$

A function in a python script, shown in Appendix 14.16, was used to solve the system of Equations for the heights h_i and the constant C in an automated way for arbitrary N_{seg}, H_{T}, and tower bottom and top lengths and widths. A corresponding bracing angle results from rearrangement of C according to

$$\alpha_{\mathrm{B}} = \arctan\left(\frac{C}{1 - C\,\tan\left(\alpha_{\mathrm{TT}}\right)}\right) \tag{7.7}$$

The constraint of constant bracing angles can not be fulfilled for both sides of the tower with possible different tower taper angles α_{TT} and if the bracings from both sides should be attached at the same joint heights h_i. From a manufacturing point of view, different attachment heights would not be a problem and would lead to reduced leg buckling lengths, but averaged resulting joint heights $h_{i,\mathrm{a}}$ and $h_{i,\mathrm{b}}$ according to

$$h_i = \frac{h_{i,\mathrm{a}} + h_{i,\mathrm{b}}}{2} \quad \forall\, i \in \left\{1, 2, \ldots, N_{\mathrm{seg}}\right\} \tag{7.8}$$

were chosen. This decision will not have so much impact to the final result and is made to keep the concept simple and to provide a more uniform look of the lattice tower. During the progress of this work, other lattice tower concepts with more complicated shapes were considered. Therefore, a more general formulation for the bracing point intersection was derived, where the point of shortest distance between two lines

$$\begin{aligned} \mathcal{L}_1 &= \underline{x}_{\mathrm{leg,left}} + \overline{n}_1\,\lambda_1 \\ \mathcal{L}_2 &= \underline{x}_{\mathrm{leg,right}} + \overline{n}_2\,\lambda_2 \end{aligned} \tag{7.9}$$

must be found. In bracing line Equations 7.9 $\underline{x}_{\mathrm{leg,left}}$ represents the position vector of a leg node on the left side and $\underline{x}_{\mathrm{leg,right}}$ represents the position vector of a leg node on the right side, respectively. $\overline{n}_1$ and $\overline{n}_2$ are unit vectors, which are aligned parallel to the bracing member lines $\mathcal{L}_1$ and $\mathcal{L}_2$. Appendix 14.17 shows the derivation of scaling scalars λ_1 and λ_2 to describe the closest point in space of the two bracing member lines. Afterwards, the tower inclination is incorporated by shifting all calculated tower node coordinates $\underline{x}_{\mathrm{node}}$ parallel to the x axis by

$$\underline{x}_{\mathrm{node,incl.},i} = \left[\underline{x}^{T}_{\mathrm{node},i} \begin{bmatrix} 1 & 0 & 0 \\ 0 & 1 & 0 \\ \tan\left(\alpha_{\mathrm{TI}}\right) & 0 & 1 \end{bmatrix}\right]^{T} \quad \forall\, i \in \left\{1, 2, \ldots, N_{\mathrm{nodes}}\right\} \tag{7.10}$$

In Equation 7.10, $\underline{x}_{\mathrm{node,incl.},i}$ is the new node position vector for node i in the inclined tower and N_{nodes} is the number of all tower nodes.

7.1.2 Nacelle Interface

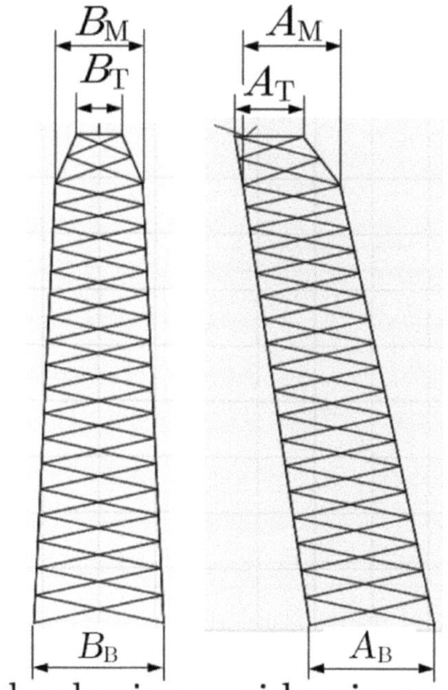

Figure 7.2: Possible adjusted tower top shape as nacelle interface

As one more advantage of this rotatable lattice tower concept, its rectangular shape at the top resembles the rectangular nacelle base area. Therefore, material may be saved in the machine foundation compared to a conventional costly transition between the circular yaw bearing and the rectangular nacelle shape. In terms of tower material efficiency, the leg distances should be as large as possible to reduce their axial loading. On the other hand, too large dimensions at the top would be adversely regarding the large area to be covered for the nacelle. Therefore, another tower wall slope can be incorporated at the two top segments, such as shown in Figure 7.2. It is a compromise to combine large leg distances over the majority of the tower height and appropriate tower top dimensions for the machine foundation. The intermediate width b_M and length a_M are assumed as

$$b_M = (b_B + b_T)/2 \tag{7.11}$$

and

$$a_M = (a_B + a_T)/2 \tag{7.12}$$

Furthermore, the tower front at the rotor side remains straight to provide the best blade to tower clearance. Note that such an adjustment is only appropriate if the leg distances reach a certain value. Otherwise one induces additional stress peaks at the sharp leg kink, which level out the previous mentioned advantage of larger leg distances. The developed rotatable inclined lattice tower in this work did not have another tower taper at the top.

7.1.3 Member Sections

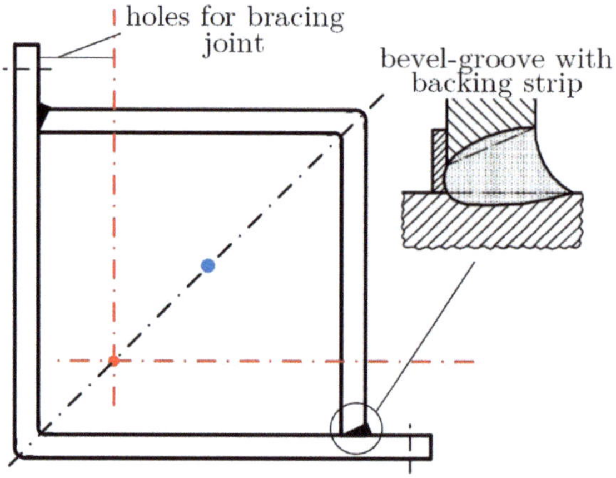

Figure 7.3: A welded lattice tower leg profile with bevel-groove welding detail according to DIN EN ISO 9692-1 [53]. The qualitative intersection of bracing member centroids are shown as a red dot and the leg centroid is shown as a blue dot. Bending radii are not shown.

A welded leg profile, assembled from two cold formed L-profiles and shown in Figure 7.3, is considered first. Welding seam preparation and welding itself are good automatable manufacturing processes and they save potential man hours compared to the alternative bolting along the legs, such as depicted in Figures 2.12 and 2.18. Moreover, lockbolts and additional filler plates may be saved to the cost of welding material and a decreased fatigue detail category. Note that Figure 6.2 may be reviewed for more information about lockbolt costs. The legs would consist out of two 90 deg folded L-profiles, which are welded together as T-joints. The proposed welding detail is a bevel-groove weld and must respect the weld preparation properties, defined in DIN EN ISO 9692-1 [53]. A bevel-groove weld is done from one side by usage of a backing strip. With such a backing strip, it can be assumed as a detail category C1 with respect to the detail categories defined in table A.3 of DNV GL's fatigue standard [62]. Note that backing strips must be continuous and if they are welded to a plate, their welding seam itself must be continuous to hinder a low detail category F for the joint [62, p. 96]. Furthermore, start and stop craters must be avoided through repair measures and non-destructive examination of the weld seam. Therefore, welded leg profiles, such as shown in Figure 7.3, thought to be an appropriate alternative. During the own performed structural fatigue analysis, it became clear that welded leg profiles without special post welding treatments, such as mentioned by Ummenhofer et al. [188] [187], would lead to a heavy and material inefficient tower. Furthermore, the potential non-intersecting neutral axes of the leg (blue dot) and the bracing members (red dot) led to eccentricities, which became a problem in terms of the bracing members fatigue despite their most high fatigue classification of detail category B1.

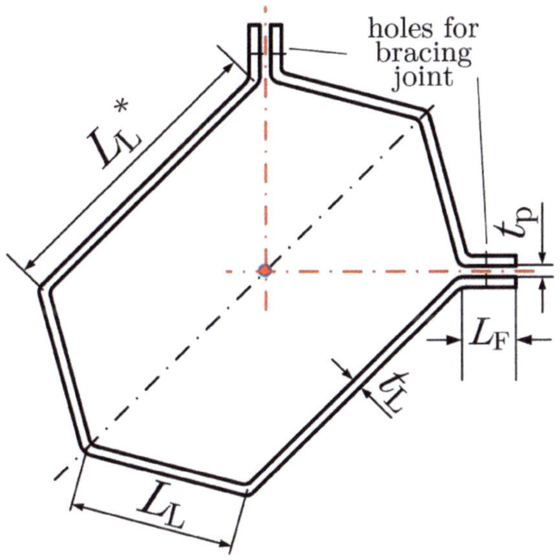

Figure 7.4: A bolted lattice tower leg profile, based on a hexagon with its dimensions. The qualitative intersection of bracing member centroids are shown as a red dot and the legs centroid is shown as blue dot. Bending radii are not shown.

Therefore, another leg profile concept, such as shown in Figure 7.4 was developed. It is based on a hexagonal shape with side length L_L and wall thickness t_L. By scaling the profile along its mirror axis in the dimension L_L^*, the legs and the bracing neutral axes can become coincident to avoid an eccentricity orthogonal to the lattice tower wall. Furthermore, a mirrored double bracing profile is an appropriate measure to avoid the local eccentricity in the junction plate with thickness t_p. Such mirrored bracing profiles are bolted along their lengths, such as for the leg profiles and indicated in Figure 2.19. The two parts are connected over their length by means of filler plates and lockbolts. DIN EN 1993-1-1 [42] shows in its section 6.4.4, that such assembled cross sections may be seen as one cross section for stability proofs under certain conditions. In the present case a minimum distance between the filler plates axes of 15 i_{min} must be met, where i_{min} is the minimum radius of gyration of both assembled parts. The flap length of the leg profile L_F depends on the minimum folding radius of the steel sheets and the dimensions of the lockbolt. Cold folded radii cause pre-induced damage to the material and are relevant in terms of fatigue. This relation must be evaluated by experiments and nonlinear FEM analyses, which are out of scope for this work. More questions arise with respect to the local stability of such folded steel plates. Stability of polygonal cross sections was investigated for example by Reinke [159] and Migita and Fukumoto [140]. An interesting finding of the second publication shows, that the axial bearing load for folded steel sheets remains constant for bending angles between 60 deg and 160 deg. Consequently, hexagonal bending angles of 120 deg provide similar stability as bending angles of 90 deg, such as presented in the cross sectional classification in tables 5.2 of DIN EN 1993-1-1 [42]. As an assumption and simplification in terms of the preliminary nature of this work, all radii were neglected to calculate the steel sheet thickness with respect to their plate widths. Thereby, Equations 2.48 indicate the length to wall thickness ratios for the one side supported flap length with respect to

$$(L_\mathrm{F} - t)\,/t \leq 13.77\,\varepsilon \tag{7.13}$$

and the two side supported length according to

$$(L_\mathrm{L} - t)\,/t \leq 42\,\varepsilon \tag{7.14}$$

Thus, with Equations 7.13 and 7.14, the minimum wall thickness of the leg profile to be not classified as class 4 is

$$t_\mathrm{min,Leg} = \max\left\{\frac{L_\mathrm{F}}{1 + 13.77\,\varepsilon}, \; \frac{L_\mathrm{L}}{1 + 42\,\varepsilon}, \; \frac{L_\mathrm{L}^*}{1 + 42\,\varepsilon}\right\} \tag{7.15}$$

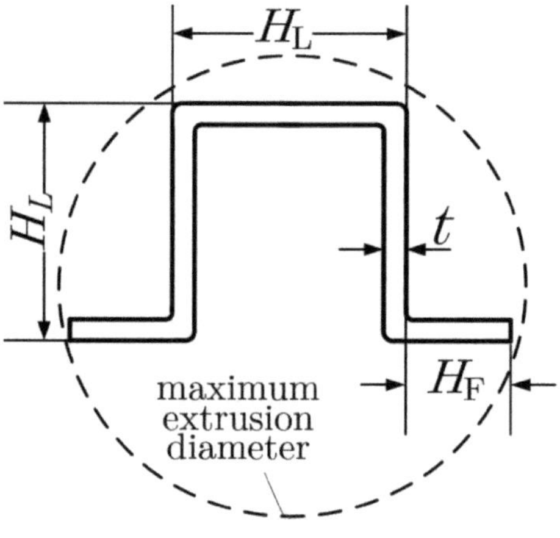

Figure 7.5: Proposed bracing hat-profile

Considering the minimum wall thickness $t_\mathrm{min,Leg}$ of Equation 7.15 means that no further local plate buckling check has to be carried out. For the bracing members hat-profiles, such as defined in Figure 7.5, were used. They take advantage of three two side supported plate lengths and are therefore relatively material efficient. Figure 2.16 indicates, that there is no reason to chose different values for the three two side supported lengths H_L. This decision achieves material efficient dimensions in terms of plate buckling and bending stiffness. The minimum bracing member wall thickness $t_\mathrm{min,Brace}$ is found as

$$t_\mathrm{min,Brace} = \max\left\{\frac{H_\mathrm{F}}{13.77\,\varepsilon}, \; \frac{H_\mathrm{L}}{2 + 42\,\varepsilon}\right\} \tag{7.16}$$

Manufacturing of relatively thick walled hat-profiles is challenging due to the small demanded folding radii. Therefore, roll forming as a cheap manufacturing alternative without any welding details for the bracing members, was chosen. A restriction to the profile dimensions is given when extrusion moulding would be an option. Through the maximum extrusion diameter, such as indicated in Figure 7.5 and prescribed by the manufacturers extruder. Common maximum extrusion diameters are in the range between 300 mm and 350 mm[1].

7.1.4 Conceptual Joint Design

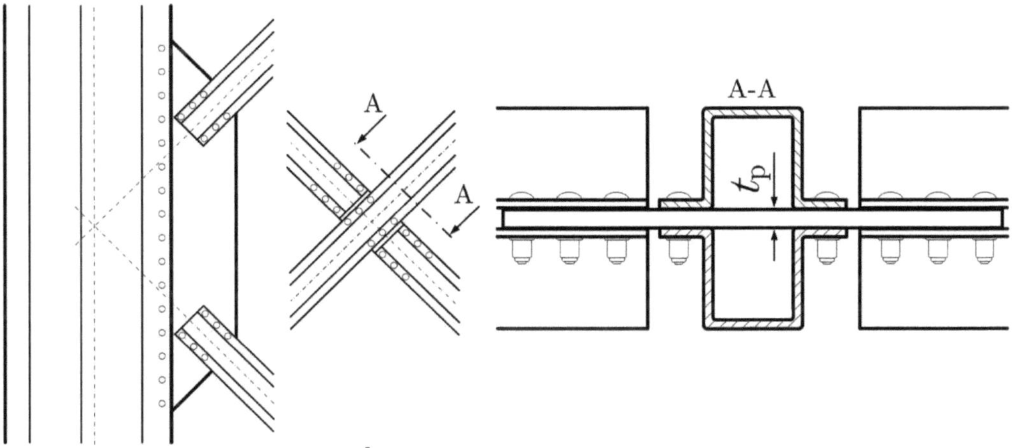

Figure 7.6: Joint details of the proposed lattice tower concept. View **a** shows the outer side of a leg profile with a gusset plate to attach the double bracing hat sections. View **b** shows the bracing members intersection joint and **c** a corresponding sectional view of it.

In this work, joint detail proofs were neglected to stay in scope of the governing task of developing and investigating a rotatable WT tower. However, a conceptual proposal for the joints are given within Figure 7.6. Note that the profile dimension proportions and amount of bolts may alternated in the real design. Figure 7.6 **a** shows the detailed brace to leg connection, which is realized by means of a junction plate. A junction plate achieves the necessary attachment space for the inclined bracing members. Otherwise the leg flap length L_F, shown in Figure 7.4 would become too large and would drive the wall thickness of the profile with respect to Equation 7.15. Although junction plates require more bolts compared to a direct attachment, they are appropriate to reduce the necessary amount of leg material. The chain lines in Figure 7.6 indicate the neutral axes of the profiles, which should intersect in one point for a

[1] According to the following three manufacturers: Meco Inc.`https://www.techpilot.de/servlets/AuctionConnector?lngCode=de&template=supplier_profile_new&ckey=aZGc7eAxVvLZfl9JZa7K#Technologien`, SMS-Group `https://www.sms-group.com/de/anlagen/alle-anlagen/strangpressen-stahl/`, and Montanstahl `https://www.montanstahl.com/de/produkte/sonderprofile/warmwalzen/`; accessed 03-April-2018

k-joint to prevent additional lever arms and bending moments. Figure 7.6 **b** shows the x-joint of the intersecting bracing members. In this joint the hat-profiles are connected by means of another junction plate. This is likewise visible in the sectional view in Figure 7.6 **c**. Here, one can see the double arranged bracing members. All bolts in Figure 7.6 are friction lockbolts to reduce the maintenance costs, such as discussed in subsection 2.7.5. Moreover, subsection 2.7.5 may be reviewed for more information about lattice tower joint proofs. Note that the junction plate thickness t_p should remain the same for all members of one tower segment.

7.2 Design as Iterative Process

The design of WT support structures is an iterative process with many steps involved. Figure 7.7 illustrates the main steps, which were applied in this work to get a final tower design.

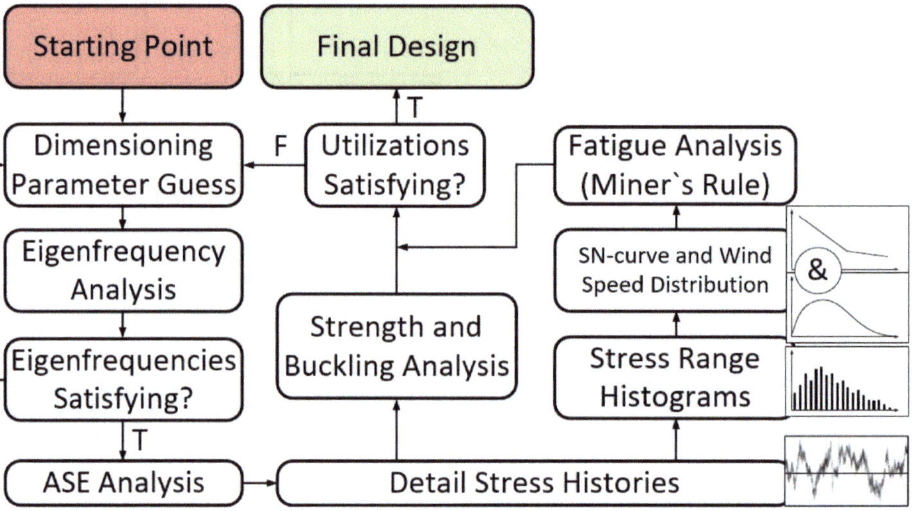

Figure 7.7: Schematic illustration of the iterative rotatable tower design procedure

A reasonable first guess of all design parameters defines the initial tower design. This design must be checked for its eigenfrequencies to ensure, that no resonance would occur in any operational conditions, such as explained in subsections 2.2.8 and 4.3.3. If this requirement is fulfilled, an ASE load analysis with all DLCs declared in Table 3.6 follows. The ASE analysis results in load timeseries for each structural member and they can be transformed to the stress timeseries of predefined member details by consideration of the corresponding member geometry. With this information, one can use the maximum stresses to perform a strength and buckling analysis according to subsection 2.2.5. After application of an appropriate stress cycle counting algorithm, the combination of the resulting stress range histograms, the detail SN-curve, the wind speed probability distribution, and the Miner's Rule the resulting fatigue utilization can be calculated. For more detailed explanations of the fatigue analysis procedure, subsection 2.2.6 may be reviewed. If the resulting utilizations of all analyses are satisfied, the design is finished. Otherwise the process repeats and starts again by guessing a new design parameter set. Note that the parameter guess may be automated based on optimization algorithms, such as multidimensional gradient searches, particle swarm algorithms, or genetic algorithms. According to the own experience, optimization with multidimensional parameter spaces and fitness functions with high computational costs is a subject on its own. Therefore, and because of the limited scope of this work, automated optimization procedures were not applied in this work. However, with some experience one can identify design driving DLCs and neglect the other DLCs to accelerate the design iteration duration. Moreover, studying the qualitative influence of parameter changes can help to reach a good utilization without automated optimization procedures.

8 Modelling the Rotatable Inclined Lattice Tower in FAST

8.1 Modification of SubDyn's Finite-Element Beam Model

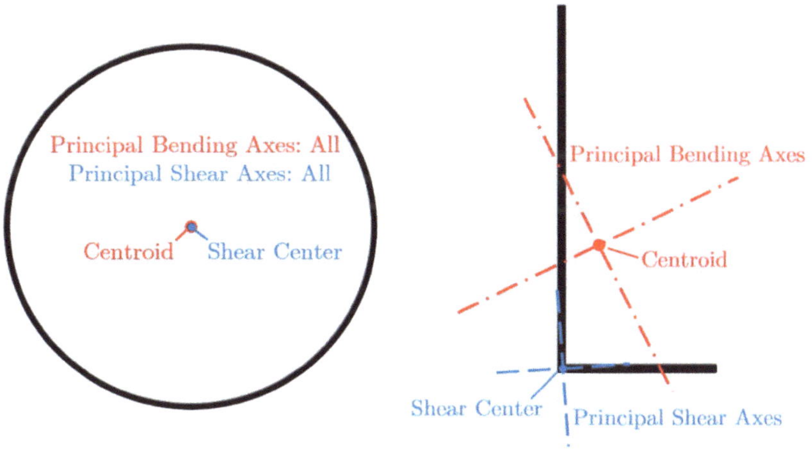

Figure 8.1: Cross sectional axes for a circular (left) and a L (right) cross section

To investigate the extreme loads and fatigue resistance of the proposed rotatable inclined lattice tower the SubDyn module of the FAST framework, presented in subsection 2.8.3, was used. In its first state it was not capable to represent other member cross sections than circular ones. This limitation is appropriate for the most offshore substructures, but not for land-based ones. In most cases offshore structures have welded member joints to reduce the assembling effort on site and because transport on open sea is less problematic for large structural pieces than ashore. Circular cross sections have a good local buckling resistance and cause less hydrodynamic drag compared to box girders and they can be attached to each other by welded joints. For resolved land-based WT substructures welded joints are adversely because of transport. As circular cross sections are difficult to attach with each other without welding, other cross sectional shapes come into consideration. Figure 8.1 shows the different principal shear and bending axes, which intersect in the shear center and the centroid for a circular and a L cross section. Thereby, a L cross section is a good representation for arbitrary cross sectional shapes, because its centroid and shear center do not coincide and their principal bending and shear axes are not parallel to each other. These properties lead to additional terms in the structural matrices and the member alignment with respect to its longitudinal axis becomes likewise relevant. The corresponding theoretical background for the changes in the structural member matrices and the direction cosine matrices in SubDyn is treated in the following two subsections 8.1.1 and 8.1.2. The validation of the changes is already published by Struve et al. [179]. It took place by means of parallel modelling between SubDyn and Abaqus. Struve et al. [179] show the general changes in SubDyn only in a superficial way, wherefore this work reveals more detail of it. However, a short comparison between the calculated eigenfrequencies in RFEM and SubDyn, provided in section 8.4, should serve as another validation of the implemented SubDyn changes as well as for the validation of the SubDyn rotatable inclined lattice tower model itself.

8.1.1 Extension of Structural Matrices

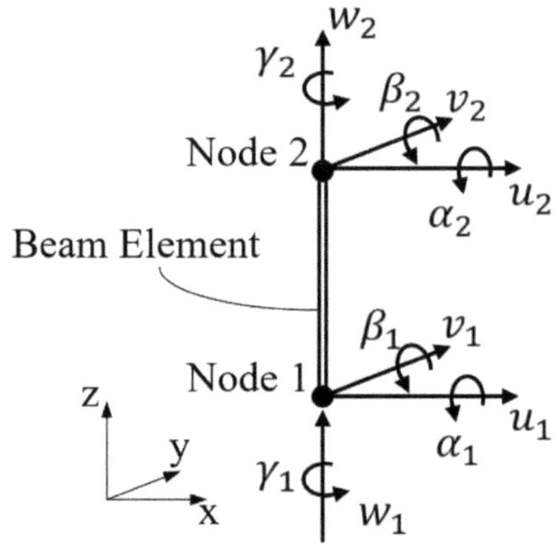

Figure 8.2: Degrees of freedom at the two nodes of a beam element

Major modifications of the SubDyn module took place in the beam element stiffness and mass matrices, $\underline{\underline{K}}_{el}^{12\times12}$ and $\underline{\underline{M}}_{el}^{12\times12}$ and the beam element direction cosine matrix $\underline{\underline{D}}_{\Phi\Theta\Psi}^{3\times3}$. The later is derived in subsection 8.1.2. In this work, a beam element was defined through two nodes, such as shown in Figure 8.2. Each node possesses six DOFs, three displacements u, v, and w and three rotations α, β, and γ. The displacements and rotations of a beam element are assembled in its displacement and rotation vector $\underline{u} = \{u_1, v_1, w_1, \alpha_1, \beta_1, \gamma_1,$
$u_2, v_2, w_2, \alpha_2, \beta_2, \gamma_2\}^T$ and are related to the corresponding static nodal loads through the beam element stiffness matrix. Analogously relates the beam element mass matrix nodal accelerations

$\underline{\ddot{u}} = \frac{\partial^2 u}{\partial t^2} = \{\ddot{u}_1, \ddot{v}_1, \ddot{w}_1, \ddot{\alpha}_1, \ddot{\beta}_1, \ddot{\gamma}_1, \ddot{u}_2, \ddot{v}_2, \ddot{w}_2, \ddot{\alpha}_2, \ddot{\beta}_2, \ddot{\gamma}_2\}^T$ to corresponding dynamic loads. Note that the symbol t represents time in this context. The nodal forces of a beam element result to

$$\underline{f} = \underbrace{\underline{\underline{K}}_{\text{el}} \, \underline{u}}_{\text{static loads}} + \underbrace{\underline{\underline{M}}_{\text{el}} \, \underline{\ddot{u}}}_{\text{dynamic loads}} \tag{8.1}$$

where damping is neglected because it is accounted for the whole system instead, such as noted in subsection 2.8.3. The resulting load vector $\underline{f} = \{F_{\text{u}1}, F_{\text{v}1}, F_{\text{w}1}, M_{\alpha 1}, M_{\beta 1}, M_{\gamma 1}, F_{\text{u}2}, F_{\text{v}2}, F_{\text{w}2}, M_{\alpha 2}, M_{\beta 2}, M_{\gamma 2}\}^T$ contains the load response to the displacements and rotations and the corresponding accelerations. The structural matrices $\underline{\underline{K}}_{\text{el}}$ and $\underline{\underline{M}}_{\text{el}}$ were changed to account for arbitrary orientations of principal bending and shear axes and for potential unequal centroid and shear center positions, such as arbitrary cross sections may have. $\underline{\underline{K}}_{\text{el}}$ and $\underline{\underline{M}}_{\text{el}}$ are calculated by the integration of the cross sectional stiffness and mass matrices $\underline{\underline{K}}_{\text{c}}^{6\times6}$ and $\underline{\underline{M}}_{\text{c}}^{6\times6}$ over the beam length. For arbitrary cross sections these cross sectional structural matrices contain additional off-diagonal terms compared to the circular cross section formulation. Equations 8.2 and 8.3 show these matrices for the case that they are defined with respect to the centroid of the cross section. A review of Bauchau's work [7] shows how to derive these cross sectional structural matrices also for arbitrary reference points.

$$\underline{\underline{K}}_{\text{c}} = \begin{Bmatrix} G\,A_{\text{x}} & -G\,A_{\text{xy}} & 0 & 0 & 0 & -G\,A_{\text{x}}\,y_{\text{s}} - G\,A_{\text{xy}}\,x_{\text{s}} \\ & G\,A_{\text{y}} & 0 & 0 & 0 & G\,A_{\text{y}}\,x_{\text{s}} + G\,A_{\text{xy}}\,y_{\text{s}} \\ & & E\,A & 0 & 0 & 0 \\ & \text{sym.} & & E\,I_{\text{x}} & -E\,I_{\text{xy}} & 0 \\ & & & & E\,I_{\text{y}} & 0 \\ & & & & & G\,A_{\text{x}}\,y_{\text{s}}^2 + G\,A_{\text{y}}\,x_{\text{s}}^2 + 2\,G\,A_{\text{xy}}\,x_{\text{s}}\,y_{\text{s}} + G\,I_{\text{t}} \end{Bmatrix} \tag{8.2}$$

$$\underline{\underline{M}}_{\text{c}} = \rho \begin{Bmatrix} A & 0 & 0 & 0 & 0 & 0 \\ & A & 0 & 0 & 0 & 0 \\ & & A & 0 & 0 & 0 \\ & \text{sym.} & & I_{\text{x}} & -I_{\text{xy}} & 0 \\ & & & & I_{\text{y}} & 0 \\ & & & & & I_{\text{p}} \end{Bmatrix} \tag{8.3}$$

Besides the material properties, shear modulus G, Young's modulus E, and density ρ, Equations 8.2 and 8.3 contain the engineering constants: cross sectional area A, corrected shear areas A_{x}, A_{y}, and A_{xy}, second area moments of inertia I_{x} and I_{y}, product of inertia I_{xy}, the torsional constant I_{t}, and the polar moment of inertia I_{p}. x_{s} and y_{s} are the shear centre coordinates. The calculation of these engineering constants for different kinds of cross sections, such as open or closed ones is documented in many textbooks, such as in Dankert [35], Bauchau [9], and Cook and Young [30]. A $A_{\text{xy}} = I_{\text{xy}} = 0$ implies, that the principal bending and shear axes are parallel to each other. This is for example the case for circular cross sections. Note that x_{s} and y_{s} is equal to the centroid for circular cross sections. Taeseong, Hansen, and Branner [181] derive the beam element structural matrices out of the potential and kinetic beam energies, E_{pot}, and E_{kin} according to

$$E_{\text{pot}} = \frac{1}{2}\int_0^L \underline{\varepsilon}^T \underline{\underline{K}}_c \underline{\varepsilon}\, dz \tag{8.4}$$

$$E_{\text{kin}} = \frac{1}{2}\int_V \rho\, \underline{\dot{r}}^T \underline{\dot{r}}\, dV, \tag{8.5}$$

where L is the beam length, V is the beam volume, $\underline{\varepsilon} = \{\gamma_{\text{zx}}, \gamma_{\text{zy}}, \varepsilon_{\text{z}}, \kappa_{\text{x}}, \kappa_{\text{y}}, \kappa_{\text{z}}\}^T = \left\{\frac{\partial u}{\partial z} - \beta, \frac{\partial v}{\partial z} + \alpha, \frac{\partial w}{\partial z}, \frac{\partial \alpha}{\partial z}, \frac{\partial \beta}{\partial z}, \frac{\partial \gamma}{\partial z}\right\}^T$ is the generalized strain vector according to Timoshenko beam kinematics, and $\underline{\dot{r}}$ is the velocity vector of a material point within the beam. Integration of the cross sectional structural matrices along the beam requires knowledge about the deflection and rotation functional shapes between the nodes. Therefore, Taeseong, Hansen, and Branner [181] assume polynomial shape functions of arbitrarily high order and minimize the elastic energy of Equation 8.4 to extract the corresponding polynomial coefficients. After introducing the correct shape function matrices and boundary conditions and carrying out mathematical transformations, the structural beam matrices can be derived as follows

$$E_{\text{pot}} = \frac{1}{2}\underline{u}^T \underbrace{\underline{\underline{N}}_\alpha^T \left[\int_0^L \underline{\underline{B}}^T \underline{\underline{K}}_c \underline{\underline{B}}\, dz\right] \underline{\underline{N}}_\alpha}_{\underline{\underline{K}}_{\text{C}}} \underline{u} \tag{8.6}$$

$$E_{\text{kin}} = \frac{1}{2}\underline{\dot{u}}^T \underbrace{\underline{\underline{N}}_\alpha^T \left[\int_0^L \underline{\underline{N}}(z)^T \underline{\underline{M}}_c \underline{\underline{N}}(z)\, dz\right] \underline{\underline{N}}_\alpha}_{\underline{\underline{M}}_{\text{C}}} \underline{\dot{u}}, \tag{8.7}$$

where the transformation matrix $\underline{\underline{N}}_\alpha^{6p\times 12}$ for the generalized degrees of freedom vector is used. $\underline{\underline{B}}^{6\times 6p}$ is the strain-displacement matrix and $\underline{\underline{N}}(z)^{6\times 6p}$ is the polynomial matrix. Furthermore, $p-1$ is the highest power of the assumed polynomials. Such rigorous beam element approach provides more generality than necessary for isotropic straight beams with arbitrary cross sections, but achieves more flexibility for future research projects. The user has now the option to provide either the engineering constants in Equations 8.2 and 8.3 or to provide full 6×6 cross sectional stiffness and mass matrices, such as shown in Equations 8.8 and 8.9 for each member.

$$\underline{\underline{K}}_c = \begin{Bmatrix} K_{11} & K_{12} & K_{13} & K_{14} & K_{15} & K_{16} \\ K_{21} & K_{22} & K_{23} & K_{24} & K_{25} & K_{26} \\ K_{31} & K_{32} & K_{33} & K_{34} & K_{35} & K_{36} \\ K_{41} & K_{42} & K_{43} & K_{44} & K_{45} & K_{46} \\ K_{51} & K_{52} & K_{53} & K_{54} & K_{55} & K_{56} \\ K_{61} & K_{62} & K_{63} & K_{64} & K_{65} & K_{66} \end{Bmatrix} \tag{8.8}$$

$$\underline{\underline{M}}_{c} = \begin{Bmatrix} M_{11} & M_{12} & M_{13} & M_{14} & M_{15} & M_{16} \\ M_{21} & M_{22} & M_{23} & M_{24} & M_{25} & M_{26} \\ M_{31} & M_{32} & M_{33} & M_{34} & M_{35} & M_{36} \\ M_{41} & M_{42} & M_{43} & M_{44} & M_{45} & M_{46} \\ M_{51} & M_{52} & M_{53} & M_{54} & M_{55} & M_{56} \\ M_{61} & M_{62} & M_{63} & M_{64} & M_{65} & M_{66} \end{Bmatrix} \tag{8.9}$$

The generality in this rigorous beam element allows also for anisotropic material properties and prepares the SubDyn module thereby for other future research and development projects.

8.1.2 Adaptation of Direction Cosine Matrices

Another important aspect concerns the direction cosine matrix, because the orientation of the beam element around its axis is relevant for arbitrary cross sections. Therefore, Figure 8.3 helps to identify the major parameters of a beam element alignment in space. It is defined through its start and end points **S** and **E** with its length $||\mathbf{E} - \mathbf{S}|| = L_e$, which can be projected to the $\mathcal{P}_{XY} = \left(\mathbf{O}, \overline{K}\right)$-plane, to become L_{eXY}. Note that the beam elements origin **S** is placed to the origin **O** of the global reference frame for convenience.

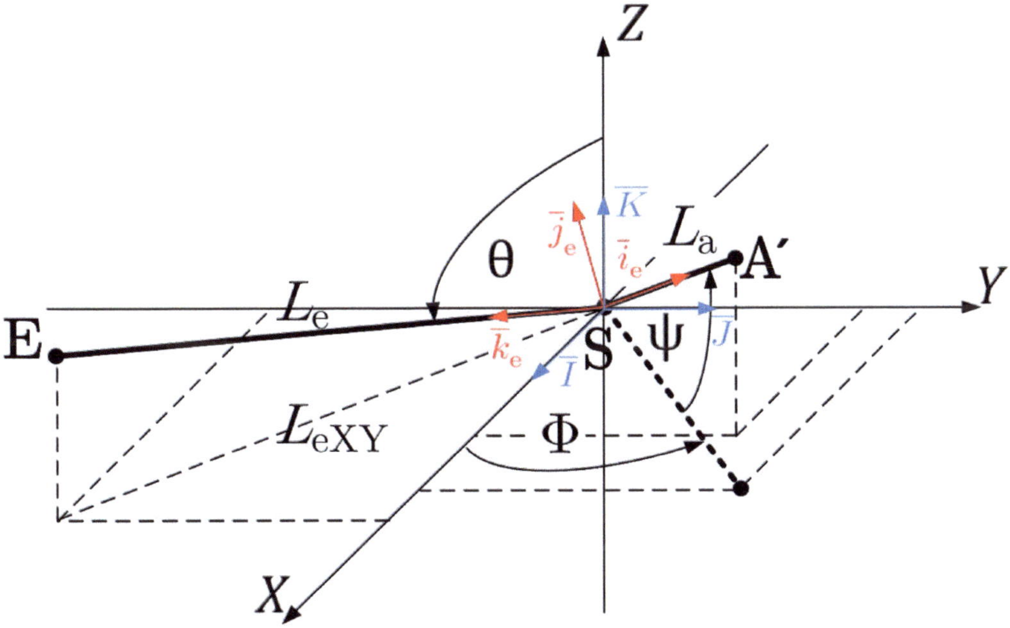

Figure 8.3: Parametrised beam element alignment in space

The restriction of the original SubDyn version for the local member axis $\bar{i}_e$ to be parallel to the $\mathcal{P}_{XY}$-plane prohibits arbitrary cross section orientations in space. Euler angles can bring any orthonormal basis to an arbitrary other one by viewing its overall rotation as a succession of three planar rotations. Thereby, the rotations happen around different axes, which means that no rotation will occur twice in a row around the same axis [8]. By allowing a third rotation around the member axis instead of the previous mentioned restriction, arbitrary cross section orientations are achievable. In this case, the direction cosine matrix is defined to describe the local beam element orthonormal basis $\mathcal{E}_e = \left(\bar{i}_e, \bar{j}_e, \bar{k}_e\right)$ with respect to the global orthonormal basis $\mathcal{I} = \left(\bar{I}, \bar{J}, \bar{K}\right)$. The order of rotations is defined by a certain sequence. Here, it is the rotation from $\mathcal{I}^{\mathcal{I}}$ to $\mathcal{E}_1^{\mathcal{I}}$ around unit vector $\bar{K}^{\mathcal{I}}$ by angle Φ, from $\mathcal{E}_1^{\mathcal{E}_1}$ to $\mathcal{E}_2^{\mathcal{E}_1}$ around unit vector $\bar{i}_1^{\mathcal{E}_1}$ by angle Θ, and from $\mathcal{E}_2^{\mathcal{E}_2}$ to $\mathcal{E}_e^{\mathcal{E}_2}$ around unit vector $\bar{k}_2^{\mathcal{E}_2}$ by angle Ψ. In this notation, the upper indices indicate the orthonormal basis in which a vector is resolved. Note that all rotations taking place in the mathematical positive counter clockwise convention when looking along the negative direction of the rotation axis.

$$\begin{bmatrix} \bar{i}_1^{\mathcal{I}} \\ \bar{j}_1^{\mathcal{I}} \\ \bar{k}_1^{\mathcal{I}} \end{bmatrix}^T = \underbrace{\begin{bmatrix} C_\Phi & -S_\Phi & 0 \\ S_\Phi & C_\Phi & 0 \\ 0 & 0 & 1 \end{bmatrix}}_{\underline{\underline{D}}_\Phi^{\mathcal{I}}} \begin{bmatrix} \bar{I}^{\mathcal{I}} \\ \bar{J}^{\mathcal{I}} \\ \bar{K}^{\mathcal{I}} \end{bmatrix}^T \tag{8.10}$$

$$\begin{bmatrix} \bar{i}_2^{\mathcal{E}_1} \\ \bar{j}_2^{\mathcal{E}_1} \\ \bar{k}_2^{\mathcal{E}_1} \end{bmatrix}^T = \underbrace{\begin{bmatrix} 1 & 0 & 0 \\ 0 & C_\Theta & -S_\Theta \\ 0 & S_\Theta & C_\Theta \end{bmatrix}}_{\underline{\underline{D}}_\Theta^{\mathcal{E}_1}} \begin{bmatrix} \bar{i}_1^{\mathcal{E}_1} \\ \bar{j}_1^{\mathcal{E}_1} \\ \bar{k}_1^{\mathcal{E}_1} \end{bmatrix}^T \tag{8.11}$$

$$\begin{bmatrix} \bar{i}_e^{\mathcal{E}_2} \\ \bar{j}_e^{\mathcal{E}_2} \\ \bar{k}_e^{\mathcal{E}_2} \end{bmatrix}^T = \underbrace{\begin{bmatrix} C_\Psi & -S_\Psi & 0 \\ S_\Psi & C_\Psi & 0 \\ 0 & 0 & 1 \end{bmatrix}}_{\underline{\underline{D}}_\Psi^{\mathcal{E}_2}} \begin{bmatrix} \bar{i}_2^{\mathcal{E}_2} \\ \bar{j}_2^{\mathcal{E}_2} \\ \bar{k}_2^{\mathcal{E}_2} \end{bmatrix}^T \tag{8.12}$$

The abbreviations S and C are representative for the geometric sine and cosine functions, and its indices show its corresponding Euler angle arguments. Note that Equations 8.10 to 8.12 describe only rotations in their own orthonormal bases, but a description of all tensors according to the global orthonormal basis $\mathcal{I}$ is required to get $\mathcal{E}_e^{\mathcal{I}}$. Therefore, the direction cosine matrices have to be transformed with respect to $\mathcal{I}$ by

$$\underline{\underline{D}}_\Theta^{\mathcal{I}} = \underline{\underline{D}}_\Phi^{\mathcal{I}} \, \underline{\underline{D}}_\Theta^{\mathcal{E}_1} \, \underline{\underline{D}}_\Phi^{\mathcal{I}\,T} \tag{8.13}$$

$$\underline{\underline{D}}_{\Psi}^{\mathcal{E}_1} = \underline{\underline{D}}_{\Theta}^{\mathcal{E}_1}\, \underline{\underline{D}}_{\Psi}^{\mathcal{E}_2}\, \underline{\underline{D}}_{\Theta}^{\mathcal{E}_1\,T} \tag{8.14}$$

$$\underline{\underline{D}}_{\Psi}^{\mathcal{I}} = \underline{\underline{D}}_{\Phi}^{\mathcal{I}}\, \underline{\underline{D}}_{\Psi}^{\mathcal{E}_1}\, \underline{\underline{D}}_{\Phi}^{\mathcal{I}\,T} \tag{8.15}$$

Using Equations 8.13 to 8.15 in 8.10 to 8.12, whereby all tensors of Equations 8.10 to 8.12 are described with respect to $\mathcal{I}$ delivers

$$\underbrace{\begin{bmatrix} \bar{i}_{\mathrm{e}}^{\mathcal{I}} \\ \bar{j}_{\mathrm{e}}^{\mathcal{I}} \\ \bar{k}_{\mathrm{e}}^{\mathcal{I}} \end{bmatrix}^T}_{\mathcal{E}_{\mathrm{e}}^{\mathcal{I}}} = \underline{\underline{D}}_{\Psi}^{\mathcal{I}}\, \underline{\underline{D}}_{\Theta}^{\mathcal{I}}\, \underline{\underline{D}}_{\Phi}^{\mathcal{I}} \begin{bmatrix} \bar{I}^{\mathcal{I}} \\ \bar{J}^{\mathcal{I}} \\ \bar{K}^{\mathcal{I}} \end{bmatrix}^T = \underline{\underline{D}}_{\Phi}^{\mathcal{I}}\, \underline{\underline{D}}_{\Theta}^{\mathcal{E}_1}\, \underline{\underline{D}}_{\Psi}^{\mathcal{E}_2} \begin{bmatrix} \bar{I}^{\mathcal{I}} \\ \bar{J}^{\mathcal{I}} \\ \bar{K}^{\mathcal{I}} \end{bmatrix}^T = \underline{\underline{D}}_{\Phi\Theta\Psi}^{\mathcal{I}} \underbrace{\begin{bmatrix} \bar{I}^{\mathcal{I}} \\ \bar{J}^{\mathcal{I}} \\ \bar{K}^{\mathcal{I}} \end{bmatrix}^T}_{\mathcal{I}^{\mathcal{I}}} \tag{8.16}$$

with

$$\underline{\underline{D}}_{\Phi\Theta\Psi}^{\mathcal{I}} = \begin{bmatrix} C_\Phi\, C_\Psi - S_\Phi\, C_\Theta\, S_\Psi & -C_\Phi\, S_\Psi - S_\Phi\, C_\Theta\, C_\Psi & S_\Phi\, S_\Theta \\ S_\Phi\, C_\Psi + C_\Phi\, C_\Theta\, S_\Psi & -S_\Phi\, S_\Psi + C_\Phi\, C_\Theta\, C_\Psi & -C_\Phi\, S_\Theta \\ S_\Theta\, S_\Psi & S_\Theta\, C_\Psi & C_\Theta \end{bmatrix} = \begin{bmatrix} \bar{i}_{\mathrm{e}}^{\mathcal{I}} & \bar{j}_{\mathrm{e}}^{\mathcal{I}} & \bar{k}_{\mathrm{e}}^{\mathcal{I}} \end{bmatrix} \tag{8.17}$$

Note that the unit vector containing vectors, resolved in their own orthonormal basis are identity matrices, such as indicated by Equation 8.18.

$$\begin{bmatrix} \bar{I}^{\mathcal{I}} \\ \bar{J}^{\mathcal{I}} \\ \bar{K}^{\mathcal{I}} \end{bmatrix}^T = \begin{bmatrix} \bar{i}_1^{\mathcal{E}_1} \\ \bar{j}_1^{\mathcal{E}_1} \\ \bar{k}_1^{\mathcal{E}_1} \end{bmatrix}^T = \begin{bmatrix} \bar{i}_2^{\mathcal{E}_2} \\ \bar{j}_2^{\mathcal{E}_2} \\ \bar{k}_2^{\mathcal{E}_2} \end{bmatrix}^T = \underline{\underline{I}}^{3\times 3} \tag{8.18}$$

The sines and cosines of Euler angles Φ and Θ are defined through the beam axis, given by $\overline{\mathbf{SE}}$ according to

$$S_\Phi = \frac{E_1 - S_1}{L_{\mathrm{eXY}}} \qquad S_\Theta = \frac{L_{\mathrm{eXY}}}{L_{\mathrm{e}}}$$
$$C_\Phi = \frac{E_2 - S_2}{L_{\mathrm{eXY}}} \qquad C_\Theta = \frac{E_3 - S_3}{L_{\mathrm{eXY}}} \tag{8.19}$$

Additional information is necessary to define the cross sectional orientation, either by setting Ψ directly or by defining a projected point $\mathbf{A}'$ in space. The projected point $\mathbf{A}'$ can be used to calculate the local element unit vector $\bar{i}_{\mathrm{e}}$ according to

$$\bar{i}_{\mathrm{e}}^{\mathcal{I}} = \frac{\mathbf{A}' - \mathbf{S}}{||\mathbf{A}' - \mathbf{S}||} \tag{8.20}$$

and the other axes are

$$\bar{k}_{\mathrm{e}}^{\mathcal{I}} = \frac{\mathbf{E} - \mathbf{S}}{||\mathbf{E} - \mathbf{S}||} \tag{8.21}$$

$$\bar{j}_{\mathrm{e}}^{\mathcal{I}} = \bar{k}_{\mathrm{e}}^{\mathcal{I}} \times \bar{i}_{\mathrm{e}}^{\mathcal{I}} \tag{8.22}$$

Because orthogonality between $\bar{i}_e$, $\bar{j}_e$, and $\bar{k}_e$ is required, one defines $\mathbf{A}'$ indirectly by another point $\mathbf{A}$, which will be projected on the shortest distance to the cross sectional plane, which intersects point $\mathbf{S}$ and to which $\bar{k}_e$ is orthogonal $\left(\mathcal{P}_{\text{proj.}} = \left(\mathbf{S}, \bar{k}_e\right)\right)$. Thus, $\mathbf{A}' \in \mathcal{P}_{\text{proj.}}$ and $\overline{\mathbf{AA}'} \perp \mathcal{P}_{\text{proj.}}$. All these proposed changes were validated through a parallel modelling approach by Struve et al. [179] and are also beneficial for non-rotatable lattice tower concepts.

8.2 Modelling the Tower in SubDyn

After the implementation of all proposed theoretical changes from section 8.1 in the SubDyn source code, written in FORTRAN 90, and the appropriate adaptation of the SubDyn input file, a way for its automatized generation had to be found. This step was necessary to get more flexibility to investigate different tower design parameter sets. A python script with the same node coordinate calculation and member connectivity functionality, as developed likewise for RFEM, was written to provide the basic geometry data for the SubDyn input file. The member list in the input file requires now a cross sectional rotation angle or a local x-axis orientation point in space, named $\mathbf{A}$ in subsection 8.1.2 for each member together with a switch OriTpe to prescribe whether the first or the later applies. This additional information is required for complete definition of the orientation of arbitrary cross sections with respect to the member axis. All cross sectional orientation points correspond to the bracing member alignment nodes, introduced in Figure 14.6. Their coordinates are calculated by definition of a new orthonormal basis, where two orthogonal unit vectors lie in a 2D-bracing-plane obtained from three points: two leg coordinates and one x-bracing intersection coordinate $\underline{P}_x$. An arbitrary orthogonal vector to this plane, added or subtracted to the bracing coordinate $\underline{P}_x$, delivers the member alignment point $\mathbf{A}$.

Calculation of the members' cross sectional properties was performed through the cross sectional analysis tool, BECAS [36]. Usage of this academic free tool, requires the citation of the following works: Blasques and Stolpe [19], Blasques [16], Blasques and Bitsche [17], and Blasques et al. [18]. It is capable to generate full 6×6 cross sectional stiffness and mass matrices $\underline{\underline{K}}_c$ and $\underline{\underline{M}}_c$ for closed and open anisotropic beam cross sections. For the purpose of this work, an automated mesh generation was coded in python to model the leg and bracing profile meshes in BECAS. A rough mesh for a leg and bracing profile is shown in Figure 8.4 where BECAS results are plotted for the mass center, the elastic center, the shear center, and the principal bending axes.

A mesh discretization study was performed to evaluate the number of necessary elements to get converged cross sectional stiffness and mass matrix entries. The results of the study are shown in Figure 8.5 where the stiffness matrix convergence of a leg and bracing member is shown on the left side and their mass matrix convergence is shown on the right side. Thereby, only non-zero structural matrix entries are shown and related to the result with the highest considered number of elements for one cross section. With dnL as the amount of elements for one straight cross sectional part and *dnt* as the amount of elements along its thickness, the maximum considered number of cross sectional elements becomes

$$n_{\text{el,max}} = dnL \times dnt \times 12 = 64 \times 64 \times 12 = 49,152 \tag{8.23}$$

for 12 straight parts. A good match to the converged result is already reached at $dnL = dnt = 16$ with $n_{\text{el}} = 3072$, where the solution is approximately 2 % apart from the 'accurate' result. Note that the sensitivity of *dnL* compared to *dnt* with respect to the convergence is not investigated further, because the calculations with BECAS and $dnL = dnt = 16$ run with an acceptable performance in terms of necessary processing time.

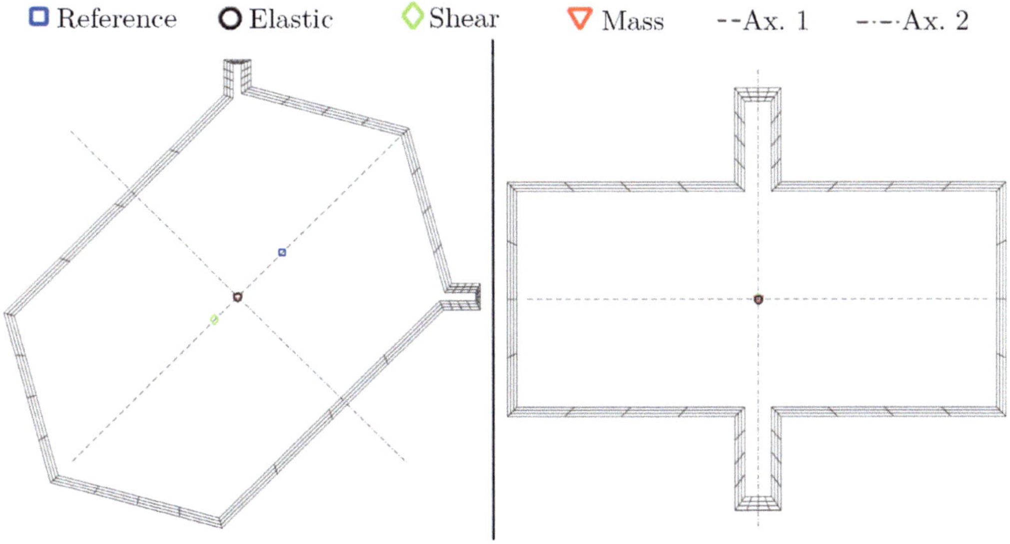

Figure 8.4: BECAS cross sectional meshes and resulting mass center, elastic center, shear center, and principal bending axes for a leg and bracing member

The cross sectional properties of all members were automatically evaluated to full 6 × 6 cross sectional stiffness and mass matrices from the generated mesh. Since they are based on the 2D finite element meshes in Figure 8.4, which represent closed cross sections rather than open cross sections, some adjustments were incorporated. They are necessary to ensure, that the torsional stiffness of the members is not overestimated because they have longitudinal parts where they are not connected by filler plates, such as explained in subsection 7.1.3. One accurate method would be to model each member in a finite element program and apply an unit torsional moment at the members tip to extract its torsional stiffness. Such a procedure would require a lot of computational time, because of the many different members to consider. Although the modifications in subsection 8.1 are rigorous enough to include all six DOF loads of a beam node, torsional loads and stiffness of lattice tower members play no relevant role in terms of their and the towers design. Therefore, a conservatively reduced torsional stiffness for each member is assumed by downscaling the cross sectional stiffness matrix $K_{c,i,6}$ $\forall i \in \{1,...,6\}$ and $K_{c,6,j}$ $\forall j \in \{1,...,6\}$ with factor 1e−3.

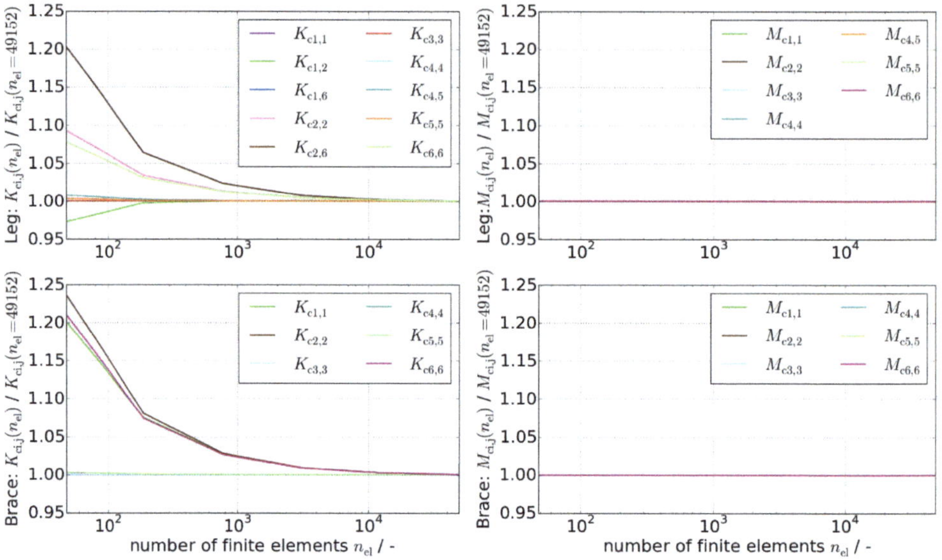

Figure 8.5: BECAS mesh study for a leg and bracing profile: Non-zero cross sectional stiffness and mass matrix entries are set into relation to their converged result

A representative SubDyn input file is shown in Appendix 14.18. It is ordered into 12 parts:

1. Input file header: Contains the SubDyn version and some fundamental model descriptions. Note that this work used an adjusted SubDyn version, which has not been declared yet.
2. Simulation control: Contains data output options, time step size, integration method, and a flag to determine, whether dynamics should be solved around the static equilibrium point, such as mentioned in subsection 2.8.3.
3. Finite-element and Craig-Bampton parameters: Contains the finite-element discretization number, a flag for enabling Craig-Bampton reduction, the amount of Craig-Bampton modes to be retained, and the damping ratio for each of these modes. A review of subsection 2.8.3 gives a brief introduction to the Craig-Bampton modal reduction method.
4. Joints: Contains the amount of joints and the declaration of each joint by means of an identification number and 3D coordinates.
5. Base reaction joints: Declares which joints are fixed in the structure. Until now, there is only full 6 DOF fixation possible.
6. Interface joints: Declares which joints are part of the interface to the rigid coupled TP point.

7. Members: Declares all structural members by setting a member identification number, listing its two joints, define the cross sectional property set of both member ends, chose the orientation type OriTpe, explained in the first paragraph of this subsection, and the corresponding information for the members rotation angle around its axis or the alignment point coordinates **A**, introduced in the end of subsection 8.1.2. The orientation type and the corresponding member alignment information were added to the original SubDyn input layout and functionality for the purposes of this work.
8. Member cross section properties: Declares member material properties, such as the Young's and shear modulus and the material density. Furthermore, it takes cross sectional geometrical properties for each property in one of three ways: (1) a circular cross section, (2) a general cross section, defined by engineering constants, such as the cross sectional area, the corrected shear areas, the shear center coordinates, the second area moments of inertia, and the torsional constant or (3) full 6 $\times$ 6 cross sectional stiffness and mass matrices, explained in subsection 8.1.1. The later two options were added to the original SubDyn input layout and functionality for the purposes of this work.
9. Joint additional mass: Declares additional concentrated masses and rotational inertias to certain joints.
10. Outputs: Contains information and flags of how and which results should be written to the output files.
11. Member outputs: Declares members from which more detailed outputs should be written to the output files, such as the loads and deflections of certain beam nodes within a member.
12. Global outputs: Declares global outputs, such as interface and boundary loads and deflections or modal parameters.

8.3 Validation of the SubDyn Model

As mentioned in subsection 8.1, the changes in SubDyn have already been validated by Struve et al. [179]. Changes were carried out to account for arbitrary member cross sections, rather than only circular member cross sections in the original version. The validation in the cited reference was done by an parallel modelling approach between SubDyn and ABAQUS. In the present validation case, the civil engineering software RFEM, was used instead of ABAQUS for two reasons: (1) Validation against another software increases the reliability of the implemented SubDyn changes. (2) In the cited reference, the option of engineering constants was used to describe the beam properties, but in this work full 6 $\times$ 6 cross sectional stiffness and mass matrices were used as input for SubDyn structural members. Appendix 14.19 may be reviewed for more detailed information about the parallel implementation in RFEM. Thereby, a certain tower design parameter set, such as discussed in section 9.1, was chosen for the tower and a natural frequency analysis was performed for comparison of both implementations.

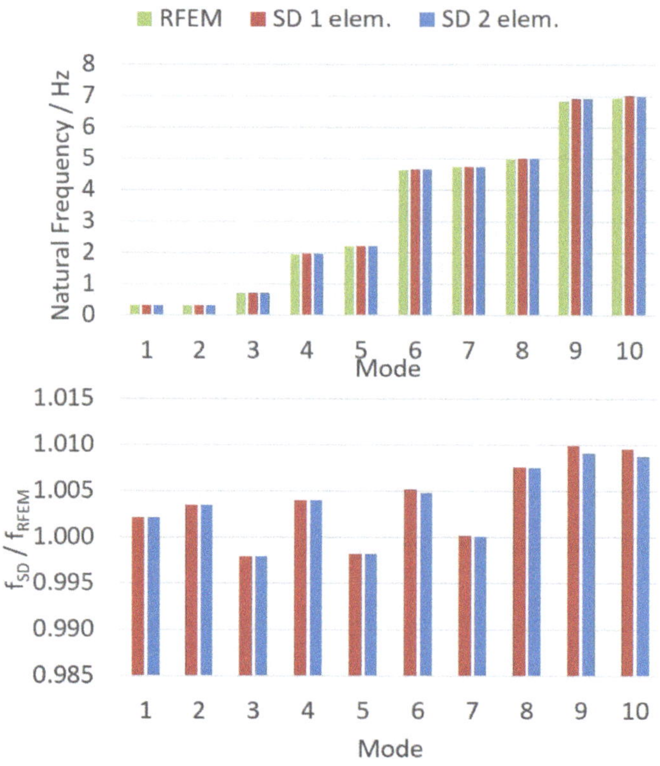

Figure 8.6: Natural frequency comparison between RFEM and SubDyn

A natural frequency comparison between both implementations is a good approach for validation of the new implemented beam element and the lattice tower in SubDyn, since it requires correct beam stiffness and mass matrices and proper member alignments in space. The later can visually be checked in the RFEM software package. Figure 8.6 shows the natural frequencies for the rotatable inclined lattice tower in RFEM and SubDyn. Note that the RFEM solution is converged with respect to the beam element discretization for element lengths ≤ 0.01 m. The calculated natural frequencies in SubDyn differ less than 1.0 % from the RFEM result for the first 10 modes by using only one beam element per member in SubDyn. Doubling the element discretization in SubDyn results in no significant improvements and affirms the good performance of the beam element formulation in subsection 8.1.1.

8.4 Implementation of the SubDyn Model in FAST

This section is divided into two subsections. The first subsection 8.4.1 treats the necessary adjustments in the FAST environment, which are focused on the structural ElastoDyn module.

The second subsection 8.4.2 discusses the necessary simulation settings in the SubDyn module, where the rotatable inclined lattice tower was modelled.

8.4.1 Adjustments to the FAST Environment

A switch, called CompSub in the main FAST input file and the provision of the corresponding SubDyn input files path is enough to incorporate SubDyn in the FAST framework. Nevertheless, a few other settings in the ElastoDyn module had to be adjusted to account for the SubDyn structure in the right way. First, the yaw compliance and ElastoDyn tower mode DOFs in the FA and SS direction, YawDOF, TwFADOF1, TwFADOF2, TwSSDOF1, and TwSSDOF2 had to be disabled. Secondly, the platform DOFs in horizontal surge and sway directions PtfmSgDOF and PtfmSwDOF, in vertical heave direction PtfmHvDOF, and around its roll PtfmRDOF, pitch PtfmPDOF, and yaw PtfmYDOF axes had to be enabled. For offshore WTs, the platform DOFs are used to connect the substructure rigidly with the tower. In case of a land-based full lattice tower, the platform becomes the connection between the tower and the RNA. It is not possible to disable the ElastoDyn tower completely, wherefore it remained in an artificial way by setting the tower height value in ElastoDyn to TowerHt = TowerBsHt + 0.001 m, where the ElastoDyn tower base height TowerBsHt = H_T equals to the full lattice tower height.

Initial Conditions
The required platform coordinates are given as

$$\begin{aligned}\mathrm{PtfmCMxt} &= \tan(\alpha_\mathrm{TI})\,H_\mathrm{T} - \frac{a_\mathrm{T}}{2} + \frac{D_\mathrm{T,NREL}}{2}\\ \mathrm{PtfmCMyt} &= 0\\ \mathrm{PtfmCMzt} &= H_\mathrm{T}\end{aligned} \tag{8.24}$$

The tower top diameter of the Baseline NREL reference WT was used to place the RNA to the same relative tower top position as for the conventional reference WT. The same relative position means, that the RNA distance to the front outer border of the rotatable inclined lattice tower was the same as to the outer tower border of the reference WT. To reach numerical stability, the definition of platform inertias > 0 was required. Here, a platform roll inertia PtfmRIner = 1/2 NacYIner and a platform pitch inertia of PtfmPIner = NacYIner were set, where NacYIner denotes the given nacelle inertia about its yaw axis. Neglecting the initialization of the self-weight-displaced platform DOFs led to numerical instabilities in the beginning of the simulation, because of high tower head accelerations. Therefore, it was important to prescribe a good approximation of the initial platform position under self weight of the structure. The SubDyn manual [32, p. 31] proposes the calculation of the initial vertical platform displacement to be

$$\mathrm{PtfmHeave} = -\frac{\left(\widetilde{M}^{(\mathrm{SD})}_{\mathrm{BB},33} + \widetilde{M}^{(\mathrm{ED})}\right) g}{\widetilde{K}^{(\mathrm{SD})}_{\mathrm{BB},33}} \tag{8.25}$$

where $\widetilde{K}^{(SD)}_{BB,33}$ and $\widetilde{M}^{(SD)}_{BB,33}$ are the 3,3 elements of the equivalent stiffness and mass matrices of the substructure, resolved with respect to the TP point. The equivalent stiffness and mass matrices are mentioned before in subsection 2.8.3 and may be evaluated with help of the stand-alone SubDyn driver. $\widetilde{M}^{(ED)}$ is the mass from ElastoDyn, which equals to the RNA mass in the present case. Since an inclined tower does not only have vertical initial displacements, but also displacements along x and a rotation about y, the whole equivalent stiffness matrix $\underline{\underline{\widetilde{K}}}_{BB}$ was used to calculate all tower head (platform) displacements. Note that the tower head, the platform and the TP are all the same for this structure. First, the stand-alone SubDyn driver was used to calculate the resulting reaction interface loads of the rotatable inclined lattice tower under RNA and self weight at the TP. The RNA weight was modelled through point masses, attached to ultra stiff and low mass members, such as likewise applied in the RFEM model, shown in Figure 14.6. Next, the initial platform displacements and rotations are calculated from Equation 8.26 as follows

$$\begin{Bmatrix} \text{PtfmSurge} \\ \text{PtfmSway} \\ \text{PtfmHeave} \\ \text{PtfmRoll}\,\dfrac{180\,\text{deg}}{\pi} \\ \text{PtfmPitch}\,\dfrac{180\,\text{deg}}{\pi} \\ \text{PtfmYaw}\,\dfrac{180\,\text{deg}}{\pi} \end{Bmatrix} = \left[\underline{\underline{\widetilde{K}}}_{BB}\right]^{-1} \begin{Bmatrix} -\text{IntfFXss} \\ -\text{IntfFYss} \\ -\text{IntfFZss} \\ -\text{IntfMXss} \\ -\text{IntfMYss} \\ -\text{IntfMZss} \end{Bmatrix} \tag{8.26}$$

where PtfmSurge, PtfmSway, PtfmHeave, PtfmRoll, PtfmPitch, and PtfmYaw are the platform surge, sway, and heave displacements and roll, pitch, and yaw rotations. IntfFXss, IntfFYss, IntfFZss, IntfMXss, IntfMYss, IntfMZss are the TP interface reaction forces and moments, whose signs must be turned. To get converged interface loads, the SubDyn stand-alone driver calculates 300 time steps under RNA and self weight. Note that an applied artificial high structural damping of 50 % increased a converged result in this analysis.

Artificial Damping through HydroDyn

Numerical problems arise in SubDyn if the Craig-Bampton[1] reduction is applied, because the interface modes experience no damping. This problem is mentioned by Damiani, Jonkman, and Hayman [32, p. 30-31] where they suggest to incorporate an augmented overall system damping by enabling the HydroDyn module [113]. HydroDyn is capable, to define a linear damper for the platform DOFs in form of a 6×6 damping matrix $\underline{\underline{C}}^{(HD)}$. In the present work, an augmented platform-heave damping of $\zeta^{(HD)} = 0.01$[2], was likewise necessary to reach numerical stability. The equation to calculate the the corresponding element in the damping matrix is

$$C^{(HD)}_{33} = 2\,\zeta^{(HD)}\sqrt{\widetilde{K}^{(SD)}_{BB,33}\left(\widetilde{M}^{(SD)}_{BB,33} + \widetilde{M}^{(ED)}\right)} \tag{8.27}$$

[1] Subsection 2.8.3 provides a brief introduction to the Craig-Bampton modal reduction method.

[2] This value is the critical damping ratio.

8.4.2 SubDyn Simulation Settings

After incorporation of all these model pre-sets, a Craig-Bampton reduction was used to retain 20 Craig-Bampton eigenvalues in the FAST embedded simulation. With this, all WT system frequencies up to 4.29 Hz may potentially couple with the rotatable inclined lattice tower natural frequencies. The value of 4.29 Hz depends on the design parameters and is calculated for the parameter set in Table 9.1. Considering the full-system linearized natural frequency analysis of the land-based Baseline NREL reference WT in [109, p. 30], no frequencies ≥ 2.02 Hz should occur for the relevant first and second blade flapwise modes, the first blade edgewise mode, and the first drivetrain torsional mode. Thus, 20 Craig-Bampton eigenvalues were sufficient. Furthermore, an appropriate member discretization should be found. The natural frequency results in Figure 8.6 justify a member discretization of NDiv = 1, because it reaches already good accordance to the converged RFEM results. This is caused by the rigorous anisotropic beam element approach, explained in subsection 8.1.1. Moreover, the dynamic solution around the static equilibrium point SttcSolve was enabled to ensure a correct self weight inclusion of the lattice structure, such as suggested in the SubDyn manual [32, p. 32]. The last important parameter was the integration step size. Since equations 3 and 4 in [32, p. 32] do not apply properly to an inclined tower, the necessary maximum time step size had to be estimated on another way. Experience has shown, that ASE simulations with a time step size of ≤ 0.0125 s in case of the implicit second-order Adams-Moulton integration method behave numerical stable. Unfortunately, this is a very small time step size, which causes long simulation times of more than 2 h to simulate 630 s of one timeseries.

9 Structural Analysis of a Rotatable Inclined Lattice Tower

9.1 Choosing Design Parameter Values

Since a material efficient design requires comparison between the performance of different input parameter sets, a flexible way for geometry and input data generation was needed. In this context, a good performance means low structural mass and a high member utilization. By means of self-written python scripts, which create the SubDyn input file automatically from the input design parameters, different designs could be tested in comparatively low amounts of time. It follows the explanation for the first design parameter set.

Parameter Assumptions

The flap lengths of the leg and bracing hat-profiles were set to $L_\mathrm{F} = 0.09$ m and $H_\mathrm{F} = 0.07$ m to ensure enough space for bolts. The material thickness of the legs and bracing members depends on Equations 7.15 and 7.16 and is therefore not a directly adjustable design parameter. Assuming a maximum extrusion moulding diameter of $D_\mathrm{extr} = 0.35$ m for the hat-profiles, restricts the two side supported length to

$$H_{\mathrm{L,max}} = \min\left\{ D_\mathrm{extr} - 2\,H_\mathrm{F}\,, \quad -\frac{H_\mathrm{F}}{2} + \sqrt{\left(\frac{H_\mathrm{F}}{2}\right)^2 + \frac{D_\mathrm{extr} - H_\mathrm{F}^2}{2}} \right\} = 0.21\ \mathrm{m} \tag{9.1}$$

For comparison, the same tower hight as for the Baseline NREL reference tower was chosen with $H_\mathrm{T} = 87.6$ m. Moreover, the steel density was increased to $\rho_\mathrm{steel} = 8500\ \mathrm{kg/m^3}$ to account for the additional weight of bolts, filler plates, paintings, cables, and structural attachments, such as the lift cage. Note that this material density is the same as assumed for the conventional reference tower and makes it therefore more comparable. Moreover, a common WT support structure steel grade with a yield strength of $355\ \mathrm{N/mm^2}$ was chosen. Cost intensive higher steel grades are uncommon, because the WT support structure design is often driven by fatigue, where higher grades are not advantageous. A bracing segment number, analogous to the Butzkies tower in Figure 2.12 was assumed. With $N_\mathrm{seg} = 20$, it is one less than for the 12.4 m higher Butzkies tower. Tower bottom length a_B and width b_B should be as large as possible to reduce the leg axial loading through the tower bending moment. On the other hand, the buckling lengths of the x-bracings should not become too big and increased leg distances increase the tower stiffness and thereby the amount of load cycles over the lifetime. Note that these load load cycles would be smaller. At this point it should be noted that the bottom tower dimensions are likewise important for the tower base yaw system. The influence is determined by the chosen tower base yaw concept, discussed in chapter 4.2. In the end it is a complex optimization problem to find optimal tower base dimensions. For this preliminary study only the material saving potential of the tower was evaluated, without accounting for a specific tower base yaw system. Therefore, the tower length and width were oriented at common lattice tower base dimensions for this hub height and rated power. Eigenfrequency analyses revealed, that first FA and SS bending frequencies reach easily the 3p excitation frequency range of the WT. A review of Figure 4.4 shows the excitation frequency ranges. Thus, the tower base length and the width were equal and relatively small with $a_{\mathrm{B}} = b_{\mathrm{B}} = 6.8$ m.

Common machine foundations have already rectangular shapes, which can be seen as tower top length and width limitation. In case of a conventional WT, the width of the machine frame would have a value, which is slightly larger than the tower top diameter. The tower top diameter of the Baseline NREL reference WT is $D_{\mathrm{T,NREL}} = 3.87$ m and a corresponding machine frame could have a width of ≈ 4 m. In the present eigenfrequency driving case for the outer dimensions, relatively small tower top dimensions of $a_{\mathrm{T}} = 4.5$ m and $b_{\mathrm{T}} = 4.5$ m were chosen. It was assumed that large machine foundations would become too heavy and reduce the material saving potential of the concept. On the other hand, a certain space is necessary to cover all nacelle mountings, such as the drivetrain and the generator. However, further investigations with respect to the machine foundation are out of scope for this work. Future research projects should consider the machine foundation weight as function of tower head dimensions for rotatable WT support structures.

For the previous explained tower dimensions, the tower inclination angle is calculated with help of Equations 6.5, 6.6, and 6.7. To avoid a cone position for the blades and a rotor shaft tilt, the corresponding parameters had to become zero: $\alpha_{\mathrm{PC}} = \alpha_{\mathrm{ST}} = 0$ deg. From this, the minimum demanded tower inclination angle results to

$$\alpha_{\mathrm{TI}} = -\arctan\left(\frac{\text{clear.} - l_{\mathrm{OH}} + d_{\mathrm{T}} - \left(\frac{1}{2}a_{\mathrm{T}} - \frac{1}{2}D_{\mathrm{T,NREL}}\right)}{H_{\mathrm{T}} - z_{\mathrm{Tip}}}\right) \qquad (9.2)$$

$$\rightarrow \alpha_{\mathrm{TI}} \frac{180\ \mathrm{deg}}{\pi} = -7.721\ \mathrm{deg}$$

where the calculated blade to tower clearance for the conventional configuration is clear. = 10.539 m. Furthermore, the parameters in the Equations 6.6 and 6.7 change for the reordered Equation 9.2: $D_{\mathrm{B}} \rightarrow a_{\mathrm{B}}$, $D_{\mathrm{T}} \rightarrow a_{\mathrm{B}}$, $z_{\mathrm{Tip}} = H_{\mathrm{T}} + l_{\mathrm{TS}} - R_{\mathrm{rot}}$. Note that the additional term $\frac{1}{2}a_{\mathrm{T}} - \frac{1}{2}D_{\mathrm{T,NREL}}$ in Equation 9.2 was necessary to shift the nacelle from the tower centreline to a value, that corresponds to the same distance between the rotor and the tower edge, as for the conventional configuration. With these settings, energy harvest from the wind is increased, such as discussed in subsection 6.2.3. Table 9.1 shows the assumed parameter set, where L_{L} and H_{L} were determined to reach a good material utilization. L_{L} and H_{L} were given for the bottom of the tower and for the top of the tower, marked by indexes $_{\mathrm{B}}$ and $_{\mathrm{T}}$, respectively. All members in-between had linear interpolated values. The constant filler plate thickness and width was set to be $t_p = 20$ mm and $w_p = 1$ mm for all members. Note that the filler plate width was conservatively set to that small value instead of a real value, such as 70 mm to account for the reduced axial, bending, and torsional stiffness, due to discontinuous filler plate occurrences along a member. The cross sectional torsional stiffness was additionally downscaled about factor 1e−3 to include the significant stiffness reduction within the open cross sectional areas along the members and is more explained at the end of section 8.2. Note that the proposed parameter set was not optimized, but should serve as a first guess in the context of this preliminary study.

Parameter	Value	Unit	Parameter	Value	Unit	Parameter	Value	Unit
H_T	87.6	m	N_{seg}	20	–	ρ_{steel}	8500	kg/m^3
f_y	355	N/mm^2	a_B	6.8	m	b_B	6.8	m
a_T	4.5	m	b_T	4.5	m	$L_{L,B}$	360	mm
$L_{L,T}$	270	mm	L_F	90	mm	$H_{L,B}$	90	mm
$H_{L,T}$	210	mm	H_F	70	mm	α_{TI}	-7.721	deg
t_p	20	mm	w_p	1	mm			

Table 9.1: Rotatable inclined lattice tower design parameter set for a tower height of 87.6 m

9.2 Modal Analysis

Some fundamentals of modal analysis are introduced in subsection 4.3.3. This section discusses the different modes and shows them in the context of WT operational conditions within the first two subsections. Furthermore, a sensitivity analysis is provided for the aspect of tower inclination with respect to the different modes. Knödel states that eigenfrequencies of structures are difficult to define as they are changing throughout vibration cycles, especially if buckling failure of the pressure x-bracing diagonal is considered [122]. Since deformations are small and buckling is not allowed for all members, this aspect is not investigated in this work.

9.2.1 Mode Shapes

Figure 9.1 depicts the first 10 mode shapes as qualitative deflections with respect to the original structure in RFEM. Note that the global x-direction is marked with a capital X in the figure. The first two mode shapes are the typical first bending modes in the FA and SS direction and the third is the torsional mode. During the evaluation of the parameter set for the rotatable inclined lattice tower for 90 m hub height, these first three eigenfrequencies were the most important ones to consider, because intersection of the excitation frequencies with these close tower eigenfrequencies led to significant fatigue. The following mode shapes from 4 to 8 are the higher global bending mode shapes, while the last two show a combined global torsion and local leg bending deformation. Table 9.2 shows the corresponding exact values of the important tower modes. As leg distances and bracing lengths are small, the first pure bracing mode shape occurred at higher frequencies of $\geq f_{12} = 8.802$ Hz. A pure bracing mode is considered as a mode if mainly bracing members deflect, without significant participation of global tower deflections.

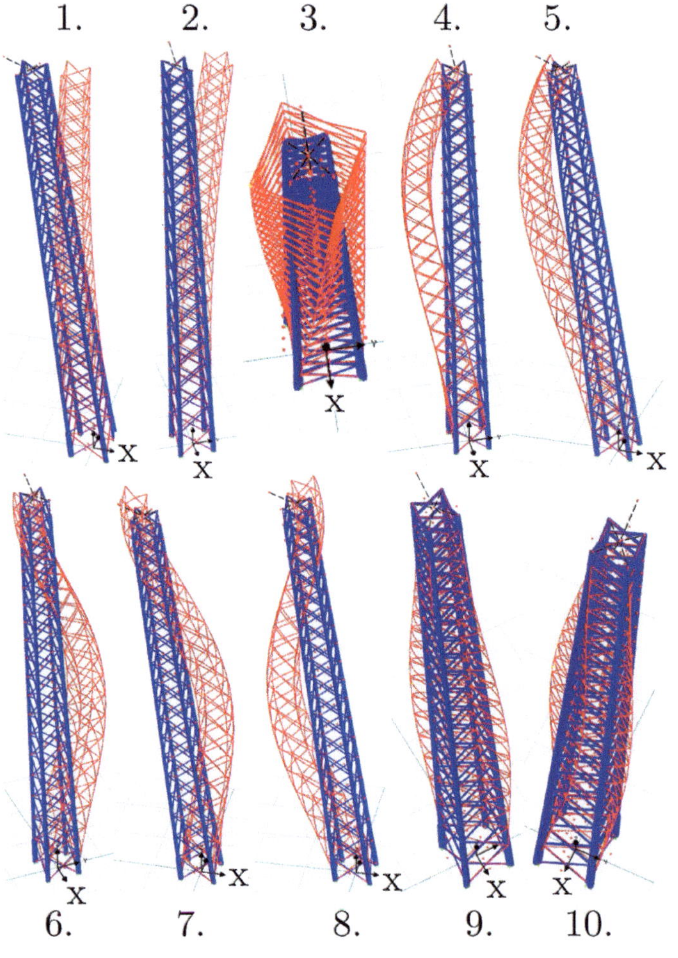

Figure 9.1: First 10 mode shapes of the rotatable inclined lattice tower for 90 m hub height

9.2.2 Campbell Diagram

In WT technology, structural eigenfrequencies should not coincide relevant excitation frequencies during operation. A good way to check for this requirement is to plot the Campbell diagram of the WT, such as introduced in subsection 4.3.3 for the reference configuration. In contrast to the previous, Figure 9.2 shows the Campbell diagram for the rotatable inclined lattice tower

(RILT) with its first 5 eigenfrequencies and some multiples of the excitation frequencies. The later were denoted as 1p, 3p, 6p, 9p, and 12p. The first multiples are more relevant, because they contain much more energy than the higher harmonics.

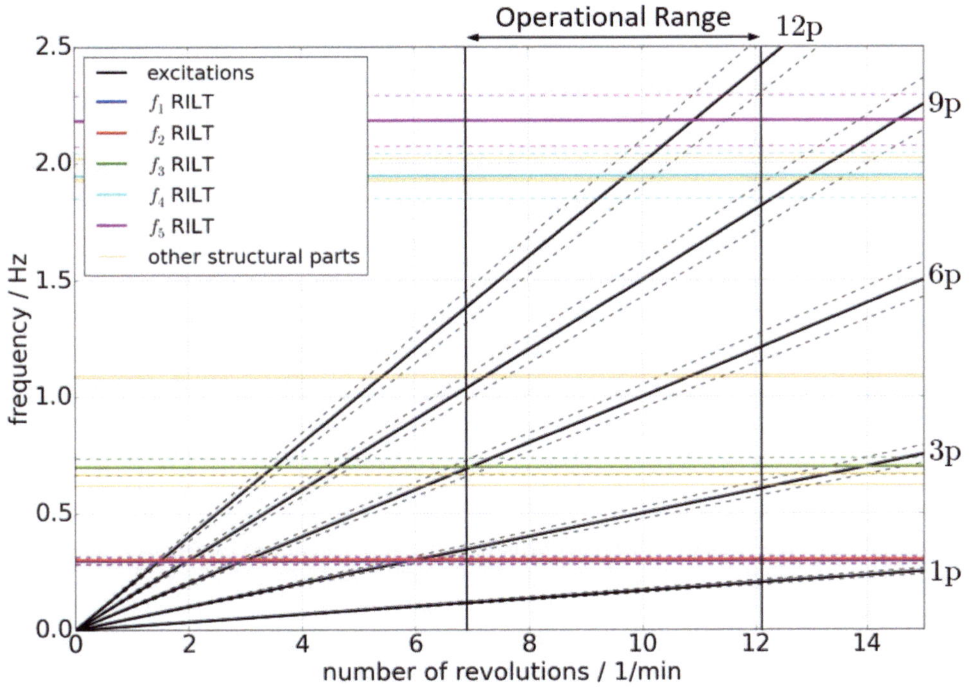

Figure 9.2: Campbell diagram of the 90 m hub height rotatable inclined lattice tower for the 5 MW NREL reference WT. All frequencies are plotted with ±5 % safety margins, except the other structural part frequencies.

Figure 9.2 illustrates, that no intersections between the important 1p and 3p excitations and the RILT eigenfrequencies exist within the operational range. The first two tower modes lie with f_1 = 0.297 Hz and f_2 = 0.301 Hz between the 1p and 3p excitations in the operational range. All other eigenfrequencies are above the 3p excitation frequencies. However, the next multiple (6p) intersects with the torsional tower eigenfrequency of f_3 = 0.654 Hz at the cut-in rotational speed. Other intersections within the operational range occur for the 12p harmonic excitation and the fourth (f_4) and fifth (f_5) tower eigenfrequencies. Interestingly, eigenfrequencies of other structural components, such as the drivetrain and the blades, are likewise hit by harmonic excitations within the operational range. These other structural parts are shown in orange colour and are part of the reference WT configuration. It reveals that the upper harmonics caused no problems with respect to these components in the past and a critical influence to the higher tower eigenfrequencies is likewise not expectable. How intersections with upper harmonics affect the loads, manifested at the full ASE analysis and in the following extreme load and fatigue analyses. It was thereby covered within the assessment procedure.

Mode	Description	Value
1	1. Tower FA Bending	0.297 Hz
2	1. Tower SS Bending	0.301 Hz
3	1. Drivetrain Torsion	0.621 Hz
4	1. Tower Torsion	0.654 Hz
5	1. Blade Asymmetric Flapwise Yaw	0.666 Hz
6	1. Blade Asymmetric Flapwise Pitch	0.668 Hz
7	1. Blade Collective Flap	0.699 Hz
8	1. Blade Asymmetric Edgewise Pitch	1.079 Hz
9	1. Blade Asymmetric Edgewise Yaw	1.090 Hz
10	2. Tower SS Bending	1.834 Hz
11	2. Blade Asymmetric Flapwise Pitch	1.922 Hz
12	2. Blade Asymmetric Flapwise Yaw	1.934 Hz
13	2. Blade Collective Flap	2.021 Hz
14	2. Tower FA Bending	2.282 Hz

Table 9.2: Wind turbine eigenfrequencies with rotatable inclined lattice tower and 90 m hub height. RNA frequencies are taken from Jonkman et al. [109, p. 30]

Table 9.2 summarizes all available structural component eigenfrequencies with their exact values. It should be noted that the tower eigenfrequencies were calculated by means of the SubDyn module where RNA masses and inertias represented at certain points in space, which are rigidly coupled to the tower top joints. More information about the modelling in SubDyn is provided in Chapter 8. Other structural part eigenfrequencies were obtained through an eigenanalyis on the first-order state matrix, which was generated within a linearization of the conventional reference FAST model and given by Jonkman et al. [109, p. 30]. The added denotation yaw and pitch in Table 9.2 corresponds to coupled asymmetric blade eigenfrequencies with nacelle yaw and pitch motions.

9.2.3 Modes Sensitivity to the Tower Inclination

This subsection provides a short sensitivity analysis of the tower modes with respect to its inclination angle α_{TI}. Recalling subsections 6.2.3 and 9.1 shows, that the tower inclination α_{TI} was considered to be dependent on the blade precone angle α_{PC} and the rotor shaft tilt angle α_{ST}. This dependency was introduced to obtain the same blade to tower clearance as for the reference configuration in cases where α_{PC} and α_{ST} were changed. Table 9.3 lists six different cases **C1** to **C6**, where these two angles were changed between their original values and zero. A combination of these options resulted into four different tower inclination angles α_{TI} plus two additional increased inclinations, **C1** and **C2**. The tower inclination influenced the eigenfrequencies in the different modes. Note that these values are different for other tower dimensions and the present ones were based on the parameter set in Table 9.1.

Parameter \ Case	C1	C2	C3	C4	C5	C6	C1/C6 in %	C3/C6 in %
α_{PC} / deg	0.0	0.0	0.0	2.5	0.0	2.5	-	-
α_{ST} / deg	0.0	0.0	0.0	0.0	5.0	5.0	-	-
$-\alpha_{TI}$ / deg	20	15	7.721	5.179	2.651	0.056	-	-
clearance / m	24.48	18.62	10.54	10.54	10.54	10.54	-	-
f_1 / Hz	0.260	0.279	0.297	0.301	0.303	0.304	86	98
f_2 / Hz	0.278	0.290	0.301	0.304	0.305	0.306	91	99
f_3 / Hz	0.580	0.616	0.654	0.662	0.667	0.670	87	98
f_4 / Hz	1.856	1.853	1.835	1.827	1.820	1.813	102	101
f_5 / Hz	2.049	2.169	2.282	2.304	2.317	2.320	88	98

Table 9.3: Sensitivity analysis of the rotatable inclined lattice tower modes with respect to the tower inclination

Table 9.3 reveals that increasing tower inclination causes decreased tower eigenfrequencies. This result was expectable, because the length of the tower rose likewise since the overall tower hight remained always the same. However, the decrease is not more than 2.3 % between the case **C3** where no precone and no shaft tilt case was applied and the almost vertical configuration **C6** for the first five tower modes. The mode frequency decrease changed up to 14 % if larger tower inclinations of $-\alpha_{TI} = 20$ deg were incorporated, such as in case **C1**.

9.3 ASE Simulation Results of IEC 61400-1 Load Cases

This section shows the results of the full RILT ASE analysis within the FAST framework and the adjusted SubDyn version. All modelling aspects and specifications are covered by the previous chapter 8. A large subset of the required DLCs according to IEC 61400-1[1] was chosen to be investigated in this section. The subset was the same subset as the one, which was analysed for the reference configuration in subsection 3.5.3 to ensure comparability. The amount of generated data in this work was so high[2], that it was impossible to show it for every single load sensor of every random seed and every DLC. Instead, a certain wind speed and random seed of DLC 1.1 was used to explain how the results looked like in general and how tower inclination affected the loads. After that, further interesting DLCs are presented, especially if some special things occurred during the analysis.

[1] An introduction to the different DLCs is given in subsection 2.2.2.

[2] The size of all ASE calculated sensor output files amounts to ≈ 470 GB. Note that these are text files and not memory optimized binary files.

9.3.1 Normal Power Production

In the normal power production DLC 1.1 with NTM, mean wind speeds between the cut-in and cut-out wind speed were considered for 6 different random seeds and a mean wind direction of −8 deg, 0 deg and 8 deg. DLC 1.1 accounts for no transient wind or WT events. In this subsection, a RILT according to case **C3** in Table 9.3 was used.

The first (top) diagram in Figure 9.3 shows the stochastic wind with a mean value, equal to the rated wind speed of 11.4 m/s and with a mean wind direction of 0 deg over the time period of 600 s[3]. In this case, the main wind direction was parallel to the WTs rotor axis, which was parallel to the global x-axis. The blue coloured sensor Wind1VelY is the wind speed in y-direction and the red sensor Wind1VelX represents the wind speed in the x-direction, respectively.

In the second diagram the blade 1 pitch angle BldPitch1 is shown in blue. The WT has a collective pitch drive, wherefore a given pitch angle of one blade can be seen as equivalent to the other two blades. Since the wind speed was around the rated wind speed, the pitch actuator had to increase the blades pitch angle from 0 deg in cases where the wind speed became more than the rated wind speed. This happens to restrict the generator power to ≈ 5 MW and to limit the loads in the system. The blade to tower clearance TipClrnc1 is shown in red and it was calculated with respect to its perpendicular distance to the tower wall in the case, that the blade tip pointed vertically downwards. In cases where the blade tip was above the yaw bearing, TipClrnc1 represented the absolute distance to it[4]. Therefore, only the minima of this sensor were interesting. Note that the blade to tower clearance was already transformed into a percentage value of its unloaded state and accounts for tower inclination as well as for a PSF of $\gamma_f\ \gamma_m\ \gamma_n = 1.35 \times 1.1 \times 1.0 = 1.485$[5]. In subsection 2.2.8, the minimum allowed clearance for different blade manufacturing testing procedures and operational conditions is discussed. Assuming the reference WT has blades with continuous bending stiffness control during the blade series production, the minimum allowed clearance is 20 % of its unloaded state. In the present case, the clearances became not less than 34.2 %.

The third diagram shows the rotor thrust RotThrust in blue and the rotor torque RotTorq in red. It can be seen, that the rotor torque rose only to about 4,500 kNm, because the blades pitched out of the wind to reduce it at higher wind speeds. At time periods with less wind speed, the rotor thrust and torque may fall to lower values. Then, the blades are fully pitched in, to 0 deg. A corresponding dip in the main wind speed Wind1VelX led consequently to a rotor thrust RotThrust and rotor torque RotTorq dip, such as shown at the time mark around 390 s. Note that this was just a dip of the wind speed in the main wind direction. Since the wind speed rose simultaneously in the transverse direction, this was rather a wind direction change than a wind speed dip.

[3] The diagram shows a 600 s time period, which starts at 30 s to avoid the initialization influences. This common practice was likewise mentioned in subsection 3.5.1

[4] Jonkman and Marshall [112, p. 95] explain how TipClrnc1 is calculated.

[5] The partial safety factors are chosen according to IEC 61400-1 [50, p. 48]

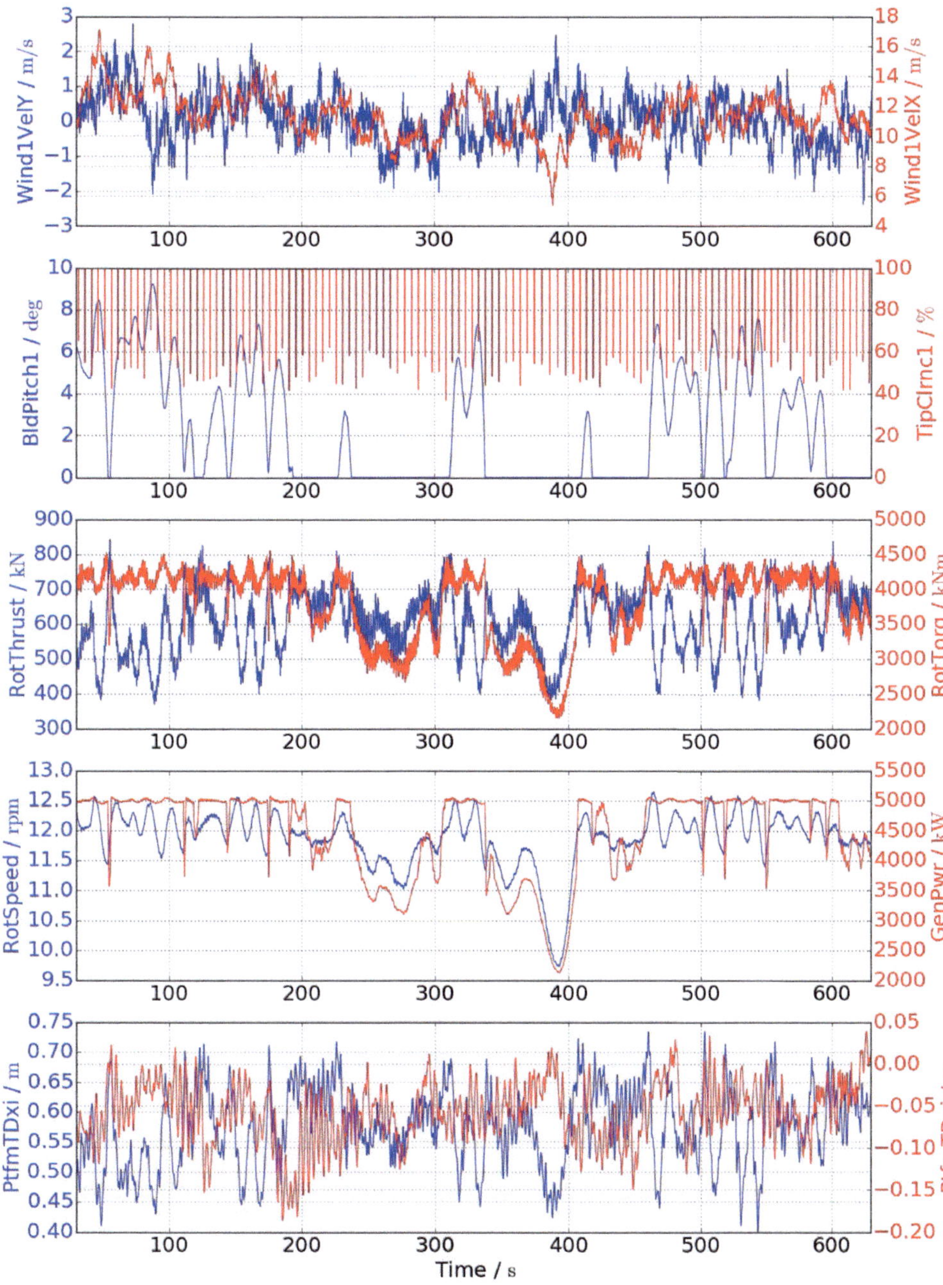

Figure 9.3: ASE simulation results for different wind turbine sensors and DLC 1.1 at rated wind speed on average. The rotatable inclined lattice tower, based on the parameter set in Table 9.1, is used.

The fourth diagram shows the rotor speed RotSpeed in blue and the generator power GenPwr in red. Both sensors had a visible correlation to each other, to the rotor torque, and to the pitch angle. A power restriction occurred as a result of increased pitch angles. Obviously, the rotor speed fluctuated around its rated value of 12.1 rpm at time periods with equal or higher than rated wind speed.

In the last diagram, tower head displacements are shown in X- and Y-direction. They are measured from the initial platform interface point at the tower top and are denoted as PtfmTDxi and PtfmTDyi for the different horizontal directions. Since the wind blew along the global x-direction, a tower top displacement in the same (positive) direction was expectable as shown. Furthermore, a good correlation between the rotor thrust in the third diagram and the tower top displacement in x-direction is visible. The transverse wind speed had a mean value of 0.04 m/s in this timeseries. However, the transverse tower top displacement PtfmTDyi fluctuated around a negative mean value. Since the rotor rotates clockwise, a sidewise tower top deflection in the negative y-direction was induced by the generator moment. This is also confirmed through analogous peaks of both sensors PtfmTDyi and RotTorq.

In Figure 9.5, load components of certain members of the RILT are depicted for the same timeseries as in Figure 9.3. Thereby, the upper three diagrams show loads of the four bottom tower legs. In the first diagram in Figure 9.5, the normal forces at the bottom of all four legs are given. The left side in Figure 9.4 helps to identify the leg enumeration with respect to the global coordinate system and the corresponding mean wind direction in this considered timeseries.

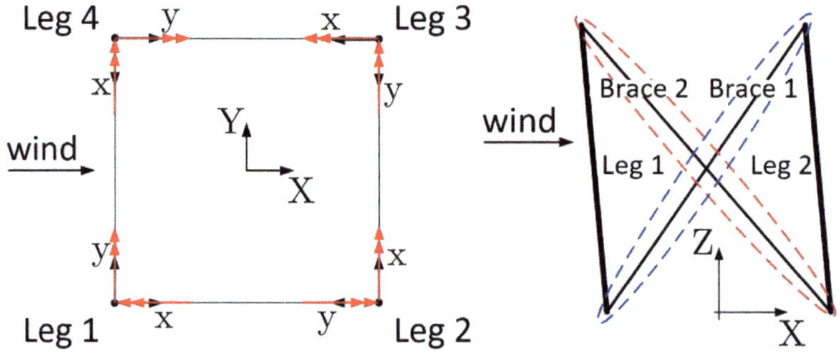

Figure 9.4: Bottom tower leg coordinate systems and mean local bending moment directions (left) and bracing member designations (right)

The front legs 1 and 4 are under tension, while the back legs 2 and 3 are under pressure. This result is interesting, because the incorporated tower inclination induced pressure to the front and tension to the back legs. This happened through the self weight induced bending moment in the tower, such as mentioned in subsection 6.2.1. Therefore, Figure 9.5 reveals that the tower inclination was not enough to equalize the normal forces in the front and back legs for this specific DLC and wind speed. Note that the wind speed drop in the main wind direction at 390 s is still detectable in a visual way for the leg normal forces. Thereby, the tension in the front and the pressure in the back legs decreased.

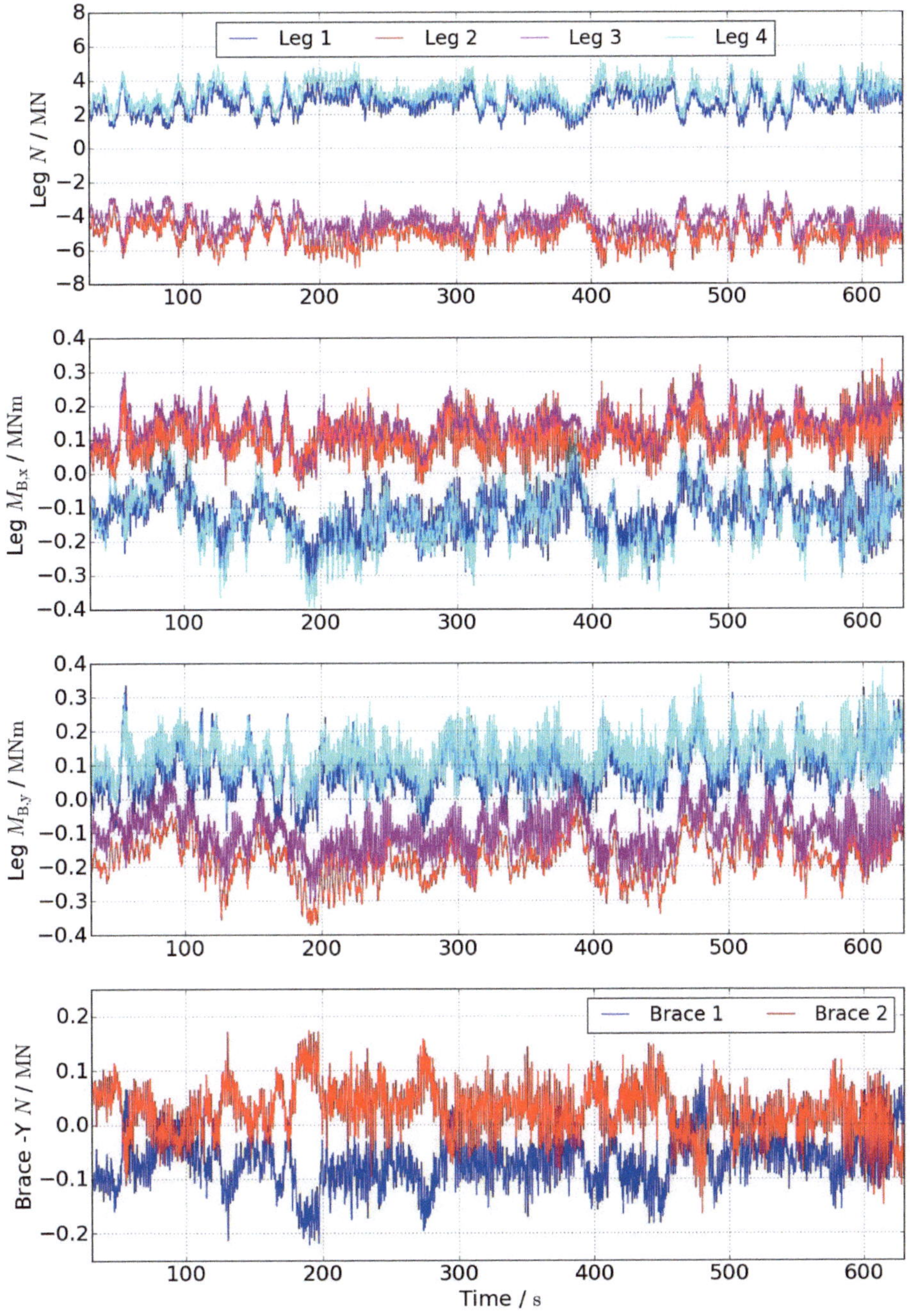

Figure 9.5: ASE simulation results for different load components in the bottom tower legs and bracing for DLC 1.1 at rated wind speed on average. The rotatable inclined lattice tower, based on the parameter set in Table 9.1, is used.

The second and third diagrams show the four legs at the bottom of the tower with their bending moments around their local x- and y-axes. Here, the left side in Figure 9.4 helps to find the local coordinate orientation of the members. Bending moment directions were entered into the Figure by red double arrows and are oriented at the mean values of diagrams two and three. Comparison with the first diagram shows that the local bending moments, which are parallel to the main tower bending axis, pointed in the same direction as the global bottom tower bending moment.

The fourth diagram shows normal forces within the diagonal bracing members on one side of the tower. It is the main wind direction and global X-axis parallel tower side on the -Y half[6]. The right side in Figure 9.4 indicates which colour corresponds to which bracing. Thereby, this is in agreement with the first diagram findings where tension in leg 1 and pressure in leg 2 led to pressure in brace 1 and tension in brace 2.

9.3.2 Loads Sensitivity to the Tower Inclination

In this section, the influence of tower inclination with respect to the tower member loads is presented. The first diagram in Figure 9.6 shows the mean leg normal force results for the same conditions as in the previous subsection 9.3.1, but for different tower inclinations, according to the three different cases **C1**, **C3**, and **C6** from Table 9.3. Each of these cases corresponds to one point in the plotted lines. Interestingly, leg normal forces did not significantly change for different tower inclinations $\alpha_{\mathrm{TI}} \leq 20$ deg. The reason for that is found in the following two diagrams. In the second diagram, all global mean bending moments are presented, which are parallel the Y-axis. One can see that the values of case **C3** ($-\alpha_{\mathrm{TI}} = 7.721$ deg) are consistent with the values from diagram two and three in Figure 9.5. However, the local mean leg bending moments, which were parallel to the global Y-axis, changed with the tower inclination. In the region of 11.9 deg $< -\alpha_{\mathrm{TI}} < 14.6$ deg occurred a minimum for all four legs in this DLC and wind speed. Assuming an exemplary tower inclination of $-\alpha_{\mathrm{TI}} = 12.5$ deg, the absolute global Y-axis parallel bending moments in the legs 1, 2, 3, and 4 could be reduced about factors 4.15, 11.85, 6.76, and 6.45, respectively.

Something similar happened to the normal forces in the two considered bracing members, marked on the right in Figure 9.4. They intersect to a minimum value at $-\alpha_{\mathrm{TI}} = 5.14$ deg. Furthermore, brace 1 is under tension and brace 2 under pressure for low tower inclinations, but this changes around the intersection point. For $-\alpha_{\mathrm{TI}} \geq 3.98$ deg brace 1 changed to be primarily under pressure and for $-\alpha_{\mathrm{TI}} \geq 6.32$ deg brace 2 changed primarily into tension. A comparison of the absolute bracing normal forces of **C6** to the local optimal intersection point at $-\alpha_{\mathrm{TI}} = 5.14$ deg shows that they can be reduced about factor 3.33 for brace 1 and about 5.24 for brace 2. These local optima changed for other segments along the tower hight and for other timeseries. Note that the presented lines are the mean values of one random seed, wind direction, and wind speed of the specific DLC 1.1. Coming back to the original statement regarding the normal forces in the legs, one identifies its bending flexibility together with the axial flexibility of the bracing members as the way to take the additional global tower

[6] On the left side in Figure 9.4, the -Y side is the X parallel side on the bottom

bending moments, which were introduced through tower inclination[7]. Since the axial stiffness of the legs were comparatively high, the loads rearranged in their bending DOF and the bracing members axial DOF.

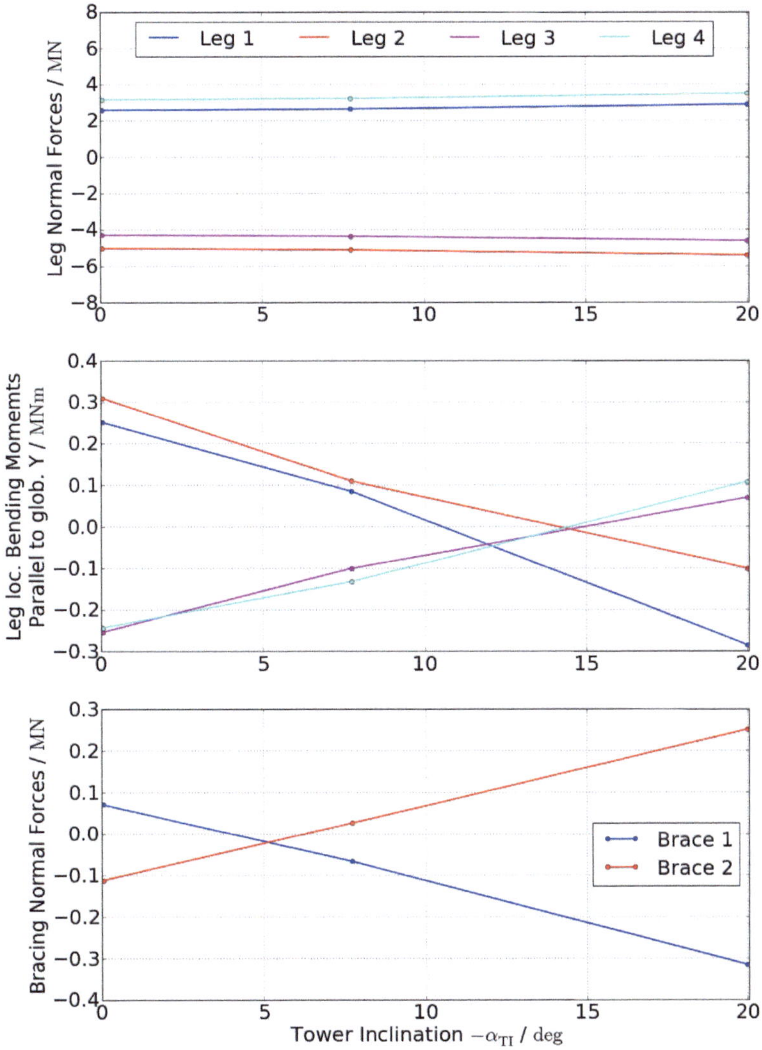

Figure 9.6: ASE simulation results for different mean load components in the bottom tower legs and bracing for DLC 1.1 at rated wind speed on average and different tower inclination angles. The rotatable inclined lattice tower, based on the parameter set in Table 9.1, is used and its inclination angle is varied according to case **C1**, **C3**, and **C6** from Table 9.3.

[7] A general discussion about tower inclination influence to loads is given in section 6.2.

At this point, one could calculate all tower inclination dependent load intersections at the different heights of the tower and for all DLCs, but it would provide no benefit regarding the material saving potential. This is because loads themselves are meaningless to the material utilization until they are applied to a certain cross sectional geometry and are transformed into stresses and thereby into utilizations. Sections 9.5 and 9.6 explain more about the material utilization and the manual optimized tower design, presented in this work.

However, Appendix 14.20 shows an exemplary plot of the member yield strength utilizations of segment 10 of a RILT as function of its inclination angle. It confirms, that inclination angles of $\alpha_{\mathrm{TI}} < 0$ can reduce not only the member loads, but consequently their utilization. A lower utilization is thereby an indicator for more material saving potential.

9.3.3 Power Production with Faults in the Control System

During the structural analysis, especially DLC 2.1 and 4.2 pointed out to be driving DLCs in terms of extreme loads. Therefore, a closer look into DLC 2.1 is given in this subsection. DLC 2.1 accounts for a fault in the control system where one blade starts to pitch to 0 deg and the measurement system detects the fault and gives the command to shut-down the WT by pitching the other two blades to 90 deg. As already explained in subsection 3.5.3, the pitch rate for this transient event was adjusted from the original pitch rate of 8 deg/s to 2 deg/s to reduce the loads[8]. For the rotatable lattice tower, the pitch rate had to be reduced once more to 0.3 deg/s for transient pitch maneuvers. Note that this adjustment has no significant impact to the annual energy harvest, because it is limited to WT shut-downs. However, a small pitch rate leads to longer periods of aerodynamic unbalance, which can cause additional fatigue damage. Since fatigue is not a design driving factor[9] and those fault events occur rarely during the wind turbines lifetime, this problem is assumed to be negligible. Moreover, start-ups are not significant in terms of extreme member utilizations, therefore the previous pitch rate of 2 deg/s remained for DLCs 3.x.

Figure 9.7 shows the same ASE sensor results as for the normal production case in subsection 9.3.1, but this time for DLC 2.1 at cut-out wind speed on average. Therefore, the pitch angle of blade one BldPitch1 was always > 0 deg until the transient event where it moved constantly to 0 deg. Because of the other two blades' pitch movements towards 90 deg, the WT shut-down and the blade tip to tower clearance curve disappears from the diagram due to rotor stop. In diagram three, one identifies a strong rotor thrust decrease at the beginning of the event. The rotor torque dropped about 30 s later from its rated value to ≈ 0 kNm. Diagram four shows how the rotor speed and the generator power decreased to zero during the event. Note that the rotational speed matches the disappearance of the rotor tip to tower clearance curve in the second diagram. The last diagram indicates where the most critical moment for the tower in this transient event occurred. It was right after its initiation and was characterised by large amplitude oscillations in the tower head FA and SS directions. The extreme cyclic ranges in both directions are remarkable. While the main wind direction parallel FA oscillation

[8] Subsection 3.5.3 explains, why this adjustment is appropriate and allowed according to the guidelines.

[9] This is the result of the later following FLS analysis in subsection 9.6.2.

had a range of 0.59 m, the corresponding extreme tower head SS oscillation range amounted to 1.29 m[10]. This extreme load relevant DLC confirmed that in some cases sidewise tower loads are larger than main wind direction parallel loads.

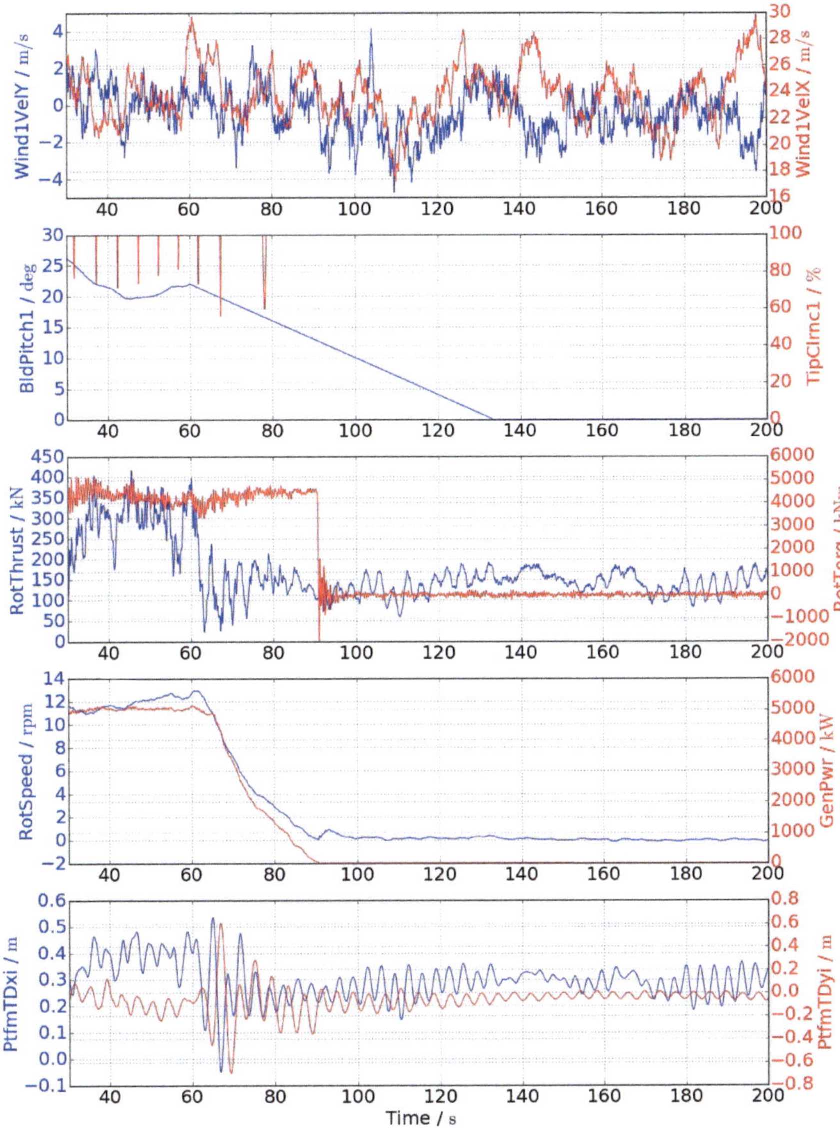

Figure 9.7: ASE simulation results for different wind turbine sensors and DLC 2.1 at cout-out wind speed on average. The rotatable inclined lattice tower, based on the parameter set in Table 9.1, is used.

[10] Note that no PSFs are applied to these deflections.

9.3.4 Shut-Down with EOG (Mexican Hat)

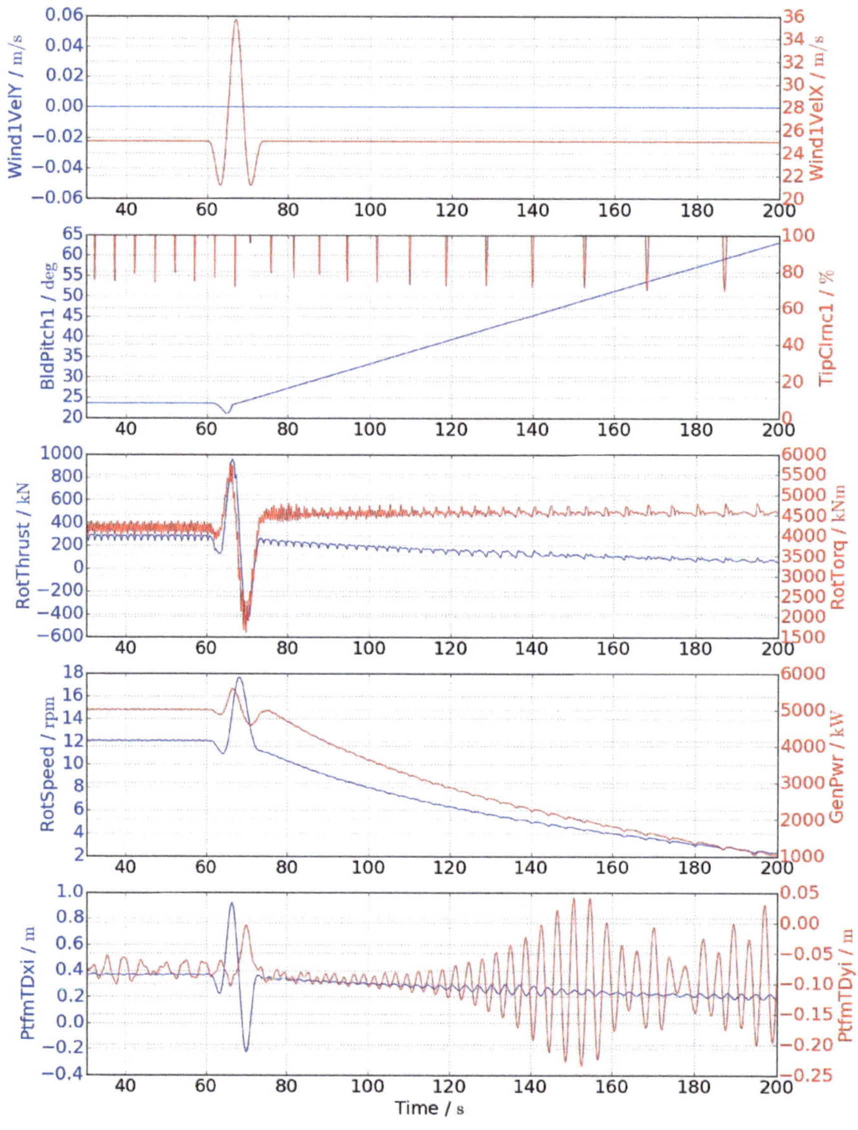

Figure 9.8: ASE simulation results for different wind turbine sensors and DLC 2.2 at cout-out wind speed on average. The rotatable inclined lattice tower, based on the parameter set in Table 9.1, is used.

The other extreme load relevant DLC is 4.2. In DLC 4.2, a normal WT shut-down is performed while the Mexican hat shaped wind speed change, called EOG, happens simultaneously. Figure 9.8 shows its results for a mainly constant cut-out wind speed and the WT shut-down starts

6.3 s after the Mexican hat initiation. The critical aspect of this DLC is the wind speed drop at the beginning of the wind event where the WT drives the pitch to lower values, such as shown in the first two diagrams. The following rapid increase in the wind speed was faster than the pitch system was able to follow it. Therefore, the rotor thrust increased drastically, such as shown in the third diagram. Since the wind speed dropped very quick after the tip of the hat and thereby the rotor thrust, a large tower head oscillation range along the main wind direction was the consequence. Its range amounted to 1.145 m in this case.

Note that the fist bending eigenfrequencies of the tower got hit by the 3p excitation frequencies during the WTs shut-down. The resonance, shown in the last diagram, started at a rotational speed of RotSpeed $\leq$ 6 rpm, such as predicted by the Campbell diagram in Figure 9.2. The resonance time period was elongated through the decreased pitch rate, such as introduced in the previous subsection 9.3.3. However, start-up and shut-down ASE timeseries were included in the fatigue analysis where this effect was covered.

9.3.5 Transport and Installation

The transport and installation DLCs are not part of the ASE analyses. Transport of single lattice members was not expected to play a crucial role in terms of their loads. However, caution should be given to their storage in the workshop and on the transport trucks. For all cases, scratches on the surfaces of lattice members should be avoided to hinder lifetime reductions. Installation of the proposed lattice tower should be carried out according to state-of-the-art procedures. Thereby, settlement in the bolted joints requires attention regarding pre-loading losses. Moreover, an inclined lattice tower would show other settlement patterns than conventional ones. At this point, some more work is required to develop an appropriate installation procedure, but it was out of scope for this work.

9.4 Post Processing Loads

In the past, WT lattice structures have been analysed by using an equivalent stiffness and mass matrix beneath the RNA within the ASE simulation. Thereby, the extracted tower head loads have been applied to the full FEM structure in another software solution to obtain all necessary proofs. FAST and the in this work adjusted SubDyn module provided timeseries of all six load components at each joint of all members within the tower for all DLCs, random seeds, wind speeds, and wind directions. Therefore, usage of another FEM software was not more necessary. Instead all load timeseries had to be transformed into stress timeseries through consideration of the members geometry. In the following subsection 9.4.1, the influence of warping stresses is estimated and afterwards the considered stresses for this work are presented in subsection 9.4.2.

9.4.1 Warping Stresses

Warping corresponds to out-of-plane deformations of a cross section due to torsional loading of a beam. Although these deformations are small, they can have a significant influence to the torsional behaviour of a beam [9, p. 362]. As indicated by subsection 8.2, the torsional stiffness of the modelled beams in the structure was conservatively downscaled. More safety was given by hindered warping, since it increases the torsional stiffness of beams, but it is associated with additional normal stresses. In this case normal stresses due to hindered warping occur especially at the connection of two cross sectional parts via filler plates and bolts. In this simplified consideration, the occurring closed cross sectional parts are assumed to be warp free, but not so for the open cross sectional sections along the assembled members. A rough calculation of expected normal stresses due to hindered warping in a leg and a bracing cross member should give clarity about its significance, especially for the later following fatigue analyses. In the first step, the exemplary calculation considered a leg cross section, which was assembled from two cross sections, such as shown in Figure 9.9.

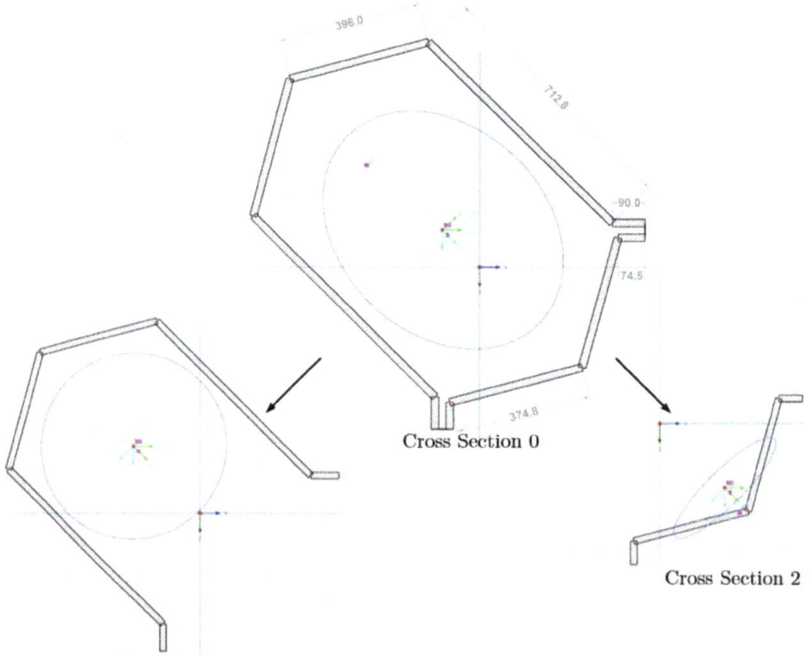

Figure 9.9: Enumeration of cross sectional parts of the leg member

Table 9.4 summarizes the necessary cross sectional parameters[11] for all three cross sections of the leg member at the bottom of the tower[12].

Parameter	Cross Section 0	Cross Section 1	Cross Section 2
Warping Constant $J_{\omega,i}$	2.892e−4 m^6	2.602e−4 m^6	4.968e−7 m^6
Torsional Constant $I_{t,i}$	7.101e−3 m^4	7.357e−6 m^4	2.733e−6 m^4
Maximum Sectorial Area $\omega_{max,i}$	1.681e−1 m^2	1.283e−1 m^2	2.151e−2 m^2

Table 9.4: Warping stress relevant cross sectional parameters of a representative leg member. Calculated in DUENQ from Dlubal [58].

For more clarification, the assumed boundary conditions of a leg member section is indicated in Figure 9.10. Thereby, the almost warp free closed cross sectional sections along the leg were considered as warping stiff.

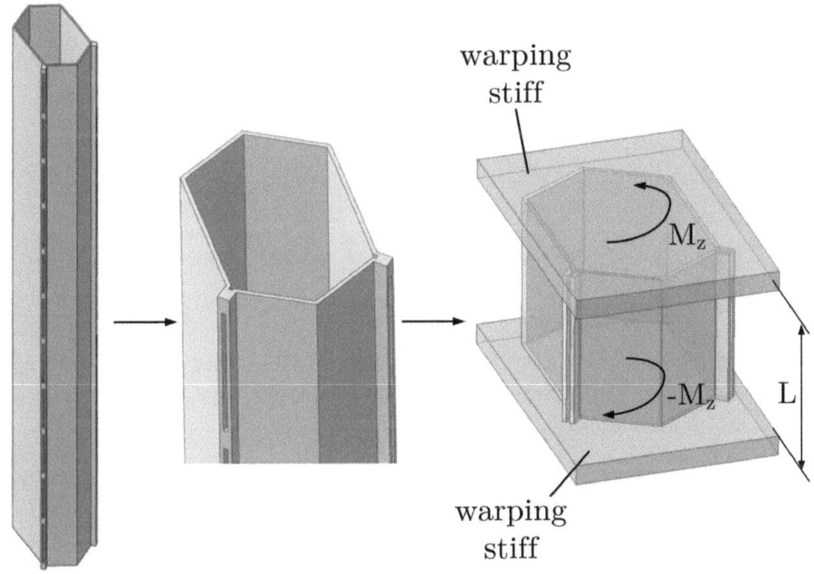

Figure 9.10: Assumed warping boundary conditions for a representative cut in a leg cross section

According to Cook and Young [30, p. 293], the torque M_z of a cantilevered member, with restrained warping at its one end is the sum of the Saint-Venant torque $M_{z,SV}$ and the Vlasov

[11] The warping constant $J_{\omega,i}$ is calculated with respect to the shear center.

[12] Note that Table 9.1 summarizes the parameters of the considered tower and the dimensions of the leg member at the bottom of the tower. However, a slight scaling of the main leg dimension L_L was necessary to ensure enough strength in the following analyses. Therefore, the dimensions of the considered leg member are $L_L = 0.36\,m \times 1.10 = 0.396\,m$, $L_F = 90$ mm, $t_p = 20$ mm, $w_p = 1$ mm, and $t = 23$ mm.

torque $M_{z,q}$. The first is associated with the cross sectional shear stresses and the second with restrained warping. Thus

$$M_z = M_{z,SV} + M_{z,q} = G\, I_t\, \kappa_z + M_{z,q} \tag{9.3}$$

Cook and Young [30, p. 281] present the general solution for the twist rate $\kappa_z(z)$ as a result of integration of the differential equation. It is given for arbitrary locations along the beam in Equation 9.4.

$$\kappa_z(z) = \frac{M_z}{G\, I_t} + C_1 \sinh(k\, z) + C_2 \cosh(k\, z) \tag{9.4}$$

C_1 and C_2 in Equation 9.4 are integration constants, which can be calculated with respect to the assumed boundary conditions. In the present case, boundary conditions were no warping at both ends of the beam, thus $\kappa_z(z = 0) = 0$ and $\kappa_z(z = L) = 0$. Using these conditions in Equation 9.4 delivers the twist rate as function of z according to

$$\kappa_z(z) = \frac{M_z}{G\, I_t}\left[1 - \cosh(k\, z) - \frac{(1 - \cosh(k\, L))}{\sinh(k\, L)} \sinh(k\, z)\right] \tag{9.5}$$

where $k = \sqrt{\frac{G\, I_t}{E\, J_\omega}}$ and $L = 10$ m was the intentionally conservative[13] assumed length of the considered member. In this case, the warping constant J_ω and torsional constant I_t were chosen to be the sum of the values of both open cross sections 1 and 2 in Figure 9.9, instead of the values of the closed cross section 0. For more clarification, this is likewise represented in Equation 9.6.

$$\begin{aligned} I_t &= I_{t,1} + I_{t,2} = 7.357\text{e}{-6}\ \text{m}^4 + 2.733\text{e}{-6}\ \text{m}^4 = 1.010\text{e}{-5}\text{m}^4 \neq I_{t,0} \\ J_\omega &= J_{\omega,1} + J_{\omega,2} = 2.602\text{e}{-4}\ \text{m}^6 + 4.968\text{e}{-7}\ \text{m}^6 = 2.607\text{e}{-4}\text{m}^6 \neq J_{\omega,0} \end{aligned} \tag{9.6}$$

As introduced, the previous Equation 9.5 led to $\kappa_z(z = 0) = \kappa_z(z = L) = 0$ deg/m and by consideration of Equation 9.3 the Saint-Venant torque became $M_{z,SV}(z = 0) = M_{z,SV}(z = L) = G\, I_t\, 0\ \text{deg/m} = 0\ \text{Nm}$ and consequently $M_z(z = 0) = M_{z,q}(z = 0) = M_{z,q}(z = L)$.

Since the cross section 0 was assembled from two other cross sections 1 and 2, the torsional moments on cross section 0 were split-up according to Equation 9.7 from PCAE [147, p. 31].

$$\begin{aligned} M_{z,SV,i} &= \frac{I_{t,i}}{I_t} M_{z,SV} \\ M_{z,q,i} &= \frac{J_{\omega,i}}{J_\omega} M_{z,q} \end{aligned} \tag{9.7}$$

[13] Since the leg members were assembled from two cross sectional parts, which were connected to each other in distances of < 1 m << 10 m, the chosen length L accounts for the leg member, as if it would not be assembled over its whole length. The critical part with respect to normal stresses appears, where the two separated open cross sectional parts were connected through bolts and filler plates.

With Equation 9.7 and 9.3, the acting torsional moment on each cross section becomes

$$M_{z,i} = M_{z,SV,i} + M_{z,q,i} \tag{9.8}$$

According to Cook and Young [30, p. 295], the normal stress under applied torsion on a beam with restrained warping for arbitrary points in the beam can be calculated by

$$\sigma_z(x,y,z) = -E\ \omega(x,y)\frac{d\ \kappa_z(z)}{d\ z} \tag{9.9}$$

On the cross sectional level, Equation 9.9 becomes maximal at maximum sectorial areas ω_{max}, such as listed in Table 9.4 for the given cross sectional parts. The derivative of the twist rate in Equation 9.5 with respect to z is

$$\frac{d\ \kappa_z(z)}{d\ z} = \frac{M_z\ k}{G\ I_t}\left[-\sinh(k\ z) - \frac{(1-\cosh(k\ L))}{\sinh(k\ L)}\cosh(k\ z)\right] \tag{9.10}$$

Usage of the corresponding cross sectional part parameters and torsional loads $M_{z,i}$, $J_{\omega,i}$, and $I_{t,i}$ and the common steel elasticity and shear moduli $E = 210{,}000\ \text{N/mm}^2$ and $G = 80{,}769\ \text{N/mm}^2$ in Equations 9.10 and 9.9 provided insight into the significance of torsional induced warping for this work. A maximum global leg torsional moment of $M_{z,max}\ \gamma_f\ \gamma_n\ \gamma_m = 200\ \text{Nm} \times 1.35 \times 1.00 \times 1.10 = 297\ \text{Nm}$[14] is extracted from the ASE analysis of DLC 2.1 for this consideration. This is one of the most relevant DLCs in terms of extreme loads, such as indicated in section 9.5. The results from this parameter set are presented in Figure 9.11. A division into the Saint-Venant and Vlasov torsional moment for the assembled and both single cross sections is shown at the top diagram. Thereby, the Vlasov part dominated at the clamped ends, but decreased with increasing z coordinate and simultaneously increasing Saint-Venant moment until the mid of the beam. From there, the values were mirrored, because of the same boundary conditions on both ends. Note that the global Vlasov torsional moment was almost completely taken by cross section 1. Moreover, maximal normal stresses within both cross sectional parts 1 and 2 are shown in the bottom diagram. Since the maximum occurring normal stress, induced by the extreme torsion on a leg member with hindered warping at one end was $0.8\ \text{N/mm}^2 > \sigma_{z,i} > -0.8\ \text{N/mm}^2$, no extreme load significant influence could be observed for the leg members.

For the bracing members, the same procedure as for the leg members was applied. Thereby, Table 9.5 shows the corresponding cross sectional parameters[15]. Note that the bracing member consisted out of two equal hat profiles, so that cross section 1 and 2 were the same.

[14] Note that the resistance PSFs are applied to the action side instead to the resisting side to get independence from the used material.

[15] Considered hat profile dimensions are $H_L = 90\ \text{mm} \times 1.10 = 99\ \text{mm}$, $H_F = 70\ \text{mm}$, $t_p = 20\ \text{mm}$, $w_p = 1\ \text{mm}$, and $t = 7\ \text{mm}$.

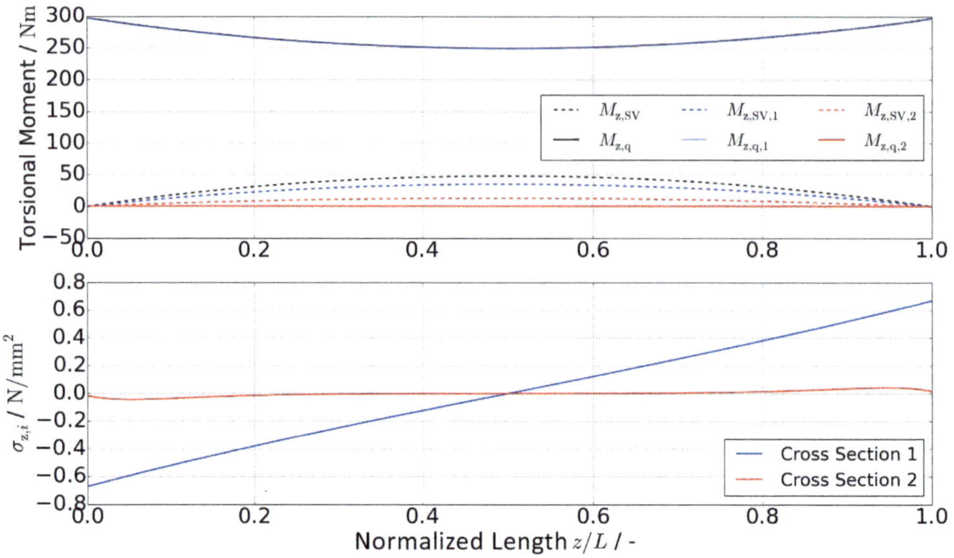

Figure 9.11: Normal stresses due to restrained warping along a representative leg member at extreme torsional loading

Parameter	Cross Section 0	Cross Section 1,2
Warping Constant $J_{\omega,i}$	5.175e−8 m^6	9.650e−9 m^6
Torsional Constant $I_{t,i}$	1.259e−5 m^4	4.770e−8 m^4
Maximum Sectorial Area $\omega_{\max,i}$	3.810e−3 m^2	4.326e−3 m^2

Table 9.5: Warping stress relevant cross sectional parameters of a representative bracing member. Calculated in DUENQ from Dlubal [58].

Application of the same member length and boundary conditions together with a maximum evaluated bracing torsional moment of $M_{z,\max}\ \gamma_f\ \gamma_n\ \gamma_m = 3.02\ \text{Nm} \times 1.35 \times 1.00 \times 1.10 = 4.48\ \text{Nm}$ led to the analogous results in Figure 9.12. In this case, Saint-Venant torsional moments superseded the Vlasov torsional moments over a wide beam section. Furthermore, normal stresses caused by hindered warping were still in a low range of $0.8\ \text{N/mm}^2 > \sigma_{z,i} > -0.8\ \text{N/mm}^2$. Therefore, warping was likewise not further considered for the extreme load conditions of bracing members.

Since fatigue depends on stress cycle ranges and the previous calculated stress levels show no potential for significant stress ranges, normal stresses caused by hindered warping was likewise not considered in the fatigue limit state analysis for all tower members.

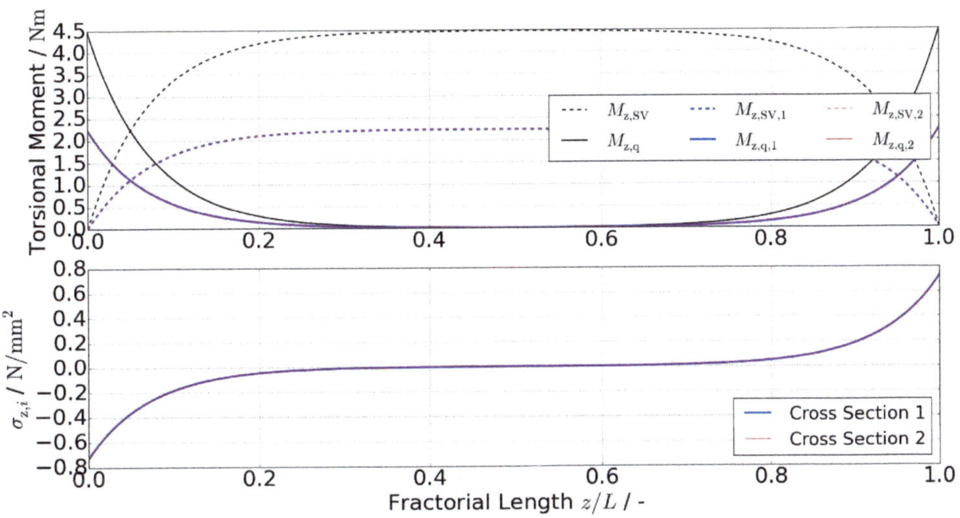

Figure 9.12: Normal stresses due to restrained warping along a representative bracing member at extreme torsional loading

9.4.2 Considered Stresses

As one simplifying step, only normal stresses through axial forces and bending moments were calculated. This convenient step was appropriate, because shear stresses and normal stresses caused by hindered warping were found to be insignificant for the necessary member dimensions. Such a shortcut is common for lattice structures. Normal stresses $\sigma_{\mathrm{N},m,j,s}$ were found to be

$$\begin{aligned}\sigma_{\mathrm{N},m,j,s} &= \frac{F_{\mathrm{z},m,j}}{A_m} + \frac{M_{\mathrm{x},m,j}}{I_{\mathrm{x},m}} s_{\mathrm{y},m,j,s} + \frac{M_{\mathrm{y},m,j}}{I_{\mathrm{y},m}} s_{\mathrm{x},m,j,s} \\ &\forall\, m \in \{1,\dots,n_\mathrm{M}\}\,;\, j \in \{1,2\}\,;\, s \in \{1,\dots,n_\mathrm{S}(m)\}\end{aligned} \tag{9.11}$$

where $F_\mathrm{z}, M_\mathrm{x}, M_\mathrm{y}$ are the normal force and bending moments around local member x- and y-axes. n_M and $n_\mathrm{S}(m)$ are the amount of members and amount of considered stress spots on the member cross section. The coordinates of a stress spot were given by s_x and s_y, respectively. All considered stress spots for the leg and bracing member cross sections are shown in Figure 9.13.

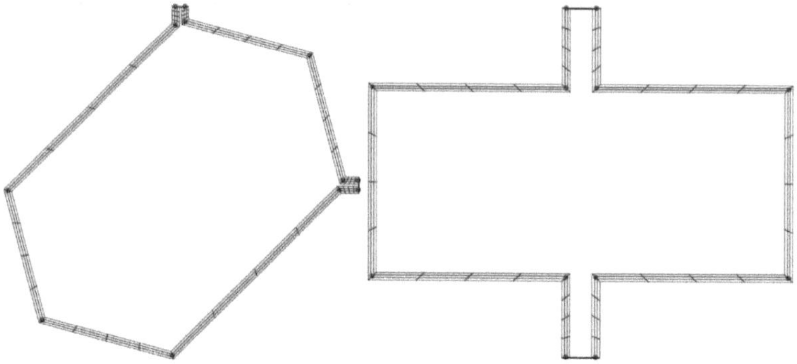

Figure 9.13: Spots, where normal stresses will be calculated for the leg (left) and bracing (right) cross sections

Counter j was used for the both member joints for each member[16]. Moreover, Equation 9.11 represents a point in time, thus it had to be evaluated for each time step in all timeseries. Cross sectional properties were calculated automatically by means of the already discretized cross sections according to

$$\begin{aligned} A_m &= \sum_{i=1}^{n_{\text{el}}} A_{\text{el},m,i} \\ I_{\text{x},m} &= \sum_{i=1}^{n_{\text{el}}} A_{\text{el},m,i}\; y_{\text{c},m,i}^2 \\ I_{\text{y},m} &= \sum_{i=1}^{n_{\text{el}}} A_{\text{el},m,i}\; x_{\text{c},m,i}^2 \end{aligned} \tag{9.12}$$

In Equation 9.12, $A_{\text{el},m,i}$ is the area of cross sectional element i of member m and $x_{\text{c},m,i}$ and $y_{\text{c},m,i}$ are its geometrical center point coordinates.

9.4.3 Buckling Analysis

Since previous stability analyses of lattice towers with larger leg distances[17] showed that failures occur especially for local members instead for the global tower, only single members were investigated. According to DIN EN 1993-1-1 [42, 6.3.2.1(2)], lateral torsional buckling is not relevant for the following cross sections

- rectangular or circular hollow
- welded tubular

[16] Note that thereby only loads at the beginning and the end of one member were considered. This convenience is appropriate because the members were assumed to have no line loads on them.

[17] In the beginning of this work, the lattice tower stability was analysed within a parallel FEM model in RFEMs [57] module RF-Stabil [55]. Thereby, only local stability failure occurred for the considered tower dimensions. Global stability failure happened, when the leg distances became very small, so that the tower itself had a high slenderness.

- box

Since all leg and bracing cross sections were assembled cross sections, which have the minimum filler plate distance of 15 i_{min}[18], they could be seen as one cross section in stability analyses. By that they built a box like shape, wherefore lateral torsional buckling should not become relevant according to the previous list. The applied analyses in this work took place according to an equivalent member method, such as suggested in the first part of the buckling analysis paragraph in subsection 2.2.5. This means that only buckling of single members subject to normal forces was considered. For the equivalent leg and bracing members, the system length between two joints was the buckling length L_{cr} and the buckling curve **b**[19] was chosen for all members and all bending axes. The choice of the buckling curve was based on the assumption that the cross sections were like hot manufactured hollow boxes. Although this would lead to a higher buckling curve **a**, the more conservative one **b** was used.

This approach was a very fundamental and rough way to get a first guess for the stability of the structure, since the proof was carried out for all members of all time steps in all computed timeseries. Procedures, such as full FEM analyses with introduced imperfections can bring more accurate results. For example they account for the influence of additional bending moments and their interaction with the axial loads or they can model the members with their assembled and discontinuous cross section more precisely, but for the scope of this work the standard hand calculation was chosen to be sufficient.

9.5 Ultimate Limit State Analysis

In this section, three different parts are considered for the ULS of the RILT and the performed ASE simulation. In the first subsection 9.5.1, the influence of tower aerodynamics on the member loads is predicted, the second subsection 9.5.2 presents the resulting extreme load utilization for the RILT with 87.6 m tower height according to case **C3**, and the last subsection 9.8.2 analyses the occurring blade tip to tower clearances.

9.5.1 Tower Aerodynamic Induced Loads

Since SubDyn has no capability to account for aerodynamic tower loads, they were not included in the ASE load analyses. Therefore, this subsection shows an estimation for loads due to aerodynamic drag on the tower itself. The general procedure how aerodynamic lattice tower loads are calculated with respect to the Eurocode 3 standard, are summarized in subsection 2.7.6. For this investigation, the tower defined in Table 9.1 was used and analytically analysed for two wind situations. Situation (1) corresponded to a 50 year extreme wind speed of 50 m/s according to DLC 6.2 with a load PSF of $\gamma_{\text{f}} = 1.1$ and power law exponent of $\alpha = 0.11$. In the second situation (2), an arbitrary DLC at cut-out wind speed of 25 m/s was considered with

[18] This aspect was explained in subsection 7.1.3.

[19] This corresponds to an imperfection coefficient of $\alpha = 0.34$ in Equation 2.33.

$\gamma_f = 1.35$ and $\alpha = 0.2$. In both wind situations, two different wind directions were investigated: (a) frontal and (b) diagonal with respect to the square tower cross section. In this investigation, the tower was seen as being vertical, but with an artificial increased height $H_{T,aero}$, namely

$$H_{T,aero} = \frac{H_T}{\cos(\alpha_{TI})} = 88.4 \text{ m} \tag{9.13}$$

The conservative tower height scaling in Equation 9.13 is applied to both wind directions. The wind force on a lattice structure is defined by Equation 2.52 where the structural coefficient became $c_s c_d = 1$, because wind gusts and major structural dynamic effects were already represented in the ASE analyses. The force coefficient c_f required two other values, namely the basic force coefficient $c_{f,0}$ and the reduction factor ψ_λ. $c_{f,0}$ could be extracted from the bottom diagrams in Figure 2.22 for the present situations. The required Reynolds number was calculated with respect to the NWP, and a kinematic viscosity of $\nu = 13.3\text{e}{-6}\ \text{m}^2/\text{s}$[20] for air. Appendix 14.21 shows the calculated Reynolds numbers for the RILT, which were in the region of $4.6\text{e}4 \leq Re \leq 2.3\text{e}5$ for all lattice tower members and considered wind speeds. The tower wall area could be calculated according to Equation 9.14

$$\begin{aligned} A_{c,front} =& H_{T,aero} \left(a_T + \frac{a_T + a_B}{2}\right) \\ A_{c,diag} =& A_{c,front} \sqrt{2} \end{aligned} \tag{9.14}$$

where the index $_{front}$ and $_{diag}$ correspond to the considered different wind directions (a) and (b), respectively. Analogously, the reference area for both situations is calculated as

$$\begin{aligned} A_{ref,front} =& SF_A \sum_{i=1}^{N_{seg}} A_{ref,front,i} = SF_A \sum_{i=1}^{N_{seg}} 2\, b_{L,p,i}\, h_i + 2\, b_{B,p,i} \sqrt{h_i^2 + a_i^2 - b_{B,p,i}^2} \\ A_{ref,diag} =& A_{ref,front} \sqrt{2} \end{aligned} \tag{9.15}$$

where $SF_A = 1.1$ is a scaling factor to account for additional tower attachments, such as a ladder, cables, and gusset plates. $b_{L,p,i}$ and $b_{B,p,i}$ in Equation 9.15 are the projected widths in air flow direction of the leg and bracing members of segment i. These projected widths are calculated as

$$\begin{aligned} b_{L,p,i} =& 2\, L_{L,i} \\ b_{B,p,i} =& \sqrt{\left(H_{L,i} + 2\, H_{F,i}\right)^2 + H_{L,i}^2} \\ & \forall\, i \in \{1, \ldots, N_{seg}\} \end{aligned} \tag{9.16}$$

where all member dimensions are linear interpolated between their tower top and bottom values in Table 9.1. Due to the following utilization study, some member dimensions had manually

[20] This value was taken from Kuchling [127, p. 622] and an air temperature of 20 °C.

been adjusted to slightly larger or smaller values. All exact used member parameters are documented in Appendix 14.22. Using Equations 9.14 and 9.15 in Equation 2.54 delivers

$$\begin{aligned}\varphi_{\text{front}} &= \frac{A_{\text{ref,front}}}{A_{\text{c,front}}} = \frac{178.26\ \text{m}^2}{499.47\ \text{m}^2} = 0.36 \\ \varphi_{\text{diag}} &= \frac{A_{\text{ref,diag}}}{A_{\text{c,diag}}} = \frac{188.80\ \text{m}^2}{706.35\ \text{m}^2} = 0.27\end{aligned} \tag{9.17}$$

These member densities together with the evaluated Reynolds number range was used to extract $c_{\text{f,0,front}} = 1.65$ and $c_{\text{f,0,diag}} = 1.95$ from the bottom diagrams in Figure 2.22. The slenderness of the structure was calculated with respect to Equation 2.56 by

$$\begin{aligned}\lambda_{\text{front}} &= \min\left\{1.4\frac{H_{\text{T,aero}}}{(a_{\text{B}} + a_{\text{T}})/2}, 70\right\} = 21.90 \\ \lambda_{\text{diag}} &= \min\left\{1.4\frac{H_{\text{T,aero}}}{(a_{\text{B}} + a_{\text{T}})/\sqrt{2}}, 70\right\} = 15.49\end{aligned} \tag{9.18}$$

and was used together with the corresponding member density to read the reduction factors $\psi_{\lambda,\text{front}} = \psi_{\lambda,\text{diag}} = 0.94$ out of Figure 2.23. By collecting all these factors, one can calculate the shear forces at each tower segment by

$$\begin{aligned}F_{\text{w,front},i} &= c_{\text{f,front}}\, V_{\text{H}} \left(\frac{z_i}{z_{\text{H}}}\right)^{\alpha} A_{\text{ref,front},i} \\ F_{\text{w,diag},i} &= c_{\text{f,diag}}\, V_{\text{H}} \left(\frac{z_i}{z_{\text{H}}}\right)^{\alpha} A_{\text{ref,diag},i}\end{aligned} \tag{9.19}$$

where z_i was conservatively assumed to be the altitude of the top of segment i. The resulting shear forces are accumulated by

$$F_{\text{acc,front},i} = \sum_{j=i}^{N_{\text{seg}}} F_{\text{w,front},j} \tag{9.20}$$

to get the accumulated shear force distribution at the first row diagrams in Figure 9.14. The second row represents the corresponding bending moments, which can be calculated by numerical integration of the accumulated shear force distribution along the tower with Equation 9.21.

$$M_{\text{w,front},i} = \sum_{j=i}^{N_{\text{seg}}-1} F_{\text{acc,front},j+1}\left(z_{j+1} - z_j\right) \tag{9.21}$$

The more interesting findings are the utilizations, which were calculated out of the axial normal forces within the leg and bracing members. In analogy to Petersen [149, p. 895], the legs were considered to take the global tower bending moments according to

$$\begin{aligned} N_{\mathrm{L,front},i} &= \pm \frac{M_{\mathrm{w,front},i}}{2\, a_i} \\ N_{\mathrm{L,diag},i} &= \pm \frac{M_{\mathrm{w,diag},i}}{\sqrt{2}\, a_i} \end{aligned} \tag{9.22}$$

The bracing members were considered to take the shear forces according to

$$\begin{aligned} N_{\mathrm{B,front},i} &= \pm \frac{F_{\mathrm{acc,front},i}}{4\, \cos\left(\alpha_{\mathrm{B}} \pm \alpha_{\mathrm{TI}}\right)} \\ N_{\mathrm{B,diag},i} &= \pm \frac{F_{\mathrm{acc,diag},i}}{4\sqrt{2}\, \cos\left(\alpha_{\mathrm{B}} \pm \alpha_{\mathrm{TI}}\right)} \end{aligned} \tag{9.23}$$

Thereby, the utilizations in terms of yield strength and buckling follow from

$$\begin{aligned} \text{a – yield strength :} \quad & \frac{\gamma_{\mathrm{f}}\, \gamma_{\mathrm{n}}\, \gamma_{\mathrm{m}}\, N_{\mathrm{front},i}}{A_{,i}\, f_{\mathrm{y}}} \\ \text{b – yield strength :} \quad & \frac{\gamma_{\mathrm{f}}\, \gamma_{\mathrm{n}}\, \gamma_{\mathrm{m}}\, N_{\mathrm{diag},i}}{A_{,i}\, f_{\mathrm{y}}} \\ \text{a – buckling :} \quad & \frac{\gamma_{\mathrm{f}}\, \gamma_{\mathrm{n}}\, \gamma_{\mathrm{m}}\, N_{\mathrm{front},i}}{A_{,i}\, \chi_{\mathrm{front}}\, f_{\mathrm{y}}} \\ \text{b – buckling :} \quad & \frac{\gamma_{\mathrm{f}}\, \gamma_{\mathrm{n}}\, \gamma_{\mathrm{m}}\, N_{\mathrm{diag},i}}{A_{,i}\, \chi_{\mathrm{diag}}\, f_{\mathrm{y}}} \end{aligned} \tag{9.24}$$

where the normal forces $N_{\mathrm{front},i}$ and $N_{\mathrm{diag},i}$ and cross sectional areas A_i correspond to the leg or bracing member loads and cross sections for segment i.

The results for both situations (1) and (2), both considered wind directions (a) and (b), and for the leg and bracing members along the tower height are summarized in the bottom four diagrams in Figure 9.14 in terms of their yield strength and buckling utilization. In general, utilizations for all members were larger in the extreme wind situation (1), than in the cut-out wind speed situation (2).

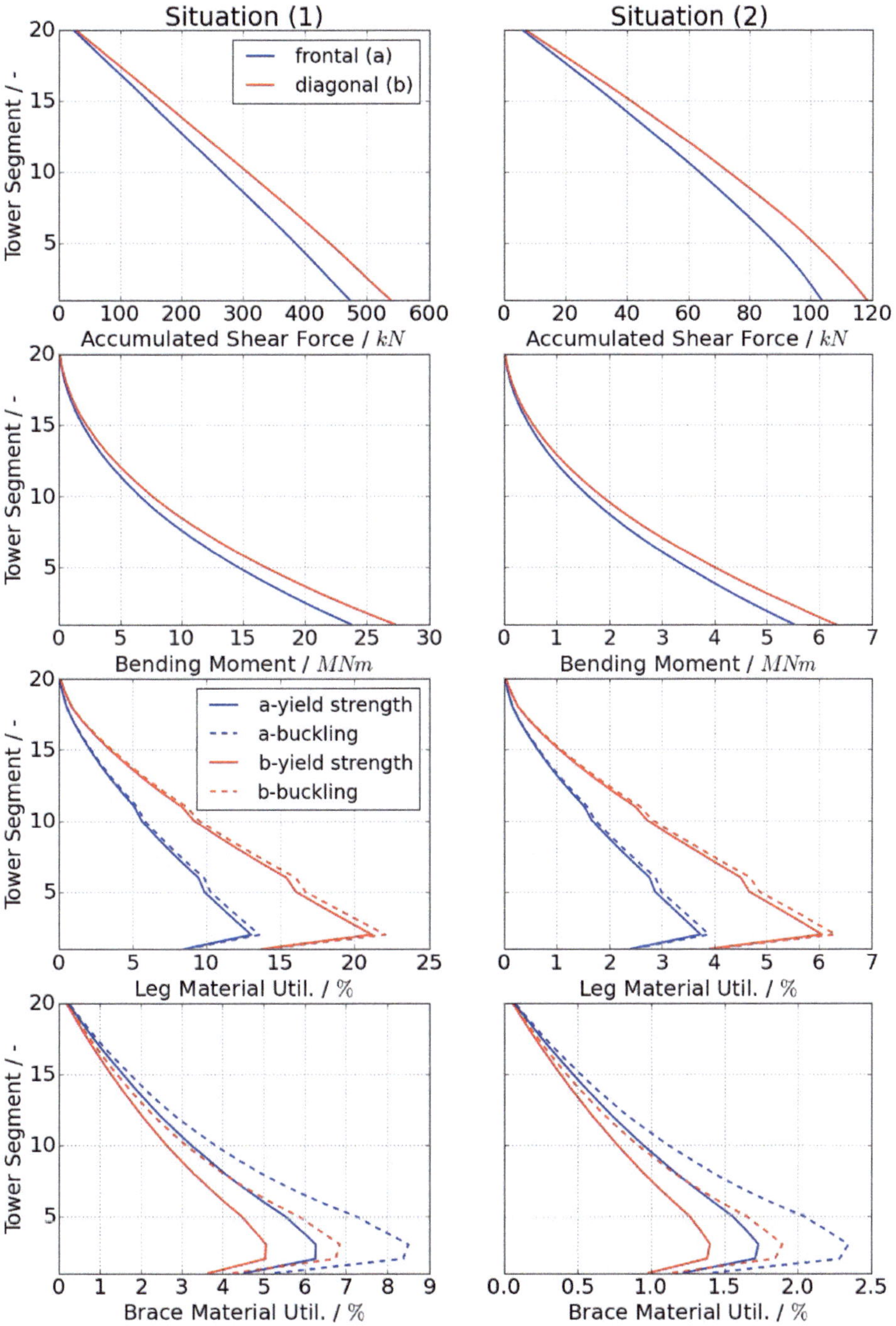

Figure 9.14: Aerodynamic loads and corresponding lattice tower member material utilization for two different situations: (1) Extreme wind speed and (2) cut-out wind speed

Furthermore, the buckling utilization was always larger than for the yield strength. This is reasonable, because χ is < 1 for all members, such as shown in Appendix 14.22. Another observation is that leg utilizations were larger in case (b) and bracing member utilizations were larger in case (a). The first results from the fact, that only two legs were activated to carry the diagonal tower loading and the second observation is because only two tower walls were activated to carry loads in the case of frontal tower wind attack. Moreover, utilizations rose at lower tower stations as consequence of increasing loads, such as shown in the first four diagrams in Figure 9.14. Note that utilizations were not continuous because of their individual dimension adjustments, mentioned earlier in this subsection. There was a special downfall of the utilizations for the bottom segment because of the relative increase of its member dimensions.

The extreme leg utilizations occurred at the second segment and were 22.12 % and 6.32 % in case (b) of situation (1) and (2). Bracing extreme utilizations in case (a) were 8.49 % and 2.35 % for (1) and (2), respectively. For the ASE analysis results, these conservative analytical evaluations were used additionally to account for wind drag on the RILT.

9.5.2 Extreme Load Utilization

Since the conversion of $\approx$ 470 GB of load timeseries to stress timeseries for several spots on the member joint cross sections produces another $\approx$ 516 GB, not every single result could be presented in this work. Therefore, only the most important results were summarized in utilization diagrams. Yield strength utilizations were calculated by means of the normal stresses, explained in subsection 9.4.2. The utilization at a certain member joint j of a cross sectional spot s of member m at time step t_i is according to Equation 2.29 calculated as

$$\text{utilization}: \frac{\gamma_\text{f}\,\gamma_\text{n}\,\gamma_\text{m}\,F_\text{k}}{f_\text{y}} \rightarrow \frac{\gamma_\text{f}\,\gamma_\text{n}\,\gamma_\text{m}\,\sigma_{\text{N},m,j,s}(t=t_i)}{f_\text{y}} \tag{9.25}$$

The buckling utilization was evaluated analogously to Equation 9.24. Note that these utilizations were calculated for all DLCs, wind speeds, wind directions, and random seeds. The considered lattice tower parameters are listed in Table 9.1 and the exact member parameters are documented in Appendix 14.22. Figure 9.15 summarizes the extreme utilizations of all members over all DLCs and thereby over all ASE timeseries. The utilizations contain likewise the previous evaluated member utilizations, caused by the wind drag on the tower. Wind drag utilizations were added to the ASE utilizations by usage of the DLC dependent load PSF γ_f and by application of the largest utilization of either case (a) or (b) for each member. This conservative step means that even though a certain DLC accounts only for frontal wind, a theoretical diagonal wind attack was applied to the tower if it caused higher utilizations. Furthermore, only two wind speeds at hub height and their following wind distributions were used for the additional wind drag on the tower: 25 m/s for all DLCs 1.x, 2.x, 3.x, 4.x, and 5.x and 50 m/s for all DLCs 6.x and 7.x.

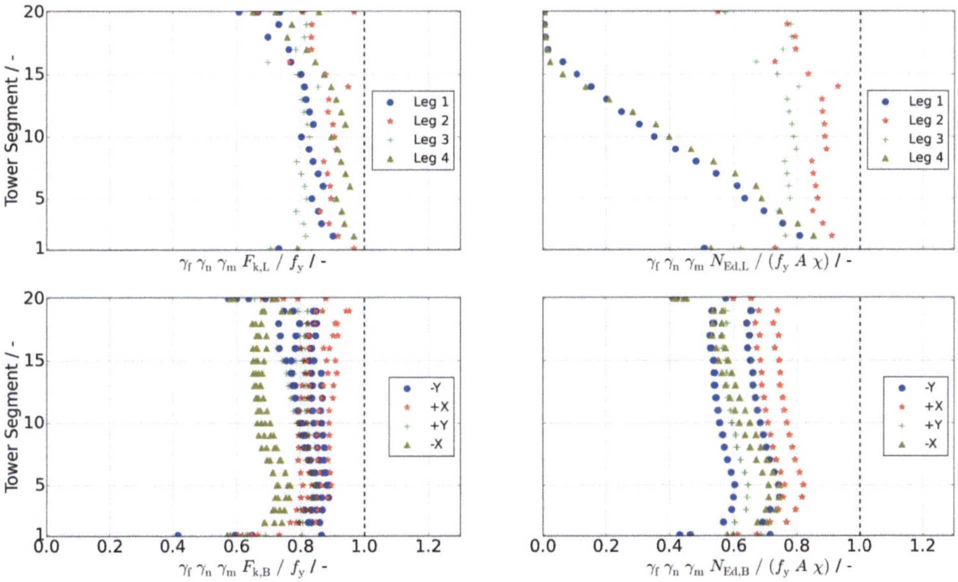

Figure 9.15: Member yield strength and buckling utilizations over all DLCs of the 87.6 m high RILT

DLC	Leg Max. YS-Util.	Leg Mean YS-Util.	Leg Max. BU-Util.	Leg Mean BU-Util.	Brace Max. YS-Util.	Brace Mean YS-Util.	Brace Max. BU-Util.	Brace Mean BU-Util.
1.1	84.0	68.4	80.9	35.9	45.2	36.0	35.7	27.2
1.3	91.2	75.1	88.0	39.4	48.7	41.2	42.3	33.0
1.4	85.5	75.1	80.6	39.5	53.3	43.6	42.1	35.3
1.5	80.8	62.6	77.7	31.9	44.5	37.3	36.4	28.6
2.1	**96.8**	73.0	**92.7**	46.7	**95.2**	**76.8**	**82.1**	**63.0**
2.3	86.0	68.8	76.1	43.5	33.1	19.6	27.3	11.5
3.1	74.4	56.9	71.6	30.5	24.4	17.0	19.0	8.4
3.2	75.2	57.6	72.5	30.9	26.3	17.7	20.6	9.1
3.3	71.4	52.5	69.7	31.0	30.1	22.1	22.7	12.9
4.1	74.9	57.0	72.1	30.5	26.7	17.2	20.6	8.3
4.2	96.7	**77.9**	88.2	**48.9**	33.9	22.1	28.1	12.7
5.1	81.3	64.4	79.7	36.8	44.3	33.0	36.8	23.0
6.1	81.9	67.0	79.4	45.9	49.9	34.2	43.3	23.6
6.2	91.8	64.6	90.9	47.1	79.6	65.7	74.5	54.9
6.3	72.7	59.2	69.8	40.6	39.2	25.9	34.8	16.6
6.4	63.4	48.8	58.8	33.4	34.5	25.2	30.7	16.3
7.1	77.0	63.8	74.3	34.6	27.3	20.9	23.6	12.6

Table 9.6: Leg and bracing member maximum and mean utilizations in terms of yield strength and buckling, given in percentage values. The maximum and minimum values over all DLCs are marked by means of bolt numbers.

The results in Figure 9.15 are divided in four different diagrams. The top left diagram shows the yield strength utilizations of the leg members along all tower segments and on the right the corresponding buckling utilizations are documented. Analogously, the bottom left diagram shows the bracing member yield strength utilizations along the tower segments and on the bottom right, bracing member buckling utilizations are depicted. Note that bracing members are distinguished with respect to their global coordinate system position. Legend entries -Y, +X, +Y, and -X indicate the respective bracing wall of the tower, according to the left drawing in Figure 9.4 or in Figure 14.6. Two examples should help for more clarification: Bracing members, denoted with +X were those members along the tower height, which were positioned on the downwind side of the tower. Bracing members, denoted with -Y were those, which could directly be seen on the left side in Figure 14.6[21].

Figure 9.15 reveals, that buckling was not as relevant as the yield strength for the member dimensions. Note that the column buckling proof considered only axial forces, but the yield strength proof considered bending moments and axial forces. Since, the scope of this work was not to optimize the tower automatically, it was optimized manually by dimension scaling factors for each segments members. For this approach, each segment got three scaling factors, one for the upwind legs, one for the downwind legs, and one for the bracing members[22]. Thereby, a maximum utilization of 96.8 % was reached for the leg 2 of the top and bottom segment, but a mean yield strength utilization of 83.3 % suggests still some more optimization potential for the legs. A closer view on the top left diagram in Figure 9.15 shows that leg 2 and 4 experience a higher utilization than leg 1 and 3. Considering only leg 2 and 4, the average utilization amounts to 87.6 %, which is a good utilization for the preliminary study and limited amount of design parameters. However, scaling of each leg dimensions, more optimization iterations, and a less conservative tower wind drag approach would certainly improve the utilization and thereby lead to further material savings. A view on the leg buckling utilization in the top right diagram indicates that the front legs 1 and 4 were always under tension at the top tower sections, while the back legs 2 and 3 experience a reasonable column buckling utilization around 80 % along the whole tower. Since self-weight induced pressure was higher at all legs of the tower bottom, the buckling utilization of the front legs reached the same level as the back legs at that section. Bracing members experienced an average yield strength utilization of 78.3 %, while the front bracing members on the -X side had the lowest and the back bracing members on the +X side had the highest utilizations. Bracing members obtain still material saving potential. Appendix 14.23 shows the corresponding utilizations of all individual DLCs. Thereby, only the maximum utilizations over all of their different wind speeds, wind directions, and random seeds are depicted. A corresponding summary of the utilizations of the different DLCs is given in Table 9.6. It shows the maximum and mean of the leg and bracing member yield strength (YS) and buckling (BU) utilizations. As mentioned in subsection 9.3.3, DLC 2.1 and 4.2 were the driving DLCs, even though adjustments in the control regimes in fault situations were implemented to decrease their influence[23].

[21] Note the global coordinate system at the bottom of the tower. It indicates, that the global Y axis points into the paper, so that the visible bracing members are the ones, denoted with -Y.

[22] A review of Appendix 14.22 shows the results of the different scaled member dimensions

[23] Subsection 9.3.3 explains more about these adjustments, which aimed to the blades pitch rates in fault situations.

9.6 Fatigue Limit State Analysis

In this section, the first subsection 9.6.1 explains more about the fatigue analyses procedure and the underlying assumptions. In the second subsection 9.6.2, the calculated tower member lifetime damages are presented and in the last subsection 9.6.3, inspection intervals for the RILT members are derived.

9.6.1 Explaining the Procedure

The fatigue analysis was performed according to the description in subsection 2.2.6 and according to the equations in the manual of the NREL fatigue assessment tool, called MLife [94]. The equations were reimplemented into a python script to gain more control over the different steps within the analysis, rather than just using the existing tool itself. However, the following brief example clarifies the the general fatigue analysis procedure.

If bolted joints are designed as slip resistant connections, the members can be checked in terms of their base material fatigue class [62, p. 95]. This led to the highest fatigue class B1 with respect to Figure 2.5 for all considered members in the RILT. The corresponding S-N curve on the resistance side was modelled by Equation 2.43 where the theoretical load ranges at the intersection between the curve and the $\log(S)$ axis are

$$S_0 = \begin{cases} \frac{1}{2}\left(N_{\mathrm{FL,B1}}\right)^{\frac{1}{m_1}} S_{\mathrm{FL,B1}} = \frac{1}{2}\left(10^7\right)^{\frac{1}{4}} 106.97 \times 10^6 \,\frac{\mathrm{N}}{\mathrm{m}^2} = 3008 \times 10^6 \frac{\mathrm{N}}{\mathrm{m}^2} & \text{for} \quad 10^4 < n(S) \leq 10^7 \\ \frac{1}{2}\left(N_{\mathrm{FL,B1}}\right)^{\frac{1}{m_2}} S_{\mathrm{FL,B1}} = \frac{1}{2}\left(10^7\right)^{\frac{1}{5}} 106.97 \times 10^6 \,\frac{\mathrm{N}}{\mathrm{m}^2} = 1343 \times 10^6 \frac{\mathrm{N}}{\mathrm{m}^2} & \text{for} \quad n(S) > 10^7 \end{cases} \tag{9.26}$$

$N_{\mathrm{FL,B1}} = 10^7$ and $S_{\mathrm{FL,B1}} = 106.97 \times 10^6\ \mathrm{N/m^2}$ are the fatigue limit lifetime cycles and its corresponding stress range for detail class B1. In Equation 9.26, two different theoretical $\log(S)$ axis intersections are calculated. One results from a S-N curve with higher slope for the finite life fatigue strength of load ranges S with life time cycles $10^4 < n \leq 10^7$ and the other corresponds to the fatigue limit S-N curve with lower slope and $> 10^7$ cycles. The inverse slopes of these two regions are denoted as $m_1 = 4$ and $m_2 = 5$. It was assumed that low cycle fatigue with $\leq 10^4$ cycles are treated as being part of the mid-term finite life fatigue strength region of $10^4 < n \leq 10^7$. Thus, the S-N curve with inverse slope m_1 was linear extrapolated to fewer cycles, such as recommended by DNV GL [62, p. 176].

Figure 9.16 shows the exemplary stress range collective of the leg 1 member of the bottom segment in the 87.6 m high RILT. One leg member had two joints and 8 spots on each of their two cross sections where stresses were calculated. Figure 9.16 shows the results of spot s_1 at joint 1 of this member. Note that the depicted stress range cycles were already scaled to lifetime cycles of a class I[24] WT with 20 years assumed lifetime. The scaling took place

[24] This is an average wind speed of $V_{\mathrm{ave}} = 10\ \mathrm{m/s}$. Furthermore, the wind speed was assumed to be Rayleigh distributed with shape factor $k = 2$.

according to subsection 2.2.6 and each cycle stress range was Goodman corrected with respect to Equation 2.44.

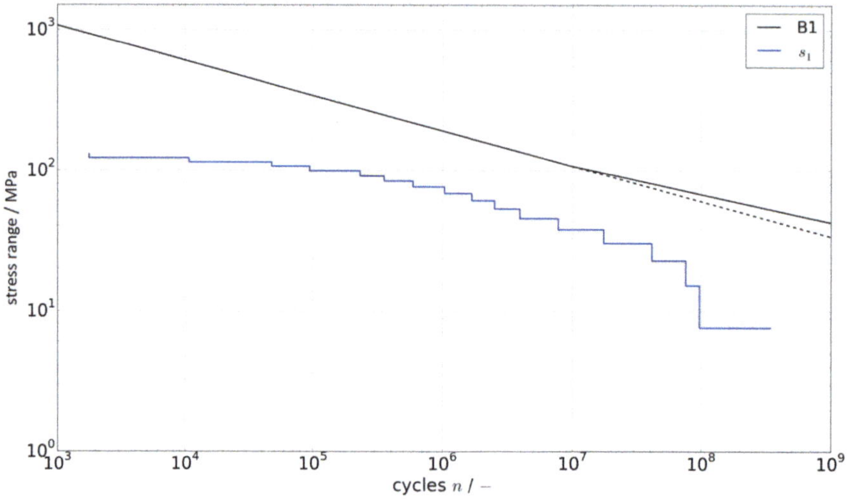

Figure 9.16: Stress range spectrum of stress calculation spot s_1 for leg 1 at the bottom segment of the C3 RILT with 87.6 m tower hight

With such stress range collectives for all tower members and their stress calculation spots and the numerical evaluation of Equation 2.40, one could calculate the lifetime damage sum at all designated positions in the tower.

9.6.2 Fatigue Utilizations of the C3 RILT

The assumed WT availability drives the influence of parking DLC 6.4 with respect to the normal operation DLC 1.2. In this work, an availability of 98 % was assumed, so that the WT was assumed to be parked at 2 % of its lifetime. Both DLCs influence to fatigue were scaled with respect to their occurring time period over the WTs lifetime and according to the wind speed probability distribution[25]. Transient start-up and shut-down events were included as proposed by the DIBt-standard [152]. The DLCs 3.1 and 4.1 required a simulation of the start-up and shut-down events with a NWP and cut-in, rated, and cut-out wind speeds. The DIBt-standard [152, p. 18] assumes that WT start-ups happen 1000 times at cut-in wind speed, 50 times at rated wind speed and 50 times at cut-out wind speed per year. For WT shut-downs, the same occurring numbers per year and wind speed are proposed.

[25] The connection between fatigue and the wind speed probability distribution was treated in subsection 2.2.6.

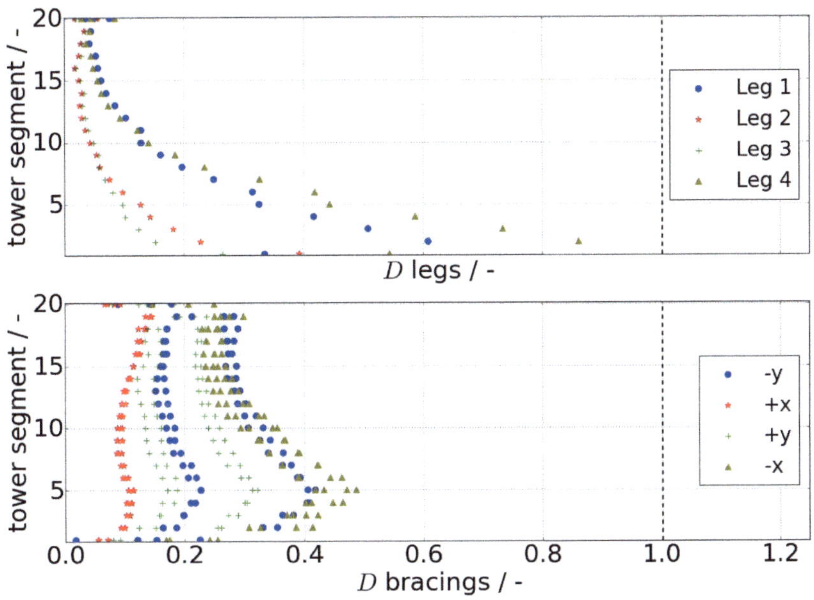

Figure 9.17: Lifetime damage of the leg and bracing members in the C3 RILT with 87.6 m tower hight

Since tower drag wind loads were not included in the ASE analysis, all load ranges had to be scaled up. The right side in Figure 9.14 shows the maximum occurring leg and bracing member utilizations under cut-out wind speed. A back translating of 6 % leg utilization corresponds to a maximum stress range of

$$\frac{1}{2}\frac{0.06 \times f_y}{\gamma_f\,\gamma_m\,\gamma_n} = \frac{1}{2}\frac{0.06 \times 355 \times 10^6\ \mathrm{N/m^2}}{1.35 \times 1.10 \times 1.00} = 7.17 \times 10^6\ \mathrm{N/m^2} \tag{9.27}$$

due to tower wind drag. Therefore, 7.17×10^6 N/m^2 was added to the stress ranges of all timeseries of all members to account conservatively for wind pressure on the tower[26]. Furthermore, a design fatigue factor of $DFF = 2.0$ was assumed, because all members were located in normal air[27] and were accessible for inspection and repair of initial fatigue and coating damages. The inner member accessibility was achieved through little holes in the joint region of members, such as shown at the bottom of the left picture in Figure 2.18. Since joint steel sheets strengthened the cross sections in the joint region, small holes caused not any strength problems. An endoscope camera can be used for visible inspections inside the structural member of interest. An inspection plan for the tower is set up in the next subsection 9.6.3.

[26] This very conservative approach may be refined by using calculated stress ranges, which are evaluated for the actual occurring wind speed in the single timeseries, rather than assuming cut-out wind speed for tower drag induced member loads for all members and timeseries. Moreover, assuming not the utilization of the member with the highest utilization for all members, but the utilization of each single member for each member increases the accuracy further.

[27] Offshore WTs may be exposed to salty air and water, whereby higher design fatigue factors are necessary.

Under the conditions, described in subsection 9.6.1 and this subsection, Figure 9.17 shows the maximum lifetime damages for the 87.6 m high C3 RILT. Since all lifetime damages were below 1[28], the WT could sustain a 20 year operation. In case of the leg members, the maximum lifetime damage of $D = DFF\ D_{\mathrm{c}} = 0.86$ occurred in the leg 4 of segment 2. Only tower bottom leg members up to the fourth segment experienced a significant lifetime damage of > 0.5. All other leg and bracing members experienced lower lifetime damages. A comparison with the extreme utilizations in Figure 9.15 reveals that fatigue was not the driving factor for this WT and tower configuration.

9.6.3 Periodical Inspection of the C3 RILT

DNV GL provides guidance notes to the required periodical inspection intervals T_{PII} as function of the chosen design fatigue factor and the calculated lifetime of a component [63, p. 131]. In Equation 9.28 it is adjusted with respect to

$$T_{\mathrm{PII},m,j,s} = T_{\mathrm{L},m,j,s}\frac{DFF}{3} = \frac{T_{\mathrm{DL}}}{D_{\mathrm{c},m,j,s}}\frac{DFF}{3} \tag{9.28}$$

where $T_{\mathrm{L},m,j,s}$ is the calculated lifetime of a stress calculation spot s on the cross section in joint j of member m and $D_{\mathrm{c},m,j,s}$ is its corresponding characteristic lifetime damage. For the leg 4 member of tower segment 2, which experienced the most damage over lifetime, a minimum inspection interval of

$$T_{\mathrm{PII,min}} = \frac{T_{\mathrm{DL}}}{0.43}\frac{DFF}{3} = \frac{20\ \text{years}}{0.43}\frac{2}{3} = 30.94\ \text{years} \tag{9.29}$$

showed that no fatigue related inspections of the RILT were necessary throughout the design lifetime of the WT.

9.7 Accidental Limit State Analysis

Although dropped objects, collision impact, fire, or explosion are accidental cases which may happen to the RILT, they were not analysed in this work to limit its scope. However, for a full assessment, these aspects have to be taken into account.

[28] Thus, they met the requirement in Equation 2.39

9.8 Serviceability Limit State Analysis

This section shows, how the serviceability limit state is hold for the given RILT. The first subsection 9.8.1 refers to the natural frequency requirements and the second subsection 9.8.2 to the blade tip to tower clearances.

9.8.1 Natural Frequency Requirements

The Campbell diagram in Figure 9.2 shows, that the requirement in Equation 2.45 is hold for the relevant 1p and 3p excitation frequencies, which were below and above the first FA and SS tower bending eigenfrequencies, respectively.

9.8.2 Blade Tip to Tower Clearance

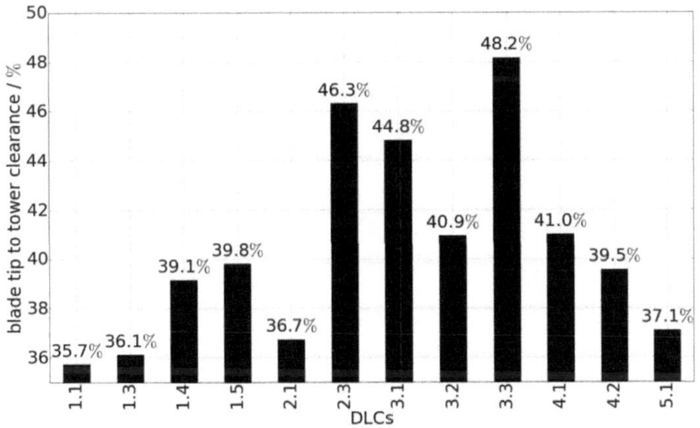

Figure 9.18: Blade tip to tower clearances in percentage of the unloaded state for the C3 RILT with 87.6 m tower hight

Figure 9.18 shows the calculated minimum occurring blade to tower clearances for the C3 RILT with 87.6 m tower hight. Since all clearances were > 30 % of the unloaded state, no additional deflection measurements were necessary for the blades throughout their manufacturing[29]. DLCs 6.x and 7.x were not depicted in Figure 9.18 because no blades pass the tower while WT parking, where one of the three blades pointing vertically upwards. Note that all DLC dependent PSFs were applied to the clearances, whereby DLC 1.1 experiences the lowest clearance from all DLCs. It amounted to 35.7 %.

[29] A review of subsection 2.2.8 indicates the relation between the minimum allowed blade tip to tower clearance and the manufacturing of the blades.

10 Potential of Rotatable Support Structures

10.1 Savings through the Material

The material usage of WT towers is the major influence factor for their costs and caused CO_2-emissions throughout the manufacturing. Since cost calculations involve many more uncertainties compared to mass calculations, a comparison of different tower concept masses was used as one indicator for the potential of a concept. In the first subsection 10.1.1, tower masses of the previously analysed RILT concept are presented and the second subsection 10.1.2 compares the RILT to other tower concepts in terms of their masses, mass related CO_2-emissions, and mass related costs.

10.1.1 Material Usage of Rotatable Inclined Lattice Towers

Parameter	Value	Unit	Parameter	Value	Unit	Parameter	Value	Unit
H_T	147.6	m	N_{seg}	24	–	ρ_{steel}	8500	kg/m^3
f_y	355	N/mm^2	a_B	20.0	m	b_B	20.0	m
a_T	8.0	m	b_T	5.0	m	$L_{L,B}$	310	mm
$L_{L,T}$	280	mm	L_F	90	mm	$H_{L,B}$	78	mm
$H_{L,T}$	190	mm	H_F	70	mm	α_{TI}	-9.534	deg
t_p	20	mm	w_p	1	mm			

Table 10.1: Rotatable inclined lattice tower design parameter set for a tower height of 147.6 m

The previous chapter 9 presented the ASE analysis and the assessment of a 87.6 m high RILT, which was mainly defined through the parameters in Table 9.1 and Appendix 14.22. SubDyn calculated the RILT overall mass to 207.403 t where an artificial high material density of 8500 kg/m^3 was assumed. This is a common[1] steel density for ASE simulations to account for the additional weight of bolts, filler plates, paintings, cables, a lift cage, and other structural

[1] Other studies, such as from Jonkman [109] used the same steel density in their ASE simulations of conventional tubular steel towers. Although a conventional tubular steel tower does not have so many bolts and filler plates as lattice towers, they have heavy flanges, and platforms inside the tower, wherefore the same overall tower material density for lattice towers is justified.

attachments, such as already mentioned in subsection 9.1. Another tower mass related aspect is the intersection of members at joints in the finite-element SubDyn model, but this was assumed to be a compensation for additional joint sheets to attach the bracing to the leg members.

The RILT was developed for a height of 87.6 m, but at least one other RILT height was necessary to calculate the relation between the tower height and mass. Therefore, another ASE simulation for a 147.6 m high tower with $\approx$ 150 m hub height was carried out in the same way as for the lower one. The fundamental parameters of that higher tower are listed in Table 10.1 and the more detailed member parameters are listed in Appendix 14.24. The analogous tower drag loads and corresponding utilizations are depicted in Appendix 14.25, followed by the different utilizations of the ULS DLCs and by all member lifetime damages from the FLS. Its overall mass was 357.403 t with $\rho_{\text{steel}} = 8500\ \text{kg/m}^3$.

10.1.2 Comparison of the Material Usage

Insight into tower masses of conventional tower concepts is given in Figure 4.7 for a 3 MW WT. One of the most suitable references for tower comparisons is the study from Engström et al. [68], especially because it was carried out for the 5 MW NREL reference WT and a scaled 3 MW version of it. In both cases, an IB WT class was assumed[2]. The 5 MW WT version had likewise been used in this work, but the database in [68] is larger for the 3 MW version. In this work, the 5 MW was more relevant and the study [68] provides data for a welded tubular steel, friction joint tubular steel, slip formed concrete, and a hybrid concrete/steel tower in that power class. Additional data for concepts, such as the normal lattice, wooden, and the guyed tubular steel tower are only available for the 3 MW version. Since these concepts were likewise interesting in the comparison, a scaling factor had to be found to scale the calculated tower masses of the 3 MW towers to a 5 MW version. Figersh, Hand, and Laxson [73, p. 20] proposed to scale the tower masses with respect to the swept rotor area and with respect to the tower height between power classes. Here, only the rotor swept area is used because tower heights remained constant for each data point. The rotor area was a good choice, because the rotor thrust scales linear with it, thus a tower mass scaling factor

$$SF_{\text{tm}} = \frac{R^2_{\text{rot,5MW}}}{R^2_{\text{rot,3MW}}} = \frac{63^2\ \text{m}^2}{50^2\ \text{m}^2} = 1.588 \tag{10.1}$$

was used to up-scale the tower masses of the 3 MW WT to the 5 MW one in cases where no 5 MW tower mass data was available. Note that possible excitation and structural eigenfrequency intersections in the operational rotational speed range of these up-scaled WTs were not checked for. If such resonance causing intersections would happen, one of the solutions presented in subsection 5.2.5 could be applied. Since this work identifies the material and cost saving potential for the RILT compared to other concepts, neglecting further investigations in this direction is an assumption on the safe side.

[2] Such as for the RILTs in this work.

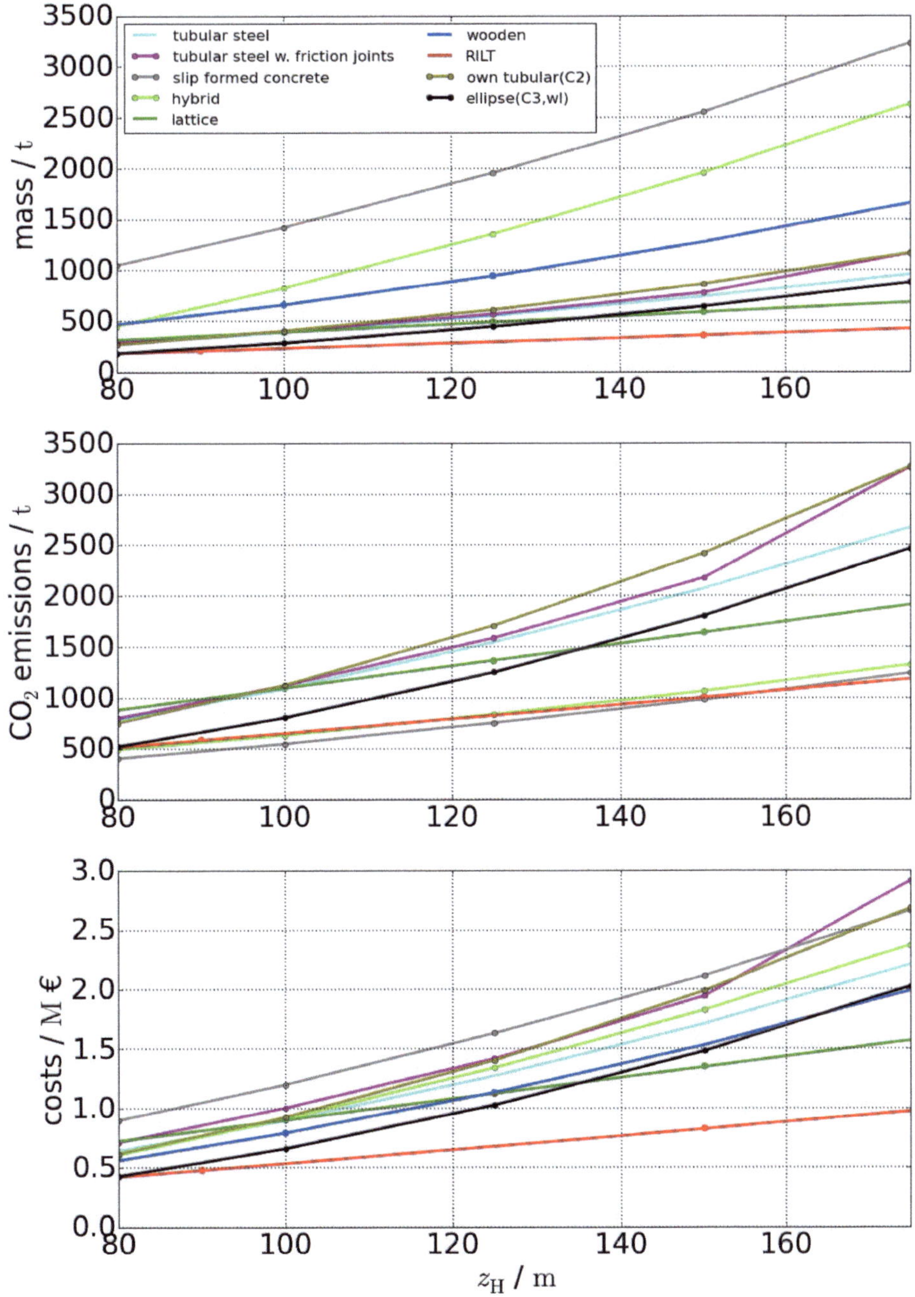

Figure 10.1: Mass, caused CO_2-emission, and cost comparison of different tower concepts with the rotatable inclined lattice tower under transport constraints

The top diagram in Figure 10.1 shows the mass data of the Engström et al. study [68] together with its up-scaled values for the normal lattice and wooden towers along rising hub heights. Furthermore, the own optimized tubular steel tower under land-based transport constraints as case **C2** of subsection 5.2.5 was entered together with the rotatable elliptical shell tower **C3,wl**, presented in the same subsection. Finally, the mass of the RILT is shown with its two available data points and a curve fit[3]. Note that all presented towers met the transport constraints, presented in subsection 2.5 along all hub heights. A first result was that concrete using concepts, such as the slip formed concrete and hybrid were the heaviest concepts, followed by the wooden tower. At a hub height of $z_H = 175$ m the slip formed concrete tower reached an enormous mass of 3,228 t and the hybrid concept at the same height reached 2,629 t. According to the large diagram scale, pure steel using concepts were relatively close to each other in terms of material usage, but as already investigated in subsection 5.2.5, the elliptical tower required less material than the conventional tubular steel concepts along all hub heights. At larger hub heights of > 95 m, the non-rotatable lattice tower required less material than the conventional tubular ones, and at $z_H > 135$ m it required even less material than the rotatable elliptical tower. The tower concept with the least material mass over all hub heights was the RILT. At low hub heights of 80 m it had with 183 t nearly the same mass as the corresponding rotatable elliptical tower with 186 t. More material was saved at larger hub heights since its mass at $z_H = 150$ m was only 61 % of the non-rotatable lattice tower, which had the lowest mass of all other concepts at that hub height.

The mass related CO_2-emissions were estimated by linear scaling factors for the different materials. Thereby, factors of 0.13 $kg_{CO_2}/kg_{concrete}$[4], 2.8 kg_{CO_2}/kg_{steel} [148], and 0.0 kg_{CO_2}/kg_{wood} were assumed. In these values, no CO_2-emissions, caused by the transport, assembly, dismantling, and disposal after lifetime of the towers were considered. Note that the CO_2-emissions in Figure 4.7 had apparently calculated with other scaling factors for concrete and their reference was not traceable. In this consideration, conventional tubular steel towers caused the most CO_2-emissions and concrete using towers together with the RILT the least.

In the third diagram in Figure 10.1, mass related costs of the different tower concepts are shown. The cost evaluation took place by linear scaling factors according to the Engström et al. reference [68]. Scaling factors are 2.3 €/$kg_{tubular\ steel}$, 2.3 €/$kg_{lattice\ steel}$, and 1.2 €/kg_{wood} [68]. Note that the values were not pure material costs, but tower costs, which involved material, labour, plates, flanges, bolts, nuts, and painting costs [68, p. 14]. Steel material prices fluctuated between ≈ 525 \$/$t_{rolled\ steel}$ to ≈ 700 \$/$t_{rolled\ steel}$ in year 2010 where the study [68] was made and between ≈ 660 \$/$t_{rolled\ steel}$ to ≈ 935 \$/$t_{rolled\ steel}$ in year 2018[5]. Since WT tower mass to cost scaling factors were not differentiated with respect to their influencing aspects in the study [68], the scaling factors were taken over for this work without adjustments for comparability. From these assumptions, the rotatable elliptical tower was cheaper than all

[3] Since three data points were available by including the origin of the graph, a quadratic polynomial $m_T = a_2\ H_T^2 + a_1\ H_T + a_0$ with $a_2 = 1.30$, $a_1 = 2.19e3$, and $a_0 = 0.00$ represented the tower mass in kg as function of the tower height.

[4] This small value results from the fact, that concrete consists only about 13 % out of CO_2 releasing cement, which causes about the same mass of CO_2-emissions than its own mass. These values originate from NRMCA [145].

[5] Prices were read from the 10 year chart of hot rolled steel, presented on `https://finanzen.handelsblatt.com/rohstoffpreise`

conventional concepts up to hub heights of $z_H \leq 135$ m. For all hub heights $z_H > 80$ m, the RILT pointed out as most cost effective tower concept. At $z_H = 100$ m the RILT saved costs of 0.661 M€ – 0.533 M€ = 0.128 M€ (19.36 %) compared to the rotatable elliptical (**C3,wl**), 0.791 M€ – 0.533 M€ = 0.258 M€ (32.62 %) compared to the wooden, 0.901 M€ – 0.533 M€ = 0.368 M€ (40.84 %) compared to the non-rotatable lattice, and 0.924 M€ – 0.533 M€ = 0.391 M€ (42.32 %) compared to the conventional own optimized tubular steel (**C2**) tower. At larger hub heights, such as $z_H = 150$ m, the RILT saved even more costs, namely 1.483 M€ – 0.822 M€ = 0.661 M€ (44.57 %) compared to the rotatable elliptical (**C3,wl**), 1.347 M€ – 0.822 M€ = 0.525 M€ (38.98 %) compared to the non-rotatable lattice, and 1.987 M€ – 0.822 M€ = 1.165 M€ (58.63 %) compared to the conventional own optimized tubular steel (**C2**) tower. However, the increased steel prices of 2018 would increase the cost savings even more, but were neglected as a conservative assumption.

10.2 Increased Energy Yield through Tower Inclination

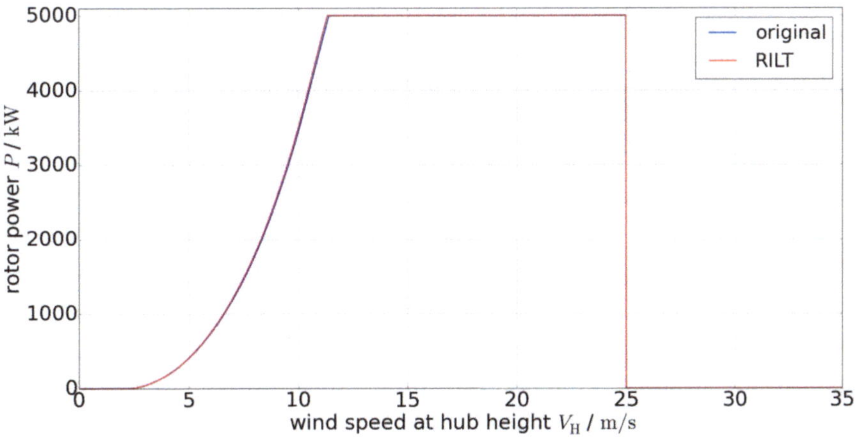

Figure 10.2: Power curves of the original 5 MW NREL reference WT and the same WT with vertical rotor alignment, such as used for the RILT

According to subsection 6.2.3, the inclined tower offered the opportunity to avoid a rotor shaft angle and a blade precone angle, which is unequal to zero. Such a vertical setting led to an 1.7 % larger effective rotor area than the reference effective rotor area from the 5 MW NREL reference WT. A corresponding minimum tower inclination angle of the RILT was necessary to ensure the same blade to tower clearance as for the reference WT. In the next step, the increased rotor area was applied to an energy production estimation for different WT classes. The WT classes may be reviewed in Table 2.3. Figure 10.2 shows the power curves for the reference rotor and the adjusted vertical rotor alignment for the RILT. The diagram itself indicates no significant difference between both WT versions. Therefore, a closer look into the lifetime generated energy and profit was necessary.

The WT rotor power as function of the wind speed at hub height $P(V_H)$ was averaged with respect to a given wind speed probability density $p(V_H)$ for a certain WT class by

$$\overline{P} = \int_0^\infty P(V_H)\, p(V_H)\, dV_H = \int_0^\infty P(V_H)\, dP_W \tag{10.2}$$

In Equation 10.2, a Weibull wind speed distribution was assumed, indicated by index $_W$ of P_W. Through discretization of wind speeds into N_B bins, one can express Equation 10.2 as

$$\overline{P} = \sum_{i=1}^{N_B} \left\{ \exp\left[-\left(\frac{V_{H,i-1}}{C} \right)^k \right] - \exp\left[-\left(\frac{V_{H,i}}{C} \right)^k \right] \right\} P\left(\frac{V_{H,i-1} + V_{H,i}}{2} \right) \tag{10.3}$$

according to Manwell, McGowan, and Rogers [137, p. 65]. Consequently, the lifetime average energy yield of a WT is calculated by

$$\overline{E}_{Life} = \overline{P}\, T_{Life}\, \overline{A}_{Life} \tag{10.4}$$

where $\overline{A}_{Life}$ is the WTs average lifetime availability.

WT Class -	Energy Yield Ref. / GWh	Energy Yield RILT / GWh	Income Ref. / M€	Income RILT / M€	Abs. Diff. / M€
1	469.1	472.0	34.105	34.319	0.214
2	387.9	391.1	28.207	28.436	0.229
3	319.9	323.0	23.260	23.488	0.228

Table 10.2: Energy and remuneration income comparison through WTs energy yield of the original 5 MW NREL reference WT and the WT with RILT and corresponding vertical rotor alignment

Taking German remunerations from the BMU [21] as basis for a WT with T_{Life} = 20 years, $\overline{A}_{Life}$ = 0.98, and a realized energy yield of 100 % of the reference energy yield resulted in an average payment of 0.0727 €/kWh[6] over lifetime. Table 10.2 lists the resulting calculated lifetime energy yield and remuneration for the 5 MW NREL reference WT and the one with RILT and vertical rotor alignment. It documents that the monetary lifetime income of the WT with RILT is only 0.63 %, 0.81 %, and 0.98 % larger than for the reference WT in WT classes

[6] The calculation of this this average lifetime remuneration is as follows: $(5\ \text{year}/T_{Life})0.0791\ \text{€/kWh} + \left(2\frac{150\,\% - 100\,\%}{0.75} \frac{1}{12\frac{\text{month}}{\text{year}}} \frac{1}{T_{Life}} \right) 0.0791\ \text{€/kWh} + \left(T_{Life} - 5\ \text{year} - (2\frac{150\,\% - 100\,\%}{0.75} \frac{1}{12\frac{\text{month}}{\text{year}}}) \frac{1}{T_{Life}} \right) 0.0432\ \text{€/kWh} = 0.0727\ \text{€/kWh}$. It was based on the equations and explanations from the BMU [21, p. 12-13]. The year of first payment is set to 2020.

1, 2, and 3, respectively. An expression in absolute values of 214 k€, 229 k€, and 228 k€ shows that it was a not negligible amount of additional money, which was generated by using the RILT instead of a conventional WT tower.

10.3 Yaw System Costs

Reliable costs for yaw systems at WT tower bases are difficult to estimate without deeper investigations. Table 2.6 reveals a wide range of 180 k€ to 450 k€ for a yaw bearing at the base of a 2.5 MW WT with 79.43 m hub height. According to Malcolm and Hansen [136], the yaw system costs are twice as high as the yaw bearing costs alone. Considering this and a discount of 10 % for mass production of these bearings leads to yaw system costs of 324 k€ to 810 k€ for a conventional roller bearing with outer diameters of 6000 mm and 5500 mm, respectively. Table 2.6 suggests, that the bearing costs may be reduced by increasing diameters, which were necessary for the RILT with larger hub heights[7]. One counter acting cost factor is the omission of the yaw system at the tower top. Malcolm and Hansen [136, p. 21] propose a table with costs for conventional tower top yaw bearings where a mass production discount is already included. In the curve fit of the data [136, p. 21], the maximum overturning moment on the yaw bearing and its diameter is included to calculate its mass. That curve fit is described as

$$m_{\text{yaw bear.}} = 0.0152\left(\frac{M_{\max}}{D_{\text{yaw bear.}}} - 36\right)^{1.489} \tag{10.5}$$

where the maximum overturning moment $M_{\max}$ must be given in kNm and the yaw bearing diameter $D_{\text{yaw bear.}}$ must be given in m to get the yaw bearing mass $m_{\text{yaw bear.}}$ in kg. The mass to cost relation is curve fitted by

$$C_{\text{yaw bear.}} = \left(m_{\text{yaw bear.}}\, 6.689 + 953\right) \frac{\$_{\text{US}}}{\text{kg}} \tag{10.6}$$

Using Equation 10.5 in 10.6 with a yaw bearing diameter of 3.87 m[8] and a tower top maximum bending moment of 3.02e4 kNm[9] resulted to costs of 112 k€[10] for the whole yaw system.

Assuming the validity of Equations 10.5 and 10.6 for the RILT base yaw bearing led to costs of 296 k€ and 127 k€ for the yaw system of the 87.6 m high and 147.6 m high RILT. Note that the yaw system costs for the smaller RILT were more than twice as high as for the larger RILT. These results were based on the extreme tower base reaction moments, extracted from the ASE analyses of the 87.6 m and 147.6 m high RILT. The largest tower base reaction moments occurred for both tower heights in DLC 2.3 at rated wind speed and were 132 kNm γ_{f} = 132 kNm 1.1 = 145 kNm and 219 kNm γ_{f} = 219 kNm 1.1 = 241 kNm for H_{T} = 87.6 m and H_{T} = 147.6 m, respectively. RILT yaw bearing diameters were calculated, based on Tables 9.1 and 10.1 and result to

[7] A comparison of Tables 9.1 and 10.1 indicates, that RILTs bottom leg widths increase with larger hub heights.

[8] This value is the tower top diameter of the 5 MW reference WT from [109].

[9] This value was taken from the conventional ASE load analysis results in Appendix 14.9

[10] The yaw system costs were assumed to be twice as the yaw bearing costs alone.

$D_{\text{yaw bear.}} = \sqrt{2} \times 6.8$ m $= 9.62$ m and $D_{\text{yaw bear.}} = \sqrt{2} \times 20$ m $= 28.28$ m for the small and high tower. Although extreme tower base overturning moments of the higher RILT were larger the costs for the bearing were lower, because of its disproportionately larger yaw bearing diameter and the resulting lower local bearing loads. This results may contain truth, but larger bearings rise also other costs, such as transport, segmentation, assembly, and meeting tight tolerances over large dimensions. The RILT has four legs at the tower base, wherefore loads are induced punctually into the yaw system. This may require additional local strengthening of the bearing at the leg to bearing joints and thereby cause more costs. It must be noted that Equations 10.5 and 10.6 are very simple and rough estimates, which do not account for transport, segmentation, assembly, and tolerances for yaw bearings with very large diameters. Therefore, the above calculated values were the result of a given limited model from the literature and must be considered critically. Other yaw bearing cost models may be derived from a more comprehensive database. Some additional yaw bearing and yaw system costs are provided by Hau [92], Gasch [77], and Manwell, McGowan, and Rogers [137]. Manufacturers data sheets may also be consulted, but must be investigated carefully because of their non-WT-specific character.

10.4 Economical Potential

The final result of this work is the economical potential of rotatable support structures for WTs and specifically for the RILT. It is a preliminary end result, because more accurate tower base yaw system costs were not available for the given tower configuration at that state of research. One major aspect of the RILT economics were the cost savings, which result from the less required material for RILTs, compared to all other support structure concepts. This aspect was investigated in subsection 10.1.2. For the end result, the difference between the cheapest conventional tower concept[11] material costs and the RILT material costs was considered for hub heights 80 m $\leq z_{\text{H}} \leq$ 175 m. This means for example that at $z_{\text{H}} = 80$ m, the RILT material related cost savings $S_{\text{RILT,mat.}}$ were those, which result from the difference to a wooden tower and at $z_{\text{H}} = 175$ m those, which resulted from the difference to a conventional lattice tower.[12] An increased WT energy yield was achievable through the vertical alignment of the rotor, which was possible through the tower inclination. The resulting additional money income ΔI_{RILT}, calculated in subsection 10.2, could be added to the profitability of RILTs together with the saved money $C_{\text{yaw sys. conv.}}$ through avoiding the tower top yaw system. On the other hand, the tower base yaw system costs $C_{\text{yaw sys. RILT}}$ had to be included in the balance to achieve a rotatable tower. The underlying cost model for the yaw system was explained in subsection 10.3. Potential foundation cost increases were assumed to be $\Delta C_{\text{found.}} = C_{\text{yaw sys.}}$. This assumption was made without any reference, because of the lack of literature for foundations of WTs with yaw system at the tower base and may be adjusted throughout future research activities. If those activities find higher tower base yaw system and additional foundation costs, other concepts,

[11] This includes the tubular steel, tubular steel with friction joints, slip formed concrete, hybrid, lattice, wooden, and own tubular tower concepts (**C2** and **C3,wl**).

[12] A review of the bottom diagram in Figure 10.1 may clarify this statement.

such as discussed in subsection 4.2.2 may be considered. The overall RILT cost savings were consequently calculated as

$$S_{\text{RILT}}(z_{\text{H}}) = S_{\text{RILT,mat.}}(z_{\text{H}}) + \Delta I_{\text{RILT}} + C_{\text{yaw sys. conv.}} - C_{\text{yaw sys. RILT}}(z_{\text{H}}) - \Delta C_{\text{found.}}(z_{\text{H}}) \quad (10.7)$$

Figure 10.3 shows the result of Equation 10.7 and its different terms. Note that a WT class I was assumed and the corresponding wind speed was always acting at the hub height, wherefore ΔI_{RILT} remained constant along z_{H}. From this assumption and the assumption that the tower top yaw bearing diameter remains constant for all conventional tower concepts and hub heights, constant costs $C_{\text{yaw sys. conv.}}$ were likewise justified. The RILT base extreme overturning moment and the yaw bearing diameter was modelled by polynomials of the second order, fitted to the data points, presented in subsection 10.3 as function of the hub height. This led to the following Equations 10.8 and 10.9.

$$M_{\text{max,RILT}}(z_{\text{H}}) = -1.511\text{e}{-1}\ z_{\text{H}}^2 + 1.625\text{e}3\ z_{\text{H}} \quad (10.8)$$

where $M_{\text{max,RILT}}$ results in kNm if z_{H} is given in m and

$$D_{\text{yaw,RILT}}(z_{\text{H}}) = 1.362\text{e}{-3}\ z_{\text{H}}^2 - 1.571\text{e}{-2}\ z_{\text{H}} \quad (10.9)$$

where $D_{\text{yaw,RILT}}$ results in m if z_{H} is given in m. Each inconstant continuous curve in Figure 10.3 is modelled by a polynomial of the fourth order to smooth the plot. The polynomials are

$$C_{\text{yaw sys. RILT}} = \Delta C_{\text{found.}} = -3.424\text{e}{-3}\ z_{\text{H}}^4 + 1.596\ z_{\text{H}}^3 - 2.530\text{e}2\ z_{\text{H}}^2 + 1.403\text{e}4\ z_{\text{H}} + 2.187\text{e}1 \quad (10.10)$$

$$S_{\text{RILT,mat.}} = 2.761\text{e}{-3}\ z_{\text{H}}^4 - 1.479\ z_{\text{H}}^3 + 2.548\text{e}2\ z_{\text{H}}^2 - 1.070\text{e}4\ z_{\text{H}} + 1.254\text{e}2 \quad (10.11)$$

$$S_{\text{RILT}} = 9.609\text{e}{-3}\ z_{\text{H}}^4 - 4.670\ z_{\text{H}}^3 + 7.607\text{e}2\ z_{\text{H}}^2 - 3.876\text{e}4\ z_{\text{H}} + 2.705\text{e}5 \quad (10.12)$$

and produce results in €. Its determination coefficient

$$R^2 = \frac{\sum (\hat{y}_i - \overline{y})^2}{\sum (y_i - \overline{y})^2} \quad (10.13)$$

is given on the right hand of the diagram where $\hat{y}_i$ is the modelled value, y_i is the calculated value, and $\overline{y}$ is the arithmetic mean of all y_i. It shows a good fitting quality of $R^2 \geq 0.997$ for all three curves.

The overall cost savings of the RILT increase with rising hub heights from 45 k€ at $z_{\text{H}} = 80$ m to 768 k€ at $z_{\text{H}} = 175$ m. These cost savings amount to 8.05 % of the cheapest conventional concept, the wooden tower, at $z_{\text{H}} = 80$ m and 48.97 % of the cheapest conventional concept, the lattice tower, at $z_{\text{H}} = 175$ m. Thereby, the developed RILT shows a huge material, CO_2, and cost saving potential for large hub heights of $z_{\text{H}} \geq 100$ m for transport constrained land-based WTs.

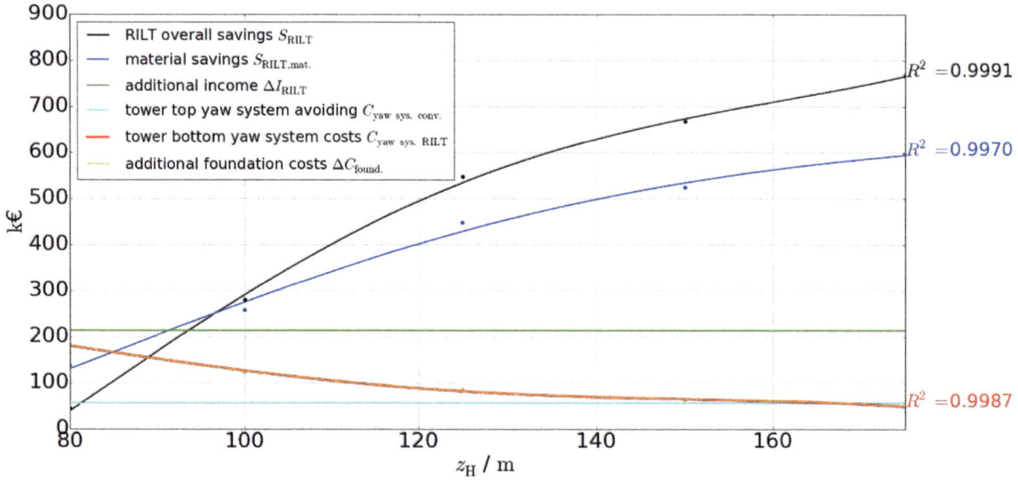

Figure 10.3: Overall RILT cost savings with respect to all other considered conventional tower concepts for the 5 MW NREL reference WT and the different monetary components, which were considered for the cost balance as function of the hub height.

11 Summary

This work presents the development and potential analysis of a new and sustainable rotatable wind turbine tower by means of aero-servo-elastic load simulations. The idea of rotatable wind turbine towers is not new, but it was never investigated on a scientific way and a meaningful depth. One obvious advantage is the permanent alignment of the tower with respect to the wind turbine rotor. Thereby, the tower can be designed less stiff lateral to the rotor axis parallel direction and less material is required. Furthermore, it can be shaped more aerodynamically to decrease its drag loads in the rotor area orthogonal direction. Such slim tower profiles achieve a better transportability under bridges and tunnels, wherefore they are appropriate for larger hub heights where conventional tubular steel towers reach their economical limits.

Another advantage is the improved accessibility of the yaw system at the tower base. A technology research revealed that yaw systems are applied for heavy structures such as railway turntables, rotatable buildings, mobile cranes, bucket wheel excavators, and rotatable shortwave antennas, which experience large axial forces and overturning moments at the rotatable joint. Thereby, a technical feasibility for yaw systems at wind turbine tower bases was given and additionally confirmed through an expertise survey. Considering the low amount of scientific research about rotatable wind turbine support structures and the scope of this work, it was focussed on the tower advantages instead of the tower base yaw system as a first research step.

The aero-servo-elastic load simulation is an appropriate tool to investigate wind turbine support structures and was carried out in the FAST environment for this work. A first aero-servo-elastic analysis with most of the design load cases of IEC 61400-1 took place with the 5 MW NREL reference wind turbine and a conventional tubular steel tower. This step was necessary to achieve comparability to the developed rotatable wind turbine tower and validity of the procedure itself. The calculated loads were used for the design of an own reference conventional tubular steel tower and an elliptical rotatable tower under extreme load conditions and transport constraints. The results were 28.94 % material savings for an 80 m high elliptical tower and 26.69% material savings for an 180 m high elliptical tower compared to the conventional tubular one with same heights. The reason for not more material savings of the elliptical tower at larger hub heights was found in the transport constraint.

Therefore, a four-legged lattice tower was used as underlying concept for the rotatable tower instead of a shell concept, such as an elliptical or aerodynamically shaped one. Lattice towers are material saving and transport constraint respecting for larger hub heights. Especially transport is no problem, because lattice tower members are small compared to huge tubular tower segments and maintenance costs remain low if slip resisting lockbolts are used for the joints. Another controversy discussed aspect, the outer appearance and acceptance of lattice towers, may be improved by usage of coverings or by application in areas with low population density. For the rotatable lattice tower the focus was not on a different stiffness in the fore-aft and side-side direction, but on increased leg distances in all directions to save as much material as possible. This aspect leads especially to more material savings at the upper part of a lattice tower, because leg distances are not restricted by the blade to tower clearance on three sides of the tower. This is because of the fixed rotor position with respect to it. An additional tower inclination against the main wind direction is another advantage for rotatable wind turbine towers. It achieves a larger blade to tower clearance on the rotor side of the tower, whereby a vertical rotor alignment is possible or/and less blade material is necessary to reach its required stiffness. Through such a vertical rotor alignment, the effective rotor area increases and the wind turbine yields more energy over its lifetime and becomes more profitable. Furthermore, additional bending moments are induced into the structure through the mass eccentricities. Since, the tower is inclined against the main wind direction, it acts as counter weight, which reduces extreme loads within the members and more material is saved. Concerning yaw movements of such eccentric high masses higher loads occur only as consequence of yaw accelerations, but not because of yaw velocities. This leads to no considerable losses in the energy yield over lifetime, because the yaw speed must not be reduced for a rotatable inclined tower.

The rotatable inclined lattice tower was modelled in the linear-beam-finite-element submodule SubDyn, which is part of the FAST framework. Since SubDyn was only capable to model circular member cross sections in a lattice structure, some changes were required. The stiffness and mass matrix formulations changed to account for arbitrary centroids, shear centers and principal bending, and shear axes on lattice member cross sections. Another adjustment was done in the direction cosine matrices to achieve control over member alignments around their length axes. After the validation of all changes, rotatable inclined lattice towers with two different heights, 87.6 m and 147.6 m were modelled and investigated via aero-servo-elastic load simulations and structural analyses. Thereby, the ultimate, fatigue, and serviceability limit states were met over the assumed wind turbines lifetime of 20 years according to IEC 61400-1 and with the support of DNV GL guidelines ST-0126 and RP-C203.

After the rotatable inclined lattice tower design, its cost savings were evaluated with respect to its saved material, the increased energy yield from the vertical rotor alignment, and the omission of the tower top yaw system. On the other side are tower base yaw system costs and assumed increased foundation costs, which reduced these cost savings. From the preliminary assumptions in this work the rotatable inclined lattice tower saved 45 k€ at 80 m to 768 k€ at 175 m hub height compared to the most cost-effective conventional concepts, such as the tubular steel, tubular steel with friction joints, slip formed concrete, hybrid, lattice, and wooden tower. This amounts to 8.05 % and 48.97 % in tower cost savings and indicates the huge cost saving potential of rotatable inclined lattice towers for larger hub heights.

This result reveals a new, sustainable, and profitable support structure concept for land-based wind turbines with large hub heights under transport constraints. Although the amount of installed offshore wind turbines increases, this concept has the potential to make land-based wind energy more attractive, especially for weak wind areas with bad accessibility. Therefore, it is a contribution to the profitability of renewable energies.

12 Outlook

The scope of this work hindered detailed proofs for the friction joints of the rotatable inclined lattice tower. Their impact on the overall structural mass and thereby on the costs was assumed to be subordinative. However, proofs are necessary to provide more comprehensive and accurate conclusions.

One main field of future research will be the tower bottom yaw system, because it was only investigated at the surface in this work. Thereby, not only conventional roller bearings, but also other concepts should be considered, such as introduced in this work. Furthermore, it is important that yaw acceleration will be limited and controllable to hinder the induction of large and periodic gyroscopic loads on the structure. The same request for more detailed analyses applies consequently to the foundation and the machinery bed interface.

Other questions rose regarding the wind turbine controls. In this work, only the standard reference NREL controller was used, while some small adjustments to the controls were implemented to react on special events, such as wind turbine shut-downs or parking design load cases with extreme wind conditions. This work showed that extreme loads depend on those control adjustments, but it was out of scope to investigate the larger influence of the adjustments with respect to the wind turbines efficiency and the loads on other structural components. Future research activities may investigate the interdependency between controls, loads, and the energy yield of wind turbines with rotatable support structure on a deeper level.

This work indicates that lattice towers are not necessarily bad in their appearance, but the visual acceptance of land-based rotatable inclined lattice towers has likewise to be investigated in more detail in the future.

13 References

[1] Abbott, I. H., and von Doenhoff, A. E. *Theory of wing sections: including a summery of airfoil data*. Dover Publ, New York, USA, 2010. ISBN 978-0-486-60586-9.

[2] Albertsen, H. Überschlägige Dimensionierung eines Azimutlagers am Turmfuß. Tech. rep., IMO Energy GmbH und Co. KG, Gremsdorf, 2012.

[3] Anderson, J. D. *Fundamentals of aerodynamics*, 3 ed. McGraw-Hill series in aeronautical and aerospace engineering. McGraw-Hill, Boston, 2001. ISBN 978-0-07-237335-6.

[4] Ashuri, T., Martins, J. R. R. A., Zaaijer, M. B., van Kuik, G. A. M., and van Bussel, G. J. W. Aeroservoelastic design definition of a 20 MW common research wind turbine model: A 20 MW common research wind turbine model. *Wind Energy* (2016). ISSN 10954244, DOI 10.1002/we.1970.

[5] Bak, C., Zahle, F., Bitsche, R., Taeseong, K., Yde, A., Henriksen, L. C., Natarajan, A., and Hansen, M. H. Description of the DTU 10 MW reference wind turbine. Tech. Rep. DTU Wind Energy Report-I-0092, Technical University of Denmark (DTU) Institut for Vindenergi, Roskilde, DK, July 2013.

[6] Bashforth, F. An attempt to test the theories of capillary action by comparing the theoretical and measured forms of drops of fluid. Tech. rep., Cambridge: at the University Press, Cambridge, USA, 1883.

[7] Bauchau, O. A. Section Builder Users Manual. Manual, School of Aerospace Engineering, Georgia Institute of Technology, Atlanta, USA, Feb. 2007.

[8] Bauchau, O. A. *Flexible multibody dynamics*. No. 176 in Solid mechanics and its applications. Springer Dordrecht, Heidelberg, London, New York, 2011. ISBN 978-94-007-0334-6.

[9] Bauchau, O. A., and Craig, J. I. *Structural analysis: with applications to aerospace structures*. No. v. 163 in Solid mechanics and its applications. Springer Dordrecht, Heidelberg, London, New York, Atlanta, USA, 2009. ISBN 978-90-481-2515-9.

[10] Bendix, H., Peschke, J., and Hofmann, K. Windenergieanlage mit horizontaler Rotorachse und mit unten liegendem Antrieb, June 2013. Patent No. DE102011119776A1.

[11] Bertagnolio, F., Rasmussen, F., Sørensen, N. N., Johansen, J., and Madsen, H. A. A stochastic model for the simulation of wind turbine blades in static stall. *Wind Energy 13*, 4 (May 2010), pp. 323–338. ISSN 10954244, 10991824, DOI 10.1002/we.342.

[12] Betz, A. *Windenergie und ihre Ausnutzung durch Windmühlen*. Ökobuch, Göttingen, 1926. ISBN 978-3-922964-11-7.

[13] Beye, T. *Konzeptstudie über die Verlegung des Azimutlagers an den Turmfuß einer Windenergieanlage*. Bachelor Thesis, Hochschule Flensburg University of Applied Sciences, Flensburg, 2012.

[14] Beye, T., Rudolf, R., and Faber, T. Konzeptstudie für Windenergieanlage mit Azimutlager am Turmfuß. *Bauingenieur Band 88* (2013), pp. 269–274.

[15] Bianculli, A. J. *Trains and technology: the American railroad in the nineteenth century*. University of Delaware Press, Newark, 2001. ISBN 978-0-87413-802-3.

[16] Blasques, J. P. Multi-material topology optimization of laminated composite beams with eigenfrequency constraints. *Composite Structures 111* (May 2014), pp. 45–55. ISSN 02638223, DOI 10.1016/j.compstruct.2013.12.021.

[17] Blasques, J. P., and Bitsche, R. D. An efficient and accurate method for computation of energy release rates in beam structures with longitudinal cracks. *Engineering Fracture Mechanics 133* (Jan. 2015), pp. 56–69. ISSN 00137944, DOI 10.1016/j.engfracmech.2014.11.002.

[18] Blasques, J. P., Bitsche, R. D., Fedorov, V., and Lazarov, B. S. Accuracy of an efficient framework for structural analysis of wind turbine blades. *Wind Energy 19*, 9 (Sept. 2016), pp. 1603–1621. ISSN 10954244, DOI 10.1002/we.1939.

[19] Blasques, J. P., and Stolpe, M. Multi-material topology optimization of laminated composite beam cross sections. *Composite Structures 94*, 11 (Nov. 2012), pp. 3278–3289. ISSN 02638223, DOI 10.1016/j.compstruct.2012.05.002.

[20] Blevins, R. D. *Flow-induced vibration*, 2 ed. Van Nostrand Reinhold, New York, 1990. ISBN 978-0-442-20651-2.

[21] BMU. Vergütungssätze, Degression und Berechnungsbeispiele nach dem neuen Erneuerbare-Energien-Gesetz (EEG) vom 04. August 2011 (EEG2012). Tech. Rep. BMU KI III 1, Bundesministerium für Umwelt, Naturschutz und nukleare Sicherheit, Berlin, May 2012.

[22] Bossanyi, E. A. GH Bladed Theory Manual. Tech. Rep. 282/BR/009, Garrad Hassan, Bristol, United Kingdom, 2009.

[23] Bošnjak, S. M., Gnjatovic, N., Savicevic, A. D., Pantelic, M. P., and L., M. I. Basic parameters of the static stability, loads and strength of the vital parts of a bucket wheel excavators slewing superstructure. *Journal of Zhejiang University Science A 17* (May 2016), pp. 353–365. ISSN 1673-565X, 1862-1775, DOI 10.1631/jzus.A1500037.

[24] Brebbia, C. A., and Dominguez, J. *Boundary elements: an introductory course*, 2 ed. Computational Mechanics Publications [u.a.], Southampton, 1991. ISBN 978-1-85312-160-9.

[25] Bruger, P., Buchmann, B., Kurrer, K.-E., and Ozimek, C. Rotatable shortwave curtain antenna operable at very high wind speeds. *IEEE Transactions on Broadcasting 42*, 1 (Mar. 1996), pp. 50–54. ISSN 00189316, DOI 10.1109/11.486075.

[26] Buhl, M., Pierce, K., Laino, D., Hansen, C., Suzuki, A., Wright, A., Jonkman, J., Hand, M., and Jonkman, B. AeroDyn Change Log, Apr. 2016.

[27] Buhl, M. L., and Manjock, A. A comparison of wind turbine aeroelastic codes used for certification. Tech. Rep. AIAA-2006-0786, National Renewable Energy Laboratory and Germanischer Lloyd WindEnergie GmbH, Colorado, USA and Hamburg, 2006.

[28] Bundesverband WindEnergie e.V. Maschinenhaus / Antriebstrang. https://www.wind-energie.de/infocenter/technik/konstruktiver-aufbau/maschinenhaus-antriebstrang, Mar. 2018. [Online; accessed 14-February-2019].

[29] Collins, J. A. *Mechanical design of machine elements and machines: a failure prevention perspective*, 2 ed. Wiley, Hoboken, USA, 2010. ISBN 978-0-470-41303-6.

[30] Cook, R. D., and Young, W. C. *Advanced mechanics of materials*, 2 ed. Prentice Hall, Upper Saddle River, USA, 1999. ISBN 978-0-13-396961-0.

[31] Corporation, T. M. MACTON®Transportation Products - Rail vehicle turntable for locomotive and railcar maintenance yards. Product Description, 2012.

[32] Damiani, R., Jonkman, J., and Hayman, G. SubDyn Users Guide and Theory Manual. Tech. Rep. NREL/TP-5000-63062, National Renewable Energy Laboratory, Golden, USA, Mar. 2015.

[33] Damiani, R., and Song, H. A Jacket Sizing Tool for Offshore Wind Turbines within the Systems Engineering Initiative. 2013. DOI 10.4043/24140-MS.

[34] Damiani, R., Song, H., Robertson, A., and Jonkman, J. Assessing the Importance of Nonlinear Structural Characteristics in the Development of a Jacket Model for the Wind Turbine CAE Tool FAST. 2013. DOI 10.1115/OMAE2013-11434.

[35] Dankert, J., and Dankert, H. *Technische Mechanik: Statik, Festigkeitslehre, Kinematik/Kinetik*, 7 ed. Springer Vieweg, Wiesbaden, 2013. ISBN 978-3-8348-1809-6.

[36] Department of Wind Energy DTU. User's Manual for BECAS A cross section analysis tool for anisotropic and inhomogeneous beam sections of arbitrary geometry. Tech. rep., Technical University of Denmark, Risø,DK, Apr. 2016.

[37] Desmond, C., Murphy, J., Blonk, L., and Haans, W. Description of an 8 MW reference wind turbine. *Journal of Physics: Conference Series 753* (Sept. 2016). ISSN 1742-6588, 1742-6596, DOI 10.1088/1742-6596/753/9/092013.

[38] Deutsches Institut für Bautechnik. Z-14.4-591: Schließringbolzen ohne Abrissteil. Allgemeine bauaufsichtliche Zulassung. Tech. Rep. | 36-1. 14.4-104/14, Zulassungsstelle für Bauprodukte und Bauarten Bautechnisches Prüfamt, Berlin, July 2015.

[39] DIN 22261-2:2015-11. Excavators, spreaders and auxiliary equipment in opencast lignite mines - Part 2: Calculation principles. Norm, Beuth Verlag GmbH, Berlin, Nov. 2015.

[40] DIN EN 13000:2014-11. Cranes - Mobile Cranes; German Version EN 13000:2010+A1:2014. Norm, Beuth Verlag GmbH, Berlin, Nov. 2014.

[41] DIN EN 1991-1-4/NA. National Annex - Nationally determined parameters - Eurocode 1: Actions on structures - Part 1-4: General actions - Wind actions. Norm, Beuth Verlag GmbH, Berlin, Dec. 2010.

[42] DIN EN 1993-1-1:2010-12. Design of steel structures - Part 1-1: General rules and rules for buildings; German Version EN 1993-1-1:2005 + AC:2009. Norm, Beuth Verlag GmbH, Berlin, Dec. 2010.

[43] DIN EN 1993-1-5:2010-12. Design of steel structures - Part 1-5: Plated structural elements; German version EN 1993-1-5:2006 + AC:2009. Norm, Beuth Verlag GmbH, Berlin, Dec. 2010.

[44] DIN EN 1993-1-6:2010-12. Design of steel structures - Part 1-6: Strength and stability of shell structures; German version EN 1993-1-6:2007 + AC:2009. Norm, Beuth Verlag GmbH, Berlin, Dec. 2010.

[45] DIN EN 1993-1-8:2010-12. Design of steel structures - Part 1-8: Design of joints; German Version EN 1993-1-8:2005 + AC:2009. Norm, Beuth Verlag GmbH, Berlin, Dec. 2010.

[46] DIN EN 1993-1-8/NA:2010-12. National Annex - Nationally determined parameters - Eurocode 3: Design of steel structures - Part 1-8: Design of joints. Norm, Beuth Verlag GmbH, Berlin, Dec. 2010.

[47] DIN EN 1993-1-9:2010-12. Design of steel structures - Part 1-9: Fatigue; German version EN 1993-1-9:2005 + AC:2009. Norm, Beuth Verlag GmbH, Berlin, Dec. 2010.

[48] DIN EN 1993-3-1:2010-12. Design of steel structures - Part 3-1: Towers, masts and chimneys - Towers and masts; German version EN 1993-3-1:2006 + AC:2009. Norm, Beuth Verlag GmbH, Berlin, Dec. 2010.

[49] DIN EN 1993-3-1/NA:2015-11. National Annex - Nationally determined parameters - Eurocode 3: Design of steel structures - Part 3-1: Towers, masts and chimneys - Towers and masts. Norm, Beuth Verlag GmbH, Berlin, Nov. 2015.

[50] DIN EN 61400-1 VDE 0127-1:2011-08. Wind turbines - Part 1: Design requirements (IEC 61400-1:2005 + A1:2010); German version En 61400-1:2005 + A1:2010. Norm, Beuth Verlag GmbH, Berlin, Aug. 2011.

[51] DIN EN 61400-1:2015-09; VDE 0127-1:2015-09. Entwurf - Windenergieanlagen - Teil 1: Auslegungsanforderungen (IEC 88/521/CD:2014). Norm, Beuth Verlag GmbH, Berlin, Sept. 2015.

[52] DIN EN ISO 4063. Welding and allied processes - Nomenclature of processes and reference numbers (ISO 4063:2009, Corrected version 2010-03-01); Trilingual version EN ISO 4063:2010. Norm, Beuth Verlag GmbH, Berlin, Mar. 2011.

[53] DIN EN ISO 9692-1. Welding and allied processes - Types of joint preparation - Part 1: Manual metal-arc welding, gas-shielded metal-arc welding, gas welding, TIG welding and beam welding of steels (ISO 9692-1:2013); German version EN ISO 9692-1:2013. Norm, Beuth Verlag GmbH, Berlin, Dec. 2013.

[54] Dlubal Software GmbH. Programm RF-DYNAM Dynamische Analyse. Program Description, Tiefenbach, 2010.

[55] Dlubal Software GmbH. Zusatzmodul RF-STABIL Verzweigungslastfaktoren, Knicklängen und Eigenformen. Program Description, Tiefenbach, May 2014.

[56] Dlubal Software GmbH. Zusatzmodul RF-IMP Generierung geometrischer Ersatzimperfektionen und verformter Ersatzmodelle. Program Description, Tiefenbach, May 2015.

[57] Dlubal Software GmbH. Programm RFEM 5 Räumliche Tragwerke nach der Finite-Elemente Methode. Program Description, Tiefenbach, 2016.

[58] Dlubal Software GmbH. Programm DUENQ 8 Querschnittswerte und Nachweise dünnwandiger Querschnitte. Program Description, Tiefenbach, 2017.

[59] Dlubal Software GmbH. Zusatzmodul RF-STAHL Allgemeine Spannungsanalyse von Flächen und Stäben. Program Description, Tiefenbach, 2017.

[60] DNV GL Group. Standard DNVGL-OS-C101 Design of offshore steel structures, general - LRFD method. Guideline, DNV GL Group, July 2015.

[61] DNV GL Group. Standard DNVGL-ST-0376 Rotor blades for wind turbines. Guideline, DNV GL Group, Dec. 2015.

[62] DNV GL Group. Standard DNVGL-RP-C203 Fatigue design of offshore steel structures. Guideline, DNV GL Group, Apr. 2016.

[63] DNV GL Group. Standard DNVGL-ST-0126: Support structures for wind turbines. Guideline, DNV GL Group, Apr. 2016.

[64] DNV GL Group. Standard DNVGL-ST-0437 Loads and site conditions for wind turbines. Guideline, DNV GL Group, Nov. 2016.

[65] DNV GL Group. Standard DNVGL-RP-C204 Design against accidental loads. Guideline, DNV GL Group, Aug. 2017.

[66] DTU. Global Wind Atlas 2.0 a free, web-based application developed, owned, and operated by the technical university of denmark (DTU) in partnership with the world bank group, utilizing data provided by vortex, with funding provided by the energy sector management assistance program (esmap). `https://globalwindatlas.info`. [Online; accessed 05-July-2018].

[67] EN 1991-1-4:2005 + A1:2010 + AC:2010. Actions on structures - Part 1-4: General actions - Wind actions; German version EN 1991-1-4:2005 + A1:2010 + AC:2010. Norm, Beuth Verlag GmbH, Berlin, Dec. 2010.

[68] Engström, S., Lyrner, T., Hassanzadeh, M., Stalin, T., and J., J. Tall towers for large wind turbines Report from Vindforsk project V-342 Höga torn för vindkraftverk. Tech. Rep. 10:48, Stockholm, July 2010.

[69] Euler, L. Découverte d'un nouveau principe de mécanique. Mémoires de l'Académie des Sciences de Berlin 6, Berlin, 1750. pp. 185–217.

[70] Euler, L. Nova methodus motum corporum rigidorum determinandi. Novi Commentari Academiae Scientiarum Imperialis Petropolitanae 20, St. Petersburg, Russia, 1775. pp. 208–238.

[71] Faber, T. *Einführung in die Windenergietechnik Kapitel 7 - Turm und Gründung*, 2 ed. Fachbuchverlag Leipzig im Carl Hanser Verlag, München, 2016. ISBN 978-3-446-44982-4.

[72] Faber, T., and Struve, A. Innovative Structures for Wind Turbines. In *WindAc Africa* (Cape Town, South Africa, Oct. 2016).

[73] Fingersh, L., Hand, M., and Laxson, A. Wind turbine design cost and scaling model. Tech. Rep. NREL/TP-500-40566, National Renewable Energy Laboratory, Colorado, USA, Dec. 2006.

[74] Forschungsgesellschaft für Strassen- und Verkehrswesen, Ed. *Richtlinien für die Anlage von Landstraßen: RAL*. No. 201 in FGSV R1, Regelwerk. FGSV-Verl, Cologne, 2012. ISBN 978-3-86446-039-5.

[75] Fröde, E., and Fröde, F. *Windmühlen*. Du Mont Buchverlag, Cologne, 1981.

[76] Gardner, L., Chan, T. M., and Abela, J. M. Structural behaviour of elliptical hollow sections under combined compression and uniaxial bending. *Advanced Steel Construction 7*, 1 (2011), pp. 86 – 112.

[77] Gasch, R., Ed. *Wind power plants: fundamentals, design, construction and operation*, 2 ed. Springer, Berlin, 2012. ISBN 978-3-642-22937-4.

[78] GE. Txchnologist - A look inside the space frame tower, a new wind turbine prototype created by GE engineers. `http://txchnologist.com/post/131761453820/a-look-inside-the-space-frame-tower-a-new-wind`, 2015. [Online; accessed 14-February-2019].

[79] Germanischer Lloyd Industrial Services GmbH. Guideline for the certification of offshore wind turbines, Dec. 2005.

[80] Germanischer Lloyd Industrial Services GmbH. Guideline for the certification of wind turbines, July 2010.

[81] Gernandt, R. Windkraftanlage mit hintereinander angeordneten Rotoren auf horizontalen Wellen und drehendem Turm zur Kraftübertragung, June 2009. Patent No. DE102007057919A1.

[82] Gillebaart, T., Bernhammer, L., Zuijlen, A., and Kuik, G. Active flap control on an aeroelastic wind turbine airfoil in gust conditions using both a cfd and an engineering model. *Journal of Physics: Conference Series 524* (06 2014), 012060. DOI 10.1088/1742-6596/524/1/012060.

[83] Glienke, R., Kupzok, A., Ebert, A., and Wanner, M.-C. Anforderungen an die mechanische Fügetechnik für Türme von WEA in Stahlbauweise für große Nabenhöhen. *Stahlbau 84*, 8 (Aug. 2015), pp. 556–570. ISSN 00389145, DOI 10.1002/stab.201510296.

[84] Golding, E. W. *The generation of electricity by wind power*. E. & F. N. Spon Ltd., New York, 1977. ISBN 978-0-470-14986-7.

[85] Goldstein, L. Theoretical analysis of an airborne wind energy conversion system with a ground generator and fast motion transfer. *Energy 55* (June 2013), pp. 987–995. ISSN 03605442, DOI 10.1016/j.energy.2013.03.087.

[86] Gröger, K. *Passivhausprojektierung des Rinn-Drehhauses*. Bachelor Thesis, Hochschule Rhein Main, Rüsselsheim, Aug. 2010.

[87] Grote, K.-H., Feldhusen, J., and Dubbel, H., Eds. *Taschenbuch für den Maschinenbau*, 24 ed. Springer Vieweg, Berlin, 2014. ISBN 978-3-642-38890-3.

[88] Gumusel, B., Kavurmacioglu, L., and Camci, C. Aerodynamic drag characteristics and shape design of a radar antenna used for airport ground traffic control. Tech. rep., Turbomachinery Aero-Heat Transfer Laboratory, Pennsylvania State University, Hawaii, USA, Feb. 2006.

[89] Haibach, E. *Betriebsfestigkeit: Verfahren und Daten zur Bauteilberechnung*, 3 ed. VDI-Buch. Springer, Berlin, 2006. ISBN 978-3-540-29363-7.

[90] HAID, L., MATHA, D., STEWART, G., LACKNER, M., JONKMAN, J., AND ROBERTSON, A. Simulation-Length Requirements in the Loads Analysis of Offshore Floating Wind Turbines. 2013. DOI 10.1115/OMAE2013-11397.

[91] HAMME, U., AND HENKEL, J. Neue Konzepte im Leichtbau - Innovativer Teleskopausleger eines Mobilkrans. *Stahlbau 82*, 4 (Apr. 2013), pp. 246–249. ISSN 00389145, DOI 10.1002/stab.201310037.

[92] HAU, E. *Wind turbines: fundamentals, technologies, application, economics*, 3 ed. Springer, Berlin, 2013. ISBN 978-3-642-27151-9.

[93] HAU, E. *Windkraftanlagen - Grundlagen, Technik, Einsatz, Wirtschaftlichkeit*, 5 ed. Springer Berlin Heidelberg, 2014. ISBN 978-3-642-28876-0, DOI 10.1007/978-3-642-28877-7.

[94] HAYMAN, G. J. MLife Theory Manual for Version 1.00. Tech. rep., National Renewable Energy Laboratory, Colorado, USA, Oct. 2012.

[95] HIMMELMANN, R. Airfoil Design for Wakeless Wind Turbine Tower Structures, June 2013. Patent No. US20130156596A1.

[96] HIMMELMANN, R., AND TONGUE, S. E. Hybrid Wind Turbine Tower with Integrated Yaw Bearing System, July 2013. Patent No. US20130171002A1.

[97] HODGES, D. H. *Nonlinear composite beam theory*. No. 213 in Progress in astronautics and aeronautics. American Institute of Aeronautics and Astronautics, Reston, USA, 2006. ISBN 978-1-56347-697-6.

[98] HOERNER, S. F., AND BORST, H. V. *Fluid-dynamic lift: Practical information on aerodynamic and hydrodynamic lift*, 2 ed. Hoerner, Albuquerque/N.M, 1992. ISBN 978-99-988-3163-6.

[99] HOFMANN, K., BENDIX, H., AND PESCHKE, J. Windenergieanlage mit horizontaler Rotorwelle und mit drehbarem Turm, Nov. 2013. Patent No. DE102012009145A1.

[100] HUCHO, W.-H., AND AHMED, S. R., Eds. *Aerodynamics of road vehicles: from fluid mechanics to vehicle engineering*, 1 ed. Butterworths, London, 1987. ISBN 978-0-408-01422-9.

[101] IGNJATOVIĆ, D., PETROVIĆ, B., JOVANČIĆ, P., AND BOŠKOVIĆ, S. Impact of the Bucket Wheel Support at Technical Parameters of the Block and Bucket Wheel Excavator Capacity. In *Proceedings of the 12th International Symposium Continuous Surface Mining - Aachen 2014*, C. Niemann-Delius, Ed. Springer International Publishing, Cham, 2015, pp. 73–81. ISBN 978-3-319-12300-4.

[102] INTERNATIONAL ELECTROTECHNICAL COMISSION. IEC 61400-3:2009 - Wind Turbines - Part 3: Design Requirements for Offshore Wind Turbines. Norm, Geneva, CH, Nov. 2009.

[103] INTERNATIONAL ELECTROTECHNICAL COMISSION. IEC 61400-1:2005 Wind turbines - Part 1: Design requirements. Norm, Geneva, CH, Aug. 2011.

[104] JONES, E., OLIPHANT, T., PETERSON, P., ET AL. SciPy: Open source scientific tools for Python. `http://www.scipy.org/`, 2001–. [Online; accessed 14-February-2019].

[105] JONKMAN, B., AND JONKMAN, J. FAST v8.16.00a-bjj. Guide to changes in FAST v8, National Renewable Energy Laboratory, Colorado, USA, July 2016.

[106] JONKMAN, B. J., AND KILCHER, L. TurbSim User's Guide: Version 1.06.00. Tech. rep., National Renewable Energy Laboratory, Golden, USA, Sept. 2012.

[107] JONKMAN, J. Overview of the FAST Servo-Elastic Module - NREL Wind Turbine Modeling Workshop with Jason Jonkman, Ph.D., Mar. 2012.

[108] JONKMAN, J. Personal communication, 2016-2018.

[109] JONKMAN, J., BUTTERFIELD, S., MUSIAL, W., AND SCOTT, G. Definition of a 5-MW Reference Wind Turbine for Offshore System Development. Tech. Rep. NREL/TP-500-38060, National Renewable Energy Laboratory, Colorado, USA, 2009.

[110] JONKMAN, J. M. Dynamics Modeling and Loads Analysis of an Offshore Floating Wind Turbine. Tech. Rep. NREL/TP-500-41958, National Renewable Energy Laboratory, Colorado, USA, Nov. 2007.

[111] JONKMAN, J. M., HAYMAN, G. J., JONKMAN, B. J., DAMIANI, R. R., AND MURRAY, R. E. AeroDyn v15 User's Guide and Theory Manual. Tech. rep., National Renewable Energy Laboratory, Golden, USA, Apr. 2017.

[112] JONKMAN, J. M., AND MARSHALL, L. B. FAST Users Guide. Tech. Rep. NREL/EL-500-38230, National Renewable Energy Laboratory, Golden, USA, Aug. 2005.

[113] JONKMAN, J. M., ROBERTSON, A. N., AND HAYMAN, G. J. HydroDyn Users Guide and Theory Manual - Draft Version. Tech. rep., National Renewable Energy Laboratory, Golden, USA, 2016.

[114] JOVAŠEVIĆ, S., SHAH MOHAMMADI, M. R., REBELO, C., PAVLOVIĆ, M., AND VELJKOVIĆ, M. New Lattice-Tubular Tower for Onshore WEC âĂŞ Part 1: Structural Optimization. *Procedia Engineering 199* (2017), pp. 3236–3241. ISSN 18777058, DOI 10.1016/j.proeng.2017.09.336.

[115] KANE, T. R., AND LEVINSON, D. A. *Dynamics: theory and applications.* McGraw Hill series in mechanical engineering. McGraw Hill, New York, USA, 1985. ISBN 978-0-07-037846-9.

[116] KARKOS, T., AND WRITER, S. Wind-turbine sections squeeze through 1935 Rumford bridge. `http://www.sunjournal.com/wind-turbine-sections-squeeze-1935-rumford-bridge/`, Aug. 2010. [Online; accessed 05-June-2020].

[117] KEINDORF, C. *Tragverhalten und Ermüdungsfestigkeit von Sandwichtürmen für Windenergieanlagen.* No. 25 in Schriftenreihe des Instituts für Stahlbau der Gottfried Wilhelm Leibniz Universität Hannover. Shaker, Aachen, 2010. ISBN 978-3-8322-8791-7.

[118] KEYSTONE TOWER SYSTEMS. Technology - Keystone Tower Systems. `http://keystonetowersystems.com/technology-2/`. [Online; accessed 14-February-2019].

[119] KIM, M.-G., AND DALHOFF, P. H. Yaw Systems for wind turbines: Overview of concepts, current challenges and design methods. *Journal of Physics: Conference Series 524* (June 2014). ISSN 1742-6596, DOI 10.1088/1742-6596/524/1/012086.

[120] Kleinhenz, F. Denkschrift 6 - Das Großwindkraftwerk MAN-Kleinhenz (Erweiterter Abdruck aus "Der Bauingenieur" 1942, Heft 23/24, Kleinhenz: "Projekt eines Großwindkraftwerkes".). Mar. 1943.

[121] Kloos, K.-H., Thomala, W., and Wiegand, H. *Schraubenverbindungen: Grundlagen, Berechnung, Eigenschaften, Handhabung*, 5 ed. Springer, Berlin, 2007. ISBN 978-3-540-21282-9.

[122] Knoedel, P. On the dynamics of steel structures with x-type bracing. *Stahlbau 80*, 8 (2011), 566–571. DOI 10.1002/stab.201101457.

[123] Kooijman, H. J. T., and et al. DOWEC 6 MW Pre-Design aero-elastic modelling of the DOWEC 6 MW pre-design in PHATAS. Report DOWEC-F1W2-HJK-01-046/9, Dutch Ministry of Economic Affairs, Sept. 2003.

[124] Koppány, S. Loss of prestress in spiral strand guy ropes in masts. *ce/papers 3*, 3-4 (2019), 785–790. DOI 10.1002/cepa.1134.

[125] Krausen, K., and Schuh, B. Antennentechnik in Stahl. *Stahlbau 66*, 3 (1997), pp. 110 – 119. ISSN 0038-9145.

[126] Kreiner, S. überschlaegige Dimensionierung eines Azimutlagers am Turmfuß von Liebherr Components Biberach GmbH. Tech. rep., Liebherr Components Biberach GmbH, Biberach an der Riss, 2012.

[127] Kuchling, H. *Taschenbuch der Physik: mit Tabellen*, 20 ed. Fachbuchverl. Leipzig im Carl Hanser Verl, München, 2011. ISBN 978-3-446-42457-9.

[128] Kutta, W. *Beitrag zur näherungsweisen Integration totaler Differentialgleichungen.* Dissertation, München, 1901.

[129] Liebherr-Werk Ehingen GmbH. Mobile crane LTM 1100-4.2 - product advantage. Product Description, 2009.

[130] Liebherr-Werk Ehingen GmbH. Mobile crane LTM 1100-4.2 - technical data. Product Description, 2012.

[131] Liebherr-Werk Ehingen GmbH. Lattice boom mobile crane LG 1750 - technical data. Product Description, 2015.

[132] Liu, K.-S. Windenergieanlage mit drehbarem Turm, Oct. 2009. Patent No. DE202009009517U1.

[133] Loth, E., and Selig, M. S. 2-D fairing for a wind turbine tower, Dec. 2014. Patent No. WO2014205348A1.

[134] Lutz, P. Windkraftanlage, Aug. 2014. Patent No. WO2014122165A1.

[135] Madsen, H. A., Johansen, J., Sørensen, N., Larsen, G., and Hansen, M. Simulation of Low Frequency Noise from a Downwind Wind Turbine Rotor. American Institute of Aeronautics and Astronautics. ISBN 978-1-62410-012-3,DOI 10.2514/6.2007-623.

[136] Malcolm, D. J., and Hansen, A. C. WindPACT Turbine Rotor Design Study. Tech. Rep. NREL/SR-500-32495, National Renewable Energy Laboratory, Golden, USA, Apr. 2006.

[137] Manwell, J. F., McGowan, J. G., and Rogers, A. L. *Wind energy explained: theory, design and application*, 2 ed. Wiley, Chichester, 2011. ISBN 978-0-470-01500-1.

[138] Matsuishi, M., and Endo, T. Fatigue of metals subjected to varying stress. Tech. rep., Fukuoka, Japan, Mar. 1968.

[139] Meslkouris, K., and Hake, E. *Statik der Stabtragwerke*, 2 ed. Springer-Lehrbuch. Springer Berlin Heidelberg, Berlin, Heidelberg, 2009. ISBN 978-3-540-88992-2,DOI 10.1007/978-3-540-88993-9.

[140] Migita, Y., and Fukumoto, Y. Local buckling behaviour of polygonal sections. *Journal of Constructional Steel Research 41*, 2-3 (Feb. 1997), pp. 221–233. ISSN 0143974X, DOI 10.1016/S0143-974X(97)00008-4.

[141] Miner, M. A. Cumulative damage in fatigue, 1945. Trans. Amer. Soc. Mech. Eng. 67, A159-164.

[142] Moriarty, P. J., and Hansen, A. C. AeroDyn theory manual. Tech. Rep. NREL/TP-500-36881, National Renewable Energy Laboratory, Colorado, USA, Jan. 2005.

[143] Moulton, F. R. Discussions: New Methods in Exterior Ballistics. *The American Mathematical Monthly 35*, 5 (May 1928), p. 246. ISSN 00029890, DOI 10.2307/2299587.

[144] Nies, J. J. Kopfteil eines vorgeneigten Turms, July 2009. Patent No. EP2075464A2.

[145] NRMCA. Concrete CO2 fact sheet. Publication, National Ready Mixed Concrete Association, Maryland, USA, 2012. No. 2PCO2.

[146] Patankar, S. V. *Numerical heat transfer and fluid flow*. Series in computational methods in mechanics and thermal sciences. Hemisphere Pub. Corp.; New York : McGraw-Hill, Washington, 1980. ISBN 978-0-07-048740-6.

[147] PCAE. Documentation of static software 4h-EC3fs for end plate butts. software documentation, Hannover, Apr. 2017.

[148] Penman, J., and et al. 2006 IPCC Guidelines for National Greenhouse Gas Inventories. Tech. Rep. volume 3 industrial processes and product use, chapter 4 Metal Industry Emissions, World Meteorological Organization, United Nations Environment Programme, 2006.

[149] Petersen, C. *Stahlbau: Grundlagen der Berechnung und baulichen Ausbildung von Stahlbauten*, 4 ed. Springer Vieweg, Wiesbaden, 2013. ISBN 978-3-8348-8610-1.

[150] Powles, S. R. J. The effects of tower shadow on the dynamics of a horizontal-axis wind turbine. *Wind Engineering 7*, 1 (1983), pp. 26–42.

[151] Prandtl, L., and Tietjens, O. J. Applied Hydro- and Aeromechanics. Dover Publications, Inc., New York, USA, 1934.

[152] Projektgruppe Windenergieanlagen des Deutschen Instituts für Bautechnik. Richtlinie für Windenergieanlagen Einwirkungen und Standsicherheitsnachweise für Turm und Gründung, 2012.

[153] Przemieniecki, J. S. *Theory of Matrix Structural Analysis*. Dover, New York, 1985. ISBN 978-0-486-64948-1.

[154] Radaj, D., and Vormwald, M. *Ermüdungsfestigkeit: Grundlagen für Ingenieure*, 3 ed. Springer, Berlin, 2007. ISBN 978-3-540-71458-3.

[155] Ransford, T. *Potential theory in the complex plane*. No. 28 in London Mathematical Society student texts. Cambridge Univ. Press, Cambridge, 1995. ISBN 978-0-521-46654-7.

[156] Rayleigh, W. S. B. J. The theory of sound. 1st ed., vol I and vol II, Macmillan, London, United Kingdom, 1887, 1888.

[157] Reddy, J. N., and Gartling, D. K. *The finite element method in heat transfer and fluid dynamics*. CRC Press, Boca Raton, Fla., 1994. ISBN 978-0-8493-9410-2.

[158] Reichsarbeitsgemeinschaft Windkraft. Denkschrift 1 über die Arbeiten der Reichsarbeitsgemeinschaft "Windkraft" im Geschäftsjahr 1939-1940. Sept. 1940.

[159] Reinke, T. *Tragverhalten von biegebeanspruchten Stahlmasten mit polygonalen Querschnitten*. Dissertation, Karlsruher Institut für Technologie, Karlsruhe, 2015.

[160] Reiso, M. *The tower shadow effect in downwind wind turbines*. Dissertation, Norwegian University of Science and Technology, Trondheim, May 2013.

[161] Richard, H. A., and Sander, M. *Technische Mechanik - Festigkeitslehre*, 2 ed. Studium. Vieweg + Teubner, Wiesbaden, 2008. ISBN 978-3-8348-0454-9.

[162] Rinn, H. W. Gebäude mit verbesserter Energieausnutzung, Nov. 1986. Patent No. EP0091626B1.

[163] Roshko, A. Experiments on the flow past a circular cylinder at very high Reynolds number. Tech. rep., Guggenheim Aeronautical Laboratory, California Institute of Technology, California, USA, Nov. 1960.

[164] Rudolf, T., R. *Taller Towers for Larger Wind Turbines: a Market Study of Support Structure Technology for Onshore Wind Turbines*. Master Thesis supervised by Prof. Dr.-Ing. Torsten Faber, Wind Energy Technology Insitute, FH Flensburg, Flensburg, Nov. 2013.

[165] Runge, C. über die numerische Auflösung von Differentialgleichungen. *Mathematische Annalen 46* (1895), pp. 167–178.

[166] SCD Technology. Data sheet SCD nezzy 8.0 MW, Jan. 2015.

[167] Schlichting, H., and Gersten, K. *Boundary-layer theory*, 8 ed. Springer, Berlin ; New York, 2000. ISBN 978-3-540-66270-9.

[168] Schulte, H., and Bruger, P. Drehbare Kurzwellen-Vorhangantenne auf der Funksendestelle Marlow. *Stahlbau 69*, 10 (Oct. 2000), pp. 781–787. ISSN 00389145, DOI 10.1002/stab.200002870.

[169] Scrouton, C., and Walshe, D. E. A means for avoiding wind-excited oscillations of structures with circular or nearly circular cross-section. *aero report: National Phsysical Laboratory (Great Britain) 335* (1957).

[170] Scruton, C. Wind effects on structures. *James Clayton Lecture Proc. Inst. of Mech. Eng*, 185/23 (1971), pp. 301–317.

[171] Shkara, Y., Werkmeister, A., Fontecha, R., Jacobs, G., and Schelenz, R. An experimental and computational investigation of wind turbine blade-tower aerodynamic interaction. In *Proceedings of 13th biennial German Wind Energy Conference DEWEK 2017* (Bremen, Oct. 2017).

[172] Skrzypinski, W. R. *Analysis and modeling of unsteady aerodynamics with application to wind turbine blade vibration at standstill conditions*. Dissertation, DTU Wind Energy, Roskilde, DK.

[173] Smith, C. On some recent applications of the coanda effect to acoustics. *The Journal of the Acoustical Society of America 128* (10 2010), 2286. DOI 10.1121/1.3508016.

[174] Smith, E., and Slocum, A. Tapered spiral welded structure, May 2014. Patent No. US8720153B2.

[175] Sockel, H. *Aerodynamik der Bauwerke*. Vieweg, Braunschweig, 1984. ISBN 978-3-528-08845-3.

[176] Song, H., Damiani, R., Robertson, A., and Jonkman, J. New Structural-Dynamics Module for Offshore Multimember Substructures within the Wind Turbine Computer-Aided Engineering Tool FAST. *Proceedings of the International Offshore and Polar Engineering Conference* (01 2013), pp. 383–391.

[177] Steel Pro Maschinenbau GmbH. ASPARAGO®Neue Türme für die Windenergie. Information Brochure, Berlin, 2016.

[178] Strømmen, E. N. *Structural Dynamics*, vol. 2 of *Springer Series in Solid and Structural Mechanics*. Springer International Publishing, Cham, 2014. ISBN 978-3-319-01801-0.

[179] Struve, A., Faber, T., Damiani, R., Jonkman, J., and Ummenhofer, T. Load-Direction-Derived Support Structures for Wind Turbines: A Lattice Tower Concept and Preparations for Future Certifications. In *Proceedings of 13th biennial German Wind Energy Conference DEWEK 2017* (Bremen, Oct. 2017).

[180] Svensson, H. *Design of foundations for wind turbines*. Master Thesis, Lund University, Lund, S, 2010.

[181] Taesong, K., Hansen, A. M., and Branner, K. Development of an anisotropic beam finite element for composite wind turbine blades in multibody system. *Renewable Energy 59* (Nov. 2013), pp. 172–183. ISSN 09601481, DOI 10.1016/j.renene.2013.03.033.

[182] Teal, D. L. E. Wind turbine tower having rotatable shroud/fairing, Mar. 2001. Patent No. GB2353825A.

[183] Team Optimus150. Presentation of WETI reference wind turbine Optimus150, Feb. 2016.

[184] Theodorsen, T. General theory of aerodynamic instability and the mechanism of flutter. Tech. Rep. 496, National Advisory Committee for Aeronautics, USA, 1934.

[185] Thompson, A. Wind turbine appartus, Apr. 2012. Patent No. EP2436924A1.

[186] Timoshenko, S. P., and Gere, J. M. *Theory of elastic stability*, 2 ed. McGraw-Hill, New York, 1961. ISBN 0-07-Y85821-7.

[187] Ummenhofer, T., and Weich, I. Hochfrequente Haemmerverfahren - Steigerung der Ermuedungsfestigkeit neuer und bestehender geschweisster Stahlkonstruktionen. *Bauingenieur 2007*, 82 (2007), pp. 237–244.

[188] Ummenhofer, T., Weich, I., Bergers, J., and Herion, S. Reduktion der Wanddicken bei Tragtürmen von Windenergieanlagen durch den Einsatz höherfester Stähle. *Stahlbau 76*, 9 (Sept. 2007), pp. 643–649. ISSN 00389145, 14371049, DOI 10.1002/stab.200710068.

[189] van der Tempel, J. *Design of Support Structures for Offshore Wind Turbines*. Dissertation, Delft University of Technology, Delft, Apr. 2006.

[190] Vestas. V80-2.0 MW Versatile megawattage brochure. `http://pdf.directindustry.com/pdf/vestas/v80-20-mw-brochure/20680-53605.html`, 2003. [Online; accessed 14-February-2019].

[191] Wallasch, A.-K., Lüers, S., and Rehfeldt, K. Wirtschaftlichkeit unterschiedlicher Nabenhöhen von Windenergieanlagen. Studie im Auftrag des Bundesministeriums für Wirtschaft und Energie, Deutsche WindGuard, Varel, Germany, June 2017. No. SP17012A1.

[192] Wang, Q., Jonkman, J., Sprague, M., and Jonkman, B. BeamDyn user's guide and theory manual. Tech. rep., Golden, USA, Mar. 216.

[193] Whitbread, R. E. Practical solutions to some windinduced vibration problems. *Nat. Phys. Lab. NPL Rep. Sci. R 124* (1975).

[194] White, F. M. *Fluid mechanics*, 5 ed. McGraw-Hill series in mechanical engineering. McGraw-Hill, Boston, 2003. ISBN 978-0-07-283180-1.

[195] Wiedemann, J. *Leichtbau 1: Elemente*, 2 ed. No. 1 in Leichtbau. Springer, Berlin, 1996. ISBN 3-540-60746-3.

[196] Wiedemann, J. *Leichtbau: Elemente und Konstruktion*, 3 ed. Klassiker der Technik. Springer, Berlin, 2007. ISBN 978-3-540-33656-3.

[197] Woodhouse, P. R., Jacobson, D., and Petersen, B. *The Everett and Monte Cristo Railway*. Oso Pub, Arlington, USA, 2000. ISBN 978-0-9647521-8-4.

[198] Zahm, A. F., Smith, R. H., and Louden, F. A. Forces on elliptic cylinders in uniform air stream. Tech. Rep. Report No. 289, National Advisory Committee for Aeronautics, Washington D.C., USA, 1928.

14 Appendix

14.1 Conversation with Clipper Windpower

Mail conversation with Clipper Windpower from the 12. June 2015

During late 2012 and through 2013 significant changes were made concerning the strategic direction of our company. Regretfully we as a company are no longer providing wind farm development services, advanced engineering design studies, engineering support activities, wind farm operation services, monitoring services, and have halted all sales and production of the company's Liberty 2.5MW wind turbine. In our go forward position Clipper remains as a much smaller entity performing as a service provider, concentrating our efforts on the existing Clipper Liberty customers and offering plant based refurbishment services for the turbine gearbox and other electro/mechanical components. Since we have concentrated our efforts in dealing with the previous Clipper customer base in a manufacturing service nature, the Sales, Marketing, Engineering, O & M Services and Wind Farm Development Departments within Clipper have been totally eliminated. Due to the circumstances described above we do not have interest or input to your query, nor is it something that we anticipate having involvement with in the near future.

Thank you,
Bob Loyd PE
President and CEO
Clipper Windpower, LLC
4601 Bowling St SW
Cedar Rapids, Iowa 52404
Tel: +1 (319) 861-7411
Fax: +1 (319) 364 2960
Email: rloyd@clipperwind.com<mailto:rloyd@clipperwind.com>
Web: http://www.clipperwind.com<http://www.clipperwind.com/>
P Think Green! Please consider the environment before printing this email

Betreff: FW: Email inquiry from Clipper website!
Von: "Robert Loyd" <RLoyd@clipperwind.com>
Datum: Fr, 12.06.2015, 20:25
An: "achim.struve@fh-flensburg.de" <achim.struve@fh-flensburg.de>
Priorität: Normal
From: Achim Struve [mailto:noreply@jotform.com]
Sent: Friday, June 12, 2015 8:40 AM
To: Info Subject: Email inquiry from Clipper website!
Email inquiry from Clipper website!
Sender Info
Your Name
Achim Struve
Your E-mail Address
achim.struve@fh-flensburg.de<mailto:achim.struve@fh-flensburg.de>
Your Message

Ladies and Gentlemen,
I am a postgraduate from the Wind Energy Technology Institute of the Flensburg University of Applied Sciences in Northern Germany. I want to do a doctorate about the theme of rotating wind energy plant towers with an aerodynamic optimized cross section shape of the tower. I found your patents with the number US20130171002A1 and US20130156596A1. These invention is exactly the theme of my doctorate and I am very interested to hear more about your research status in this. Is this an current topic of your company? I look forward to hearing from you soon.
Yours faithfully

M.Eng. Achim Struve

14.2 Considered DLCs According to Different References

The following table summarizes the considered DLCs of the ASE load calculations regarding its references.

Reference	Respected DLC list
NREL WindPACT [136]	normal turbulence model, extreme coherent gust with wind direction change, extreme coherent gust, extreme direction change, extreme operating gust, extreme vertical wind shear, extreme horizontal wind shear and emergency stops with wind speeds $\left(\frac{\mathrm{m}}{\mathrm{s}}\right)$: 8 (only for normal turbulence model), 12, 16, 20 and 24 and the extreme wind model
NREL 5 MW [109]	1.1, 1.3, 1.4, 1.5, 1.6 (own DLC), 2.1, 2.3, 6.1, 6.2, 6.3, 7.1
DTU 10 MW [5]	1.1, 1.2, 1.3, 1.5, 2.1, 2.3, 2.4, 4.1, 4.2, 5.1, 6.1, 6.2, 6.3, 6.4, 7.1

14.3 Statistical Outliers Check in Validation Procedure for RootMyc1

These are the timeseries of load sensors where the extreme out-of-plane blade root bending moment of all DLCs 1.x in an own simulation occurred. The underlying WT and support structure was the 5 MW baseline NREL reference WT with conventional circular steel tower. In the timeseries, the generator power (GenPwr) in kW, the pitch angle of blade 1 (BldPitch1) in deg, the rotor thrust (RotThrust) in kN, the rotor speed (RotSpeed) in rpm, the wind speed in global x-direction (Wind1Velx) in m/s, and the out-of-plane blade root bending moment (RootMyc1) in kNm are shown. The first 30 s were cut away because of possible numerical initialization influences. Note that the values of these load sensors are raw without post processing, such as applied partial safety factors.

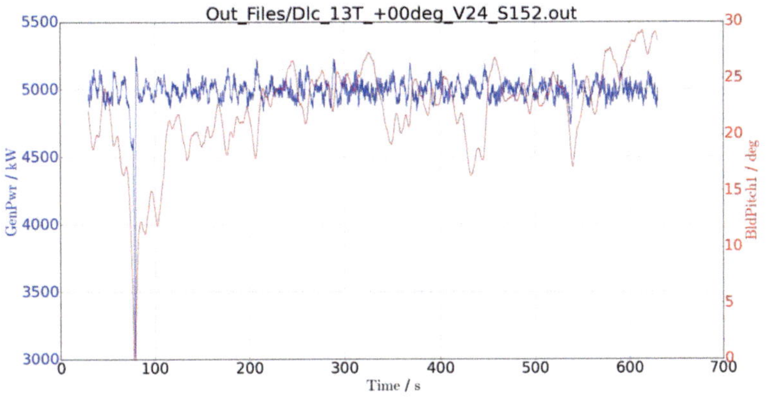
Out_Files/Dlc_13T_+00deg_V24_S152.out
GenPwr / kW
BldPitch1 / deg
Time / s
5500
5000
4500
4000
3500
3000
30
25
20
15
10
5
0
0
100
200
300
400
500
600
700

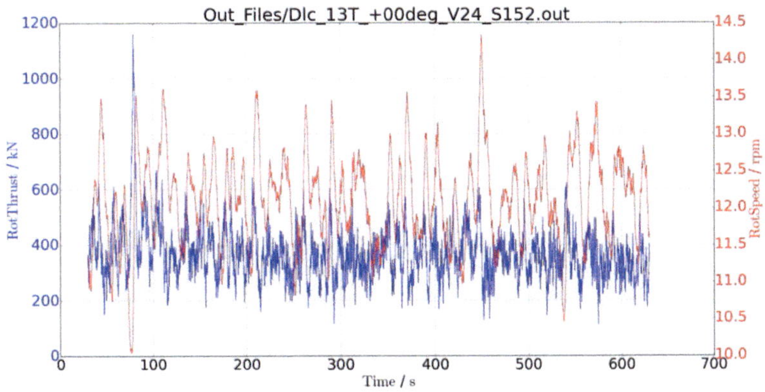
Out_Files/Dlc_13T_+00deg_V24_S152.out
RotThrust / kN
RotSpeed / rpm
Time / s
1200
1000
800
600
400
200
0
14.5
14.0
13.5
13.0
12.5
12.0
11.5
11.0
10.5
10.0
0
100
200
300
400
500
600
700

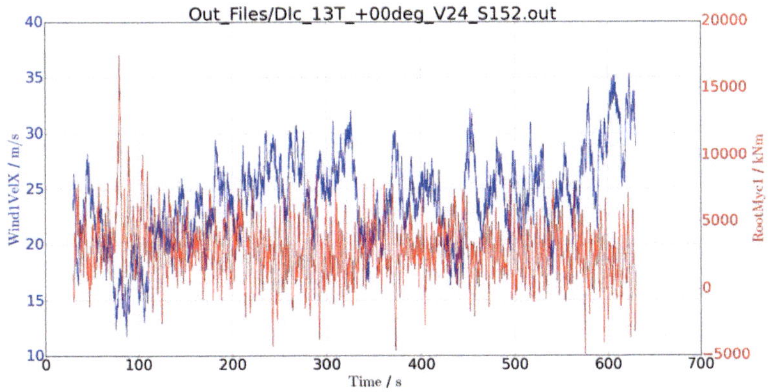
Out_Files/Dlc_13T_+00deg_V24_S152.out
Wind1VelX / m/s
RootMyc1 / kNm
Time / s
40
35
30
25
20
15
10
20000
15000
10000
5000
0
−5000
0
100
200
300
400
500
600
700

14.4 Summary of Required Simulation Time Periods in Comparison Study

Summary of the required simulation time periods and its corresponding random seeds to be used for each wind speed in turbulent wind fields. Only DLCs, considered in the comparison study of Jonkman [110] are listed.

DLC	Wind condition	Seeds/Wind Speed	Sim. Time
1.1	Turbulent	6	10 min
1.3	Turbulent	6	10 min
1.4	Deterministic	-	1 min
1.5	Deterministic	-	1 min
2.1	Turbulent	12	10 min
2.3	Deterministic	-	1 min
6.1	Turbulent	6	60 min
6.2	Turbulent	6	60 min
6.3	Turbulent	6	60 min
7.1	Turbulent	6	60 min

14.5 ASE Validation - Relevant FAST Input Files

The following pages show the relevant FAST environmental inputs, used for the validation of the ASE load simulations. They consist out of the following module input files: FAST, ElastoDyn, ElastoDyn Tower, AeroDyn, and ServoDyn. The detailed blade aerodynamic input files and the controller source code are not listed, but can be looked up in the Appendix B of the comparison study [110].

Furthermore, the InflowWind file is not listed because it has only the purpose to load the correct turbulent wind field file in this case. Note that there are three different structural blade input files to mimic the rotor mass unbalance, but only one of them is listed here.

FAST Input

```
------- FAST v8.16.* INPUT FILE ------------------------------------------------
ASE Load Validation Procedure
---------------------- SIMULATION CONTROL --------------------------------------
False         Echo            - Echo input data to <RootName>.ech (flag)
"FATAL"       AbortLevel      - Error level when simulation should abort (string) {"WARNING", "SEVERE", "FATAL"}
        630   TMax            - Total run time (s)
     0.0125   DT              - Recommended module time step (s)
          2   InterpOrder     - Interpolation order for input/output time history (-) {1=linear, 2=quadratic}
          0   NumCrctn        - Number of correction iterations (-) {0=explicit calculation, i.e., no corrections}
      99999   DT_UJac         - Time between calls to get Jacobians (s)
      1E+06   UJacSclFact     - Scaling factor used in Jacobians (-)
---------------------- FEATURE SWITCHES AND FLAGS ------------------------------
          1   CompElast       - Compute structural dynamics (switch) {1=ElastoDyn; 2=ElastoDyn + BeamDyn for blades}
          1   CompInflow      - Compute inflow wind velocities (switch) {0=still air; 1=InflowWind; 2=external from OpenFOAM}
          1   CompAero        - Compute aerodynamic loads (switch) {0=None; 1=AeroDyn v14; 2=AeroDyn v15}
          1   CompServo       - Compute control and electrical-drive dynamics (switch) {0=None; 1=ServoDyn}
          0   CompHydro       - Compute hydrodynamic loads (switch) {0=None; 1=HydroDyn}
          0   CompSub         - Compute sub-structural dynamics (switch) {0=None; 1=SubDyn}
          0   CompMooring     - Compute mooring system (switch) {0=None; 1=MAP++; 2=FEAMooring; 3=MoorDyn; 4=OrcaFlex}
          0   CompIce         - Compute ice loads (switch) {0=None; 1=IceFloe; 2=IceDyn}
---------------------- INPUT FILES ---------------------------------------------
"ElastoDyn/NREL_5MW_ElastoDyn_Dlc_13T_+00deg_V20_S170.dat"    EDFile          - Name of file containing ElastoDyn input parameters (quoted string)
"NREL_5MW_BeamDyn.dat"    BDBldFile(1)    - Name of file containing BeamDyn input parameters for blade 1 (quoted string)
"NREL_5MW_BeamDyn.dat"    BDBldFile(2)    - Name of file containing BeamDyn input parameters for blade 2 (quoted string)
"NREL_5MW_BeamDyn.dat"    BDBldFile(3)    - Name of file containing BeamDyn input parameters for blade 3 (quoted string)
"InflowWind/NREL_5MW_InflowWind_Dlc_13T_+00deg_V20_S170.dat"    InflowFile      - Name of file containing inflow wind input parameters (quoted string)
"NREL_5MW_AeroDyn.dat"    AeroFile        - Name of file containing aerodynamic input parameters (quoted string)
"ServoDyn/NREL_5MW_ServoDyn_Dlc_13T_+00deg_V20_S170.dat"    ServoFile       - Name of file containing control and electrical-drive input parameters (quoted string)
"unused"      HydroFile       - Name of file containing hydrodynamic input parameters (quoted string)
"unused"      SubFile         - Name of file containing sub-structural input parameters (quoted string)
"unused"      MooringFile     - Name of file containing mooring system input parameters (quoted string)
"unused"      IceFile         - Name of file containing ice input parameters (quoted string)
---------------------- OUTPUT --------------------------------------------------
False         SumPrint        - Print summary data to "<RootName>.sum" (flag)
         10   SttsTime        - Amount of time between screen status messages (s)
      99999   ChkptTime       - Amount of time between creating checkpoint files for potential restart (s)
       0.05   DT_Out          - Time step for tabular output (s) (or "default")
         30   TStart          - Time to begin tabular output (s)
          1   OutFileFmt      - Format for tabular (time-marching) output file (switch) {1: text file [<RootName>.out], 2: binary file [<RootName>.outb], 3: both}
True          TabDelim        - Use tab delimiters in text tabular output file? (flag) {uses spaces if false}
"ES10.3E2"    OutFmt          - Format used for text tabular output, excluding the time channel.  Resulting field should be 10 characters. (quoted string)
---------------------- LINEARIZATION -------------------------------------------
False         Linearize       - Linearization analysis (flag)
          2   NLinTimes       - Number of times to linearize (-) [>=1] [unused if Linearize=False]
         30,         60    LinTimes        - List of times at which to linearize (s) [1 to NLinTimes] [unused if Linearize=False]
          1   LinInputs       - Inputs included in linearization (switch) [0=none; 1=standard; 2=all module inputs (debug)] [unused if Linearize=False]
          1   LinOutputs      - Outputs included in linearization (switch) [0=none; 1=from OutList(s); 2=all module outputs (debug)] [unused if Linearize=False]
False         LinOutJac       - Include full Jacobians in linearization output (for debug) (flag) [unused if Linearize=False; used only if LinInputs=LinOutputs=2]
False         LinOutMod       - Write module-level linearization output files in addition to output for full system? (flag) [unused if Linearize=False]
---------------------- VISUALIZATION ------------------------------------------
          0   WrVTK           - VTK visualization data output: (switch) {0=none; 1=initialization data only; 2=animation}
          1   VTK_type        - Type of VTK visualization data: (switch) {1=surfaces; 2=basic meshes (lines/points); 3=all meshes (debug)} [unused if WrVTK=0]
true          VTK_fields      - Write mesh fields to VTK data files? (flag) {true/false} [unused if WrVTK=0]
         15   VTK_fps         - Frame rate for VTK output (frames per second){will use closest integer multiple of DT} [used only if WrVTK=2]
```

ElastoDyn Input I

```
------- ELASTODYN v1.03.* INPUT FILE -------------------------------------------
NREL 5.0 MW Baseline Land-Based Wind Turbine
---------------------- SIMULATION CONTROL --------------------------------------
False         Echo        - Echo input data to "<RootName>.ech" (flag)
          3   Method      - Integration method: {1: RK4, 2: AB4, or 3: ABM4} (-)
"DEFAULT"     DT          - Integration time step (s)
---------------------- ENVIRONMENTAL CONDITION ---------------------------------
    9.80665   Gravity     - Gravitational acceleration (m/s^2)
---------------------- DEGREES OF FREEDOM --------------------------------------
True          FlapDOF1    - First flapwise blade mode DOF (flag)
True          FlapDOF2    - Second flapwise blade mode DOF (flag)
True          EdgeDOF     - First edgewise blade mode DOF (flag)
False         TeetDOF     - Rotor-teeter DOF (flag) [unused for 3 blades]
True          DrTrDOF     - Drivetrain rotational-flexibility DOF (flag)
True          GenDOF      - Generator DOF (flag)
True          YawDOF      - Yaw DOF (flag)
True          TwFADOF1    - First fore-aft tower bending-mode DOF (flag)
True          TwFADOF2    - Second fore-aft tower bending-mode DOF (flag)
True          TwSSDOF1    - First side-to-side tower bending-mode DOF (flag)
True          TwSSDOF2    - Second side-to-side tower bending-mode DOF (flag)
False         PtfmSgDOF   - Platform horizontal surge translation DOF (flag)
False         PtfmSwDOF   - Platform horizontal sway translation DOF (flag)
False         PtfmHvDOF   - Platform vertical heave translation DOF (flag)
False         PtfmRDOF    - Platform roll tilt rotation DOF (flag)
False         PtfmPDOF    - Platform pitch tilt rotation DOF (flag)
False         PtfmYDOF    - Platform yaw rotation DOF (flag)
---------------------- INITIAL CONDITIONS --------------------------------------
          0   OoPDefl     - Initial out-of-plane blade-tip displacement (meters)
          0   IPDefl      - Initial in-plane blade-tip deflection (meters)
          0   BlPitch(1)  - Blade 1 initial pitch (degrees)
          0   BlPitch(2)  - Blade 2 initial pitch (degrees)
          0   BlPitch(3)  - Blade 3 initial pitch (degrees) [unused for 2 blades]
          0   TeetDefl    - Initial or fixed teeter angle (degrees) [unused for 3 blades]
          0   Azimuth     - Initial azimuth angle for blade 1 (degrees)
       12.1   RotSpeed    - Initial or fixed rotor speed (rpm)
          0   NacYaw      - Initial or fixed nacelle-yaw angle (degrees)
          0   TTDspFA     - Initial fore-aft tower-top displacement (meters)
          0   TTDspSS     - Initial side-to-side tower-top displacement (meters)
          0   PtfmSurge   - Initial or fixed horizontal surge translational displacement of platform (meters)
          0   PtfmSway    - Initial or fixed horizontal sway translational displacement of platform (meters)
          0   PtfmHeave   - Initial or fixed vertical heave translational displacement of platform (meters)
          0   PtfmRoll    - Initial or fixed roll tilt rotational displacement of platform (degrees)
          0   PtfmPitch   - Initial or fixed pitch tilt rotational displacement of platform (degrees)
          0   PtfmYaw     - Initial or fixed yaw rotational displacement of platform (degrees)
---------------------- TURBINE CONFIGURATION -----------------------------------
          3   NumBl       - Number of blades (-)
         63   TipRad      - The distance from the rotor apex to the blade tip (meters)
        1.5   HubRad      - The distance from the rotor apex to the blade root (meters)
       -2.5   PreCone(1)  - Blade 1 cone angle (degrees)
       -2.5   PreCone(2)  - Blade 2 cone angle (degrees)
       -2.5   PreCone(3)  - Blade 3 cone angle (degrees) [unused for 2 blades]
          0   HubCM       - Distance from rotor apex to hub mass [positive downwind] (meters)
          0   UndSling    - Undersling length [distance from teeter pin to the rotor apex] (meters) [unused for 3 blades]
          0   Delta3      - Delta-3 angle for teetering rotors (degrees) [unused for 3 blades]
          0   AzimB1Up    - Azimuth value to use for I/O when blade 1 points up (degrees)
    -5.0191   OverHang    - Distance from yaw axis to rotor apex [3 blades] or teeter pin [2 blades] (meters)
      1.912   ShftGagL    - Distance from rotor apex [3 blades] or teeter pin [2 blades] to shaft strain gages [positive for upwind rotors] (meters)
         -5   ShftTilt    - Rotor shaft tilt angle (degrees)
        1.9   NacCMxn     - Downwind distance from the tower-top to the nacelle CM (meters)
          0   NacCMyn     - Lateral  distance from the tower-top to the nacelle CM (meters)
       1.75   NacCMzn     - Vertical distance from the tower-top to the nacelle CM (meters)
  -3.09528    NcIMUxn     - Downwind distance from the tower-top to the nacelle IMU (meters)
          0   NcIMUyn     - Lateral  distance from the tower-top to the nacelle IMU (meters)
   2.23336    NcIMUzn     - Vertical distance from the tower-top to the nacelle IMU (meters)
   1.96256    Twr2Shft    - Vertical distance from the tower-top to the rotor shaft (meters)
       87.6   TowerHt     - Height of tower above ground level [onshore] or MSL [offshore] (meters)
          0   TowerBsHt   - Height of tower base above ground level [onshore] or MSL [offshore] (meters)
          0   PtfmCMxt    - Downwind distance from the ground level [onshore] or MSL [offshore] to the platform CM (meters)
          0   PtfmCMyt    - Lateral distance from the ground level [onshore] or MSL [offshore] to the platform CM (meters)
          0   PtfmCMzt    - Vertical distance from the ground level [onshore] or MSL [offshore] to the platform CM (meters)
          0   PtfmRefzt   - Vertical distance from the ground level [onshore] or MSL [offshore] to the platform reference point (meters)
---------------------- MASS AND INERTIA ----------------------------------------
          0   TipMass(1)  - Tip-brake mass, blade 1 (kg)
          0   TipMass(2)  - Tip-brake mass, blade 2 (kg)
          0   TipMass(3)  - Tip-brake mass, blade 3 (kg) [unused for 2 blades]
      56780   HubMass     - Hub mass (kg)
     115926   HubIner     - Hub inertia about rotor axis [3 blades] or teeter axis [2 blades] (kg m^2)
    534.116   GenIner     - Generator inertia about HSS (kg m^2)
     240000   NacMass     - Nacelle mass (kg)
2.60789E+06   NacYIner    - Nacelle inertia about yaw axis (kg m^2)
          0   YawBrMass   - Yaw bearing mass (kg)
          0   PtfmMass    - Platform mass (kg)
1303945.0   PtfmRIner   - Platform inertia for roll tilt rotation about the platform CM (kg m^2)
2607890   PtfmPIner   - Platform inertia for pitch tilt rotation about the platform CM (kg m^2)
          0   PtfmYIner   - Platform inertia for yaw rotation about the platform CM (kg m^2)
---------------------- BLADE ---------------------------------------------------
         17   BldNodes    - Number of blade nodes (per blade) used for analysis (-)
"Blade/NREL_5MW_Blade.dat"    BldFile(1)  - Name of file containing properties for blade 1 (quoted string)
"Blade/NREL_5MW_Blade_light.dat"    BldFile(2)  - Name of file containing properties for blade 2 (quoted string)
"Blade/NREL_5MW_Blade_heavy.dat"    BldFile(3)  - Name of file containing properties for blade 3 (quoted string) [unused for 2 blades]
---------------------- ROTOR-TEETER --------------------------------------------
          0   TeetMod     - Rotor-teeter spring/damper model {0: none, 1: standard, 2: user-defined from routine UserTeet} (switch) [unused for 3 blades]
          0   TeetDmpP    - Rotor-teeter damper position (degrees) [used only for 2 blades and when TeetMod=1]
          0   TeetDmp     - Rotor-teeter damping constant (N-m/(rad/s)) [used only for 2 blades and when TeetMod=1]
          0   TeetCDmp    - Rotor-teeter rate-independent Coulomb-damping moment (N-m) [used only for 2 blades and when TeetMod=1]
          0   TeetSStP    - Rotor-teeter soft-stop position (degrees) [used only for 2 blades and when TeetMod=1]
          0   TeetHStP    - Rotor-teeter hard-stop position (degrees) [used only for 2 blades and when TeetMod=1]
          0   TeetSSSp    - Rotor-teeter soft-stop linear-spring constant (N-m/rad) [used only for 2 blades and when TeetMod=1]
          0   TeetHSSp    - Rotor-teeter hard-stop linear-spring constant (N-m/rad) [used only for 2 blades and when TeetMod=1]
```

ElastoDyn Input II

```
---------------------- DRIVETRAIN ----------------------------------------------
          100   GBoxEff      - Gearbox efficiency (%)
           97   GBRatio      - Gearbox ratio (-)
8.67637E+08   DTTorSpr     - Drivetrain torsional spring (N-m/rad)
  6.215E+06   DTTorDmp     - Drivetrain torsional damper (N-m/(rad/s))
---------------------- FURLING -------------------------------------------------
False          Furling      - Read in additional model properties for furling turbine (flag) [must currently be FALSE)
"unused"       FurlFile     - Name of file containing furling properties (quoted string) [unused when Furling=False]
---------------------- TOWER ---------------------------------------------------
           20   TwrNodes     - Number of tower nodes used for analysis (-)
"Tower/NREL_5MW_ElastoDyn_Tower.dat"    TwrFile      - Name of file containing tower properties (quoted string)
---------------------- OUTPUT --------------------------------------------------
True           SumPrint     - Print summary data to "<RootName>.sum" (flag)
            1   OutFile      - Switch to determine where output will be placed: {1: in module output file only; 2: in glue code output file only; 3: both} (currentl
True           TabDelim     - Use tab delimiters in text tabular output file? (flag) (currently unused)
"ES10.3E2"     OutFmt       - Format used for text tabular output (except time).  Resulting field should be 10 characters. (quoted string) (currently unused)
            0   TStart       - Time to begin tabular output (s) (currently unused)
            1   DecFact      - Decimation factor for tabular output {1: output every time step} (-) (currently unused)
            9   NTwGages     - Number of tower nodes that have strain gages for output [0 to 9] (-)
           2,          4,          6,   8,   10, 12, 14, 16, 18     TwrGagNd     - List of tower nodes that have strain gages [1 to TwrNodes] (-) [unused if NTwGages=0]
            4   NBlGages     - Number of blade nodes that have strain gages for output [0 to 9] (-)
            5,          9,          13,     17     BldGagNd     - List of blade nodes that have strain gages [1 to BldNodes] (-) [unused if NBlGages=0]
                OutList      - The next line(s) contains a list of output parameters.  See OutListParameters.xlsx for a listing of available output channels, (-)
"OoPDefl1"                 - Blade 1 out-of-plane and in-plane deflections and tip twist
"IPDefl1"                  - Blade 1 out-of-plane and in-plane deflections and tip twist
"TwstDefl1"                - Blade 1 out-of-plane and in-plane deflections and tip twist
"Azimuth"                  - Blade 1 azimuth angle
"TTDspFA"                  - Tower fore-aft and side-to-side displacements and top twist
"TTDspSS"                  - Tower fore-aft and side-to-side displacements and top twist
"TTDspTwst"                 - Tower fore-aft and side-to-side displacements and top twist
"YawPzn"                   - Nacelle yaw angle (position)
"HSSBrTq"                  - High-speed shaft brake torque
"YawBrFxp"                 - Fore-aft shear, side-to-side shear, and vertical forces at the top of the tower (not rotating with nacelle yaw)
"YawBrFyp"                 - Fore-aft shear, side-to-side shear, and vertical forces at the top of the tower (not rotating with nacelle yaw)
"YawBrFzp"                 - Fore-aft shear, side-to-side shear, and vertical forces at the top of the tower (not rotating with nacelle yaw)
"YawBrMxp"                 - Side-to-side bending, fore-aft bending, and yaw moments at the top of the tower (not rotating with nacelle yaw)
"YawBrMyp"                 - Side-to-side bending, fore-aft bending, and yaw moments at the top of the tower (not rotating with nacelle yaw)
"YawBrMzp"                 - Side-to-side bending, fore-aft bending, and yaw moments at the top of the tower (not rotating with nacelle yaw)
"TwrBsFxt"                 - Fore-aft shear, side-to-side shear, and vertical forces at the base of the tower (mudline)
"TwrBsFyt"                 - Fore-aft shear, side-to-side shear, and vertical forces at the base of the tower (mudline)
"TwrBsFzt"                 - Fore-aft shear, side-to-side shear, and vertical forces at the base of the tower (mudline)
"TwrBsMxt"                 - Side-to-side bending, fore-aft bending, and yaw moments at the base of the tower (mudline)
"TwrBsMyt"                 - Side-to-side bending, fore-aft bending, and yaw moments at the base of the tower (mudline)
"TwrBsMzt"                 - Side-to-side bending, fore-aft bending, and yaw moments at the base of the tower (mudline)
"YawBrRDzt"
"YawBrTDxp"                - Tower-top / yaw bearing fore-aft (translational) deflection (relative to the undeflected position)    Directed along the xp-axis  (m)
"YawBrTDyp"                - Tower-top / yaw bearing side-to-side (translational) deflection (relative to the undeflected position)    Directed along the yp-axis  (
"YawBrTDzp"                - Tower-top / yaw bearing axial (translational) deflection (relative to the undeflected position)   Directed along the zp-axis  (m)
"YawBrTDxt"                - TTDspFA   Tower-top / yaw bearing fore-aft (translational) deflection (relative to the undeflected position)  Directed along the xt-axi
"YawBrTDyt"                - TTDspSS   Tower-top / yaw bearing side-to-side (translation) deflection (relative to the undeflected position)    Directed along the yt
"YawBrTDzt"                - TTDspAx   Tower-top / yaw bearing axial (translational) deflection (relative to the undeflected position) Directed along the zt-axis  (
"YawBrTAxp"                - Tower-top / yaw bearing fore-aft (translational) acceleration (absolute)  Directed along the xp-axis  (m/s^2)
"YawBrTAyp"                - Tower-top / yaw bearing side-to-side (translational) acceleration (absolute)  Directed along the yp-axis  (m/s^2)
"YawBrTAzp"                - Tower-top / yaw bearing axial (translational) acceleration (absolute) Directed along the zp-axis  (m/s^2)
"TipClrnc1"
"BldPitch1"                - Blade 1 pitch angle
"BldPitch2"                - Blade 1 pitch angle
"BldPitch3"                - Blade 1 pitch angle
"GenSpeed"                 - Low-speed shaft and high-speed shaft speeds
"RotThrust"                - Rotor thrust
"RotTorq"                  - Rotor torque and low-speed shaft 0- and 90-bending moments at the main bearing
"RotSpeed"                 - Low-speed shaft and high-speed shaft speeds
"RootFxc1"
"RootFyc1"
"RootFzc1"
"RootMxc1"
"RootMyc1"
"RootMzc1"
"RootFxc2"
"RootFyc2"
"RootFzc2"
"RootMxc2"
"RootMyc2"
"RootMzc2"
"RootFxc3"
"RootFyc3"
"RootFzc3"
"RootMxc3"
"RootMyc3"
"RootMzc3"
END of input file (the word "END" must appear in the first 3 columns of this last OutList line)
---------------------------------------------------------------------------------------
```

ElastoDyn Tower Input

```
------- ELASTODYN V1.00.* TOWER INPUT FILE -------------------------------------
NREL 5.0 MW offshore baseline tower input properties.
---------------------- TOWER PARAMETERS ----------------------------------------
         11   NTwInpSt     - Number of input stations to specify tower geometry
          1   TwrFADmp(1)  - Tower 1st fore-aft mode structural damping ratio (%)
          1   TwrFADmp(2)  - Tower 2nd fore-aft mode structural damping ratio (%)
          1   TwrSSDmp(1)  - Tower 1st side-to-side mode structural damping ratio (%)
          1   TwrSSDmp(2)  - Tower 2nd side-to-side mode structural damping ratio (%)
---------------------- TOWER ADJUSTMUNT FACTORS --------------------------------
          1   FAStTunr(1)  - Tower fore-aft modal stiffness tuner, 1st mode (-)
          1   FAStTunr(2)  - Tower fore-aft modal stiffness tuner, 2nd mode (-)
          1   SSStTunr(1)  - Tower side-to-side stiffness tuner, 1st mode (-)
          1   SSStTunr(2)  - Tower side-to-side stiffness tuner, 2nd mode (-)
          1   AdjTwMa      - Factor to adjust tower mass density (-)
          1   AdjFASt      - Factor to adjust tower fore-aft stiffness (-)
          1   AdjSSSt      - Factor to adjust tower side-to-side stiffness (-)
---------------------- DISTRIBUTED TOWER PROPERTIES ----------------------------
  HtFract       TMassDen         TwFAStif       TwSSStif
   (-)           (kg/m)           (Nm^2)         (Nm^2)
0.0000000E+00  5.5908700E+03  6.1434300E+11  6.1434300E+11
1.0000000E-01  5.2324300E+03  5.3482100E+11  5.3482100E+11
2.0000000E-01  4.8857600E+03  4.6326700E+11  4.6326700E+11
3.0000000E-01  4.5508700E+03  3.9913100E+11  3.9913100E+11
4.0000000E-01  4.2277500E+03  3.4188300E+11  3.4188300E+11
5.0000000E-01  3.9164100E+03  2.9101100E+11  2.9101100E+11
6.0000000E-01  3.6168300E+03  2.4602700E+11  2.4602700E+11
7.0000000E-01  3.3290300E+03  2.0645700E+11  2.0645700E+11
8.0000000E-01  3.0530100E+03  1.7185100E+11  1.7185100E+11
9.0000000E-01  2.7887500E+03  1.4177600E+11  1.4177600E+11
1.0000000E+00  2.5362700E+03  1.1582000E+11  1.1582000E+11
---------------------- TOWER FORE-AFT MODE SHAPES ------------------------------
     0.7004   TwFAM1Sh(2) - Mode 1, coefficient of x^2 term
     2.1963   TwFAM1Sh(3) -       , coefficient of x^3 term
    -5.6202   TwFAM1Sh(4) -       , coefficient of x^4 term
     6.2275   TwFAM1Sh(5) -       , coefficient of x^5 term
     -2.504   TwFAM1Sh(6) -       , coefficient of x^6 term
   -70.5319   TwFAM2Sh(2) - Mode 2, coefficient of x^2 term
   -63.7623   TwFAM2Sh(3) -       , coefficient of x^3 term
    289.737   TwFAM2Sh(4) -       , coefficient of x^4 term
   -176.513   TwFAM2Sh(5) -       , coefficient of x^5 term
    22.0706   TwFAM2Sh(6) -       , coefficient of x^6 term
---------------------- TOWER SIDE-TO-SIDE MODE SHAPES --------------------------
      1.385   TwSSM1Sh(2) - Mode 1, coefficient of x^2 term
    -1.7684   TwSSM1Sh(3) -       , coefficient of x^3 term
     3.0871   TwSSM1Sh(4) -       , coefficient of x^4 term
    -2.2395   TwSSM1Sh(5) -       , coefficient of x^5 term
     0.5357   TwSSM1Sh(6) -       , coefficient of x^6 term
    -121.21   TwSSM2Sh(2) - Mode 2, coefficient of x^2 term
    184.415   TwSSM2Sh(3) -       , coefficient of x^3 term
   -224.904   TwSSM2Sh(4) -       , coefficient of x^4 term
    298.536   TwSSM2Sh(5) -       , coefficient of x^5 term
   -135.838   TwSSM2Sh(6) -       , coefficient of x^6 term
```

ElastoDyn Blade Input

```
------- ELASTODYN V1.00.* INDIVIDUAL BLADE INPUT FILE --------------------------
NREL 5.0 MW land-based baseline blade input properties.
---------------------- BLADE PARAMETERS ----------------------------------------
        49    NBlInpSt    - Number of blade input stations (-)
   2.5        BldFlDmp(1) - Blade flap mode #1 structural damping in percent of critical (%)
   2.5        BldFlDmp(2) - Blade flap mode #2 structural damping in percent of critical (%)
   2.5        BldEdDmp(1) - Blade edge mode #1 structural damping in percent of critical (%)
---------------------- BLADE ADJUSTMENT FACTORS --------------------------------
         1    FlStTunr(1) - Blade flapwise modal stiffness tuner, 1st mode (-)
         1    FlStTunr(2) - Blade flapwise modal stiffness tuner, 2nd mode (-)
   1.04536    AdjBlMs     - Factor to adjust blade mass density (-)  !bjj: value for AD14=1.04536; value for AD15=1.057344
         1    AdjFlSt     - Factor to adjust blade flap stiffness (-)
         1    AdjEdSt     - Factor to adjust blade edge stiffness (-)
---------------------- DISTRIBUTED BLADE PROPERTIES ----------------------------
    BlFract      PitchAxis      StrcTwst      BMassDen       FlpStff       EdgStff
      (-)          (-)           (deg)         (kg/m)        (Nm^2)        (Nm^2)
0.0000000E+00  2.5000000E-01  1.3308000E+01  6.7893500E+02  1.8110000E+10  1.8113600E+10
3.2500000E-03  2.5000000E-01  1.3308000E+01  6.7893500E+02  1.8110000E+10  1.8113600E+10
1.9510000E-02  2.5049000E-01  1.3308000E+01  7.7336300E+02  1.9424900E+10  1.9558600E+10
3.5770000E-02  2.5490000E-01  1.3308000E+01  7.4055000E+02  1.7455900E+10  1.9497800E+10
5.2030000E-02  2.6716000E-01  1.3308000E+01  7.4004200E+02  1.5287400E+10  1.9788800E+10
6.8290000E-02  2.7941000E-01  1.3308000E+01  5.9249600E+02  1.0782400E+10  1.4858500E+10
8.4550000E-02  2.9167000E-01  1.3308000E+01  4.5027500E+02  7.2297200E+09  1.0220600E+10
1.0081000E-01  3.0392000E-01  1.3308000E+01  4.2405400E+02  6.3095400E+09  9.1447000E+09
1.1707000E-01  3.1618000E-01  1.3308000E+01  4.0063800E+02  5.5283600E+09  8.0631600E+09
1.3335000E-01  3.2844000E-01  1.3308000E+01  3.8206200E+02  4.9800600E+09  6.8844400E+09
1.4959000E-01  3.4069000E-01  1.3308000E+01  3.9965500E+02  4.9368400E+09  7.0091800E+09
1.6585000E-01  3.5294000E-01  1.3308000E+01  4.2632100E+02  4.6916600E+09  7.1676800E+09
1.8211000E-01  3.6519000E-01  1.3181000E+01  4.1682000E+02  3.9494600E+09  7.2716600E+09
1.9837000E-01  3.7500000E-01  1.2848000E+01  4.0618600E+02  3.3865200E+09  7.0817000E+09
2.1465000E-01  3.7500000E-01  1.2192000E+01  3.8142000E+02  2.9337400E+09  6.2445300E+09
2.3089000E-01  3.7500000E-01  1.1561000E+01  3.5282200E+02  2.5689600E+09  5.0489600E+09
2.4715000E-01  3.7500000E-01  1.1072000E+01  3.4947700E+02  2.3886500E+09  4.9484900E+09
2.6341000E-01  3.7500000E-01  1.0792000E+01  3.4653800E+02  2.2719900E+09  4.8080200E+09
2.9595000E-01  3.7500000E-01  1.0232000E+01  3.3933300E+02  2.0500500E+09  4.5014000E+09
3.2846000E-01  3.7500000E-01  9.6720000E+00  3.3000400E+02  1.8282500E+09  4.2440700E+09
3.6098000E-01  3.7500000E-01  9.1100000E+00  3.2199000E+02  1.5887100E+09  3.9952800E+09
3.9350000E-01  3.7500000E-01  8.5340000E+00  3.1382000E+02  1.3619300E+09  3.7507600E+09
4.2602000E-01  3.7500000E-01  7.9320000E+00  2.9473400E+02  1.1023800E+09  3.4471400E+09
4.5855000E-01  3.7500000E-01  7.3210000E+00  2.8712000E+02  8.7580000E+08  3.1390700E+09
4.9106000E-01  3.7500000E-01  6.7110000E+00  2.6334300E+02  6.8130000E+08  2.7342400E+09
5.2358000E-01  3.7500000E-01  6.1220000E+00  2.5320700E+02  5.3472000E+08  2.5548700E+09
5.5610000E-01  3.7500000E-01  5.5460000E+00  2.4166600E+02  4.0890000E+08  2.3340300E+09
5.8862000E-01  3.7500000E-01  4.9710000E+00  2.2063800E+02  3.1454000E+08  1.8287300E+09
6.2115000E-01  3.7500000E-01  4.4010000E+00  2.0029300E+02  2.3863000E+08  1.5841000E+09
6.5366000E-01  3.7500000E-01  3.8340000E+00  1.7940400E+02  1.7588000E+08  1.3233600E+09
6.8618000E-01  3.7500000E-01  3.3320000E+00  1.6509400E+02  1.2601000E+08  1.1836800E+09
7.1870000E-01  3.7500000E-01  2.8900000E+00  1.5441100E+02  1.0726000E+08  1.0201600E+09
7.5122000E-01  3.7500000E-01  2.5030000E+00  1.3893500E+02  9.0880000E+07  7.9781000E+08
7.8376000E-01  3.7500000E-01  2.1160000E+00  1.2955500E+02  7.6310000E+07  7.0961000E+08
8.1626000E-01  3.7500000E-01  1.7300000E+00  1.0726400E+02  6.1050000E+07  5.1819000E+08
8.4878000E-01  3.7500000E-01  1.3420000E+00  9.8776000E+01  4.9480000E+07  4.5487000E+08
8.8130000E-01  3.7500000E-01  9.5400000E-01  9.0248000E+01  3.9360000E+07  3.9512000E+08
8.9756000E-01  3.7500000E-01  7.6000000E-01  8.3001000E+01  3.4670000E+07  3.5372000E+08
9.1382000E-01  3.7500000E-01  5.7400000E-01  7.2906000E+01  3.0410000E+07  3.0473000E+08
9.3008000E-01  3.7500000E-01  4.0400000E-01  6.8772000E+01  2.6520000E+07  2.8142000E+08
9.3821000E-01  3.7500000E-01  3.1900000E-01  6.6264000E+01  2.3840000E+07  2.6171000E+08
9.4636000E-01  3.7500000E-01  2.5300000E-01  5.9340000E+01  1.9630000E+07  1.5881000E+08
9.5447000E-01  3.7500000E-01  2.1600000E-01  5.5914000E+01  1.6000000E+07  1.3788000E+08
9.6260000E-01  3.7500000E-01  1.7800000E-01  5.2484000E+01  1.2830000E+07  1.1879000E+08
9.7073000E-01  3.7500000E-01  1.4000000E-01  4.9114000E+01  1.0080000E+07  1.0163000E+08
9.7886000E-01  3.7500000E-01  1.0100000E-01  4.5818000E+01  7.5500000E+06  8.5070000E+07
9.8699000E-01  3.7500000E-01  6.2000000E-02  4.1669000E+01  4.6000000E+06  6.4260000E+07
9.9512000E-01  3.7500000E-01  2.3000000E-02  1.1453000E+01  2.5000000E+05  6.6100000E+06
1.0000000E+00  3.7500000E-01  0.0000000E+00  1.0319000E+01  1.7000000E+05  5.0100000E+06
---------------------- BLADE MODE SHAPES ---------------------------------------
     0.0622   BldFl1Sh(2) - Flap mode 1, coeff of x^2
     1.7254   BldFl1Sh(3) -            , coeff of x^3
    -3.2452   BldFl1Sh(4) -            , coeff of x^4
     4.7131   BldFl1Sh(5) -            , coeff of x^5
    -2.2555   BldFl1Sh(6) -            , coeff of x^6
    -0.5809   BldFl2Sh(2) - Flap mode 2, coeff of x^2
     1.2067   BldFl2Sh(3) -            , coeff of x^3
   -15.5349   BldFl2Sh(4) -            , coeff of x^4
    29.7347   BldFl2Sh(5) -            , coeff of x^5
   -13.8255   BldFl2Sh(6) -            , coeff of x^6
     0.3627   BldEdgSh(2) - Edge mode 1, coeff of x^2
     2.5337   BldEdgSh(3) -            , coeff of x^3
    -3.5772   BldEdgSh(4) -            , coeff of x^4
      2.376   BldEdgSh(5) -            , coeff of x^5
    -0.6952   BldEdgSh(6) -            , coeff of x^6
```

AeroDyn Input

```
--------- AeroDyn v14.04.* INPUT FILE ------------------------------------------------------------------------
NREL 5.0 MW offshore baseline aerodynamic input properties.
"BEDDOES"      StallMod     - Dynamic stall included [BEDDOES or STEADY] (unquoted string)
"USE_CM"       UseCm        - Use aerodynamic pitching moment model? [USE_CM or NO_CM] (unquoted string)
"EQUIL"        InfModel     - Inflow model [DYNIN or EQUIL] (unquoted string)
"WAKE"         IndModel     - Induction-factor model [NONE or WAKE or SWIRL] (unquoted string)
      0.005    AToler       - Induction-factor tolerance (convergence criteria) (-)
"PRANDtl"      TLModel      - Tip-loss model (EQUIL only) [PRANDtl, GTECH, or NONE] (unquoted string)
"PRANDtl"      HLModel      - Hub-loss model (EQUIL only) [PRANdtl or NONE] (unquoted string)
"NEWTOWER"     TwrShad        - INSTEAD OF:   0.0       TwrShad      - Tower-shadow velocity deficit (-)
True           TwrPotent      - Calculate tower potential flow (flag)  INSTEAD OF  9999.9      ShadHWid     - Tower-shadow half width (m)
False          TwrShadow      - Calculate tower shadow (flag)          INSTEAD OF  9999.9      T_Shad_Refpt- Tower-shadow reference point (m)
"NREL_5MW_AeroDyn_Tower.dat"    TwrFile        - Tower drag file name (quoted string)
False          CalcTwrAero    - Calculate aerodynamic drag of the tower at the ElastoDyn nodes.  TwrPotent must be true.
      1.225    AirDens        - Air density (kg/m^3)
  1.464E-05    KinVisc      - Kinematic air viscosity [CURRENTLY IGNORED] (m^2/sec)
0.0125         DTAero       - Time interval for aerodynamic calculations (sec)
          8    NumFoil      - Number of airfoil files (-)
"AeroData/Cylinder1.dat"    FoilNm      - Names of the airfoil files [NumFoil lines] (quoted strings)
"AeroData/Cylinder2.dat"
"AeroData/DU40_A17.dat"
"AeroData/DU35_A17.dat"
"AeroData/DU30_A17.dat"
"AeroData/DU25_A17.dat"
"AeroData/DU21_A17.dat"
"AeroData/NACA64_A17.dat"
         17    BldNodes    - Number of blade nodes used for analysis (-)
RNodes          AeroTwst        DRNodes         Chord           NFoil           PrnElm
2.8667000E+00   1.3308000E+01   2.7333000E+00   3.5420000E+00           1       NOPRINT
5.6000000E+00   1.3308000E+01   2.7333000E+00   3.8540000E+00           1       NOPRINT
8.3333000E+00   1.3308000E+01   2.7333000E+00   4.1670000E+00           2       NOPRINT
1.1750000E+01   1.3308000E+01   4.1000000E+00   4.5570000E+00           3       NOPRINT
1.5850000E+01   1.1480000E+01   4.1000000E+00   4.6520000E+00           4       NOPRINT
1.9950000E+01   1.0162000E+01   4.1000000E+00   4.4580000E+00           4       NOPRINT
2.4050000E+01   9.0110000E+00   4.1000000E+00   4.2490000E+00           5       NOPRINT
2.8150000E+01   7.7950000E+00   4.1000000E+00   4.0070000E+00           6       NOPRINT
3.2250000E+01   6.5440000E+00   4.1000000E+00   3.7480000E+00           6       NOPRINT
3.6350000E+01   5.3610000E+00   4.1000000E+00   3.5020000E+00           7       NOPRINT
4.0450000E+01   4.1880000E+00   4.1000000E+00   3.2560000E+00           7       NOPRINT
4.4550000E+01   3.1250000E+00   4.1000000E+00   3.0100000E+00           8       NOPRINT
4.8650000E+01   2.3190000E+00   4.1000000E+00   2.7640000E+00           8       NOPRINT
5.2750000E+01   1.5260000E+00   4.1000000E+00   2.5180000E+00           8       NOPRINT
5.6166700E+01   8.6300000E-01   2.7333000E+00   2.3130000E+00           8       NOPRINT
5.8900000E+01   3.7000000E-01   2.7333000E+00   2.0860000E+00           8       NOPRINT
6.1633300E+01   1.0600000E-01   2.7333000E+00   1.4190000E+00           8       NOPRINT
```

ServoDyn Input

```
------- SERVODYN v1.05.* INPUT FILE ------------------------------------------
NREL 5.0 MW Baseline Land-Based Wind Turbine
---------------------- SIMULATION CONTROL --------------------------------------
False         Echo         - Echo input data to <RootName>.ech (flag)
"default"     DT           - Communication interval for controllers (s) (or "default")
---------------------- PITCH CONTROL -------------------------------------------
          5   PCMode       - Pitch control mode (0: none, 3: user-defined from routine PitchCntrl, 4: user-defined from Simulink/Labview, 5: user-defined from Bla
          0   TPCOn        - Time to enable active pitch control (s) [unused when PCMode=0]
     9999.9   TPitManS(1)  - Time to start override pitch maneuver for blade 1 and end standard pitch control (s)
     9999.9   TPitManS(2)  - Time to start override pitch maneuver for blade 2 and end standard pitch control (s)
     9999.9   TPitManS(3)  - Time to start override pitch maneuver for blade 3 and end standard pitch control (s) [unused for 2 blades]
          2   PitManRat(1) - Pitch rate at which override pitch maneuver heads toward final pitch angle for blade 1 (deg/s)
          2   PitManRat(2) - Pitch rate at which override pitch maneuver heads toward final pitch angle for blade 2 (deg/s)
          2   PitManRat(3) - Pitch rate at which override pitch maneuver heads toward final pitch angle for blade 3 (deg/s) [unused for 2 blades]
          0   BlPitchF(1)  - Blade 1 final pitch for pitch maneuvers (degrees)
          0   BlPitchF(2)  - Blade 2 final pitch for pitch maneuvers (degrees)
          0   BlPitchF(3)  - Blade 3 final pitch for pitch maneuvers (degrees) [unused for 2 blades]
---------------------- GENERATOR AND TORQUE CONTROL ----------------------------
          5   VSContrl     - Variable-speed control mode (0: none, 1: simple VS, 3: user-defined from routine UserVSCont, 4: user-defined from Simulink/Labview, 5
          2   GenModel     - Generator model (1: simple, 2: Thevenin, 3: user-defined from routine UserGen) (switch) [used only when VSContrl=0]
       94.4   GenEff       - Generator efficiency [ignored by the Thevenin and user-defined generator models] (%)
True          GenTiStr     - Method to start the generator (T: timed using TimGenOn, F: generator speed using SpdGenOn) (flag)
True          GenTiStp     - Method to stop the generator (T: timed using TimGenOf, F: when generator power = 0) (flag)
     9999.9   SpdGenOn     - Generator speed to turn on the generator for a startup (HSS speed) (rpm) [used only when GenTiStr=False]
          0   TimGenOn     - Time to turn on the generator for a startup (s) [used only when GenTiStr=True]
     9999.9   TimGenOf     - Time to turn off the generator (s) [used only when GenTiStp=True]
---------------------- SIMPLE VARIABLE-SPEED TORQUE CONTROL --------------------
     9999.9   VS_RtGnSp    - Rated generator speed for simple variable-speed generator control (HSS side) (rpm) [used only when VSContrl=1]
     9999.9   VS_RtTq      - Rated generator torque/constant generator torque in Region 3 for simple variable-speed generator control (HSS side) (N-m) [used only
     9999.9   VS_Rgn2K     - Generator torque constant in Region 2 for simple variable-speed generator control (HSS side) (N-m/rpm^2) [used only when VSContrl=1]
     9999.9   VS_SlPc      - Rated generator slip percentage in Region 2 1/2 for simple variable-speed generator control (%) [used only when VSContrl=1]
---------------------- SIMPLE INDUCTION GENERATOR ------------------------------
     9999.9   SIG_SlPc     - Rated generator slip percentage (%) [used only when VSContrl=0 and GenModel=1]
     9999.9   SIG_SySp     - Synchronous (zero-torque) generator speed (rpm) [used only when VSContrl=0 and GenModel=1]
     9999.9   SIG_RtTq     - Rated torque (N-m) [used only when VSContrl=0 and GenModel=1]
     9999.9   SIG_PORt     - Pull-out ratio (Tpullout/Trated) (-) [used only when VSContrl=0 and GenModel=1]
---------------------- THEVENIN-EQUIVALENT INDUCTION GENERATOR -----------------
     9999.9   TEC_Freq     - Line frequency [50 or 60] (Hz) [used only when VSContrl=0 and GenModel=2]
       9998   TEC_NPol     - Number of poles [even integer > 0] (-) [used only when VSContrl=0 and GenModel=2]
     9999.9   TEC_SRes     - Stator resistance (ohms) [used only when VSContrl=0 and GenModel=2]
     9999.9   TEC_RRes     - Rotor resistance (ohms) [used only when VSContrl=0 and GenModel=2]
     9999.9   TEC_VLL      - Line-to-line RMS voltage (volts) [used only when VSContrl=0 and GenModel=2]
     9999.9   TEC_SLR      - Stator leakage reactance (ohms) [used only when VSContrl=0 and GenModel=2]
     9999.9   TEC_RLR      - Rotor leakage reactance (ohms) [used only when VSContrl=0 and GenModel=2]
     9999.9   TEC_MR       - Magnetizing reactance (ohms) [used only when VSContrl=0 and GenModel=2]
---------------------- HIGH-SPEED SHAFT BRAKE ----------------------------------
          0   HSSBrMode    - HSS brake model (0: none, 1: simple, 3: user-defined from routine UserHSSBr, 4: user-defined from Simulink/Labview, 5: user-defined f
     9999.9   THSSBrDp     - Time to initiate deployment of the HSS brake (s)
        0.6   HSSBrDT      - Time for HSS-brake to reach full deployment once initiated (sec) [used only when HSSBrMode=1]
    28116.2   HSSBrTqF     - Fully deployed HSS-brake torque (N-m)
---------------------- NACELLE-YAW CONTROL -------------------------------------
          0   YCMode       - Yaw control mode (0: none, 3: user-defined from routine UserYawCont, 4: user-defined from Simulink/Labview, 5: user-defined from Blad
     9999.9   TYCOn        - Time to enable active yaw control (s) [unused when YCMode=0]
          0   YawNeut      - Neutral yaw position--yaw spring force is zero at this yaw (degrees)
9.02832E+09   YawSpr       - Nacelle-yaw spring constant (N-m/rad)
  1.916E+07   YawDamp      - Nacelle-yaw damping constant (N-m/(rad/s))
     9999.9   TYawManS     - Time to start override yaw maneuver and end standard yaw control (s)
          2   YawManRat    - Yaw maneuver rate (in absolute value) (deg/s)
          0   NacYawF      - Final yaw angle for override yaw maneuvers (degrees)
---------------------- TUNED MASS DAMPER ---------------------------------------
False         CompNTMD     - Compute nacelle tuned mass damper (true/false) (flag)
"NRELOffshrBsline5MW_ServoDyn_TMD.dat"    NTMDfile     - Name of the file for nacelle tuned mass damper (quoted string) [unused when CompNTMD is false]
False         CompTTMD     - Compute tower tuned mass damper (true/false) (flag)
"NRELOffshrBsline5MW_ServoDyn_TMD.dat"    TTMDfile     - Name of the file for tower tuned mass damper (quoted string) [unused when CompTTMD is false]
---------------------- BLADED INTERFACE ---------------------------------------- [used only with Bladed Interface]
"ServoData/DISCON_x64.dll"    DLL_FileName - Name/location of the dynamic library {.dll [Windows] or .so [Linux]} in the Bladed-DLL format (-) [used only with Bla
"DISCON.IN"    DLL_InFile   - Name of input file sent to the DLL (-) [used only with Bladed Interface]
"DISCON"      DLL_ProcName - Name of procedure in DLL to be called (-) [case sensitive; used only with DLL Interface]
"default"     DLL_DT       - Communication interval for dynamic library (s) (or "default") [used only with Bladed Interface]
false         DLL_Ramp     - Whether a linear ramp should be used between DLL_DT time steps [introduces time shift when true] (flag) [used only with Bladed Interf
9999.9        BPCutoff     - Cuttoff frequency for low-pass filter on blade pitch from DLL (Hz) [used only with Bladed Interface]
          0   NacYaw_North - Reference yaw angle of the nacelle when the upwind end points due North (deg) [used only with Bladed Interface]
          0   Ptch_Cntrl   - Record 28: Use individual pitch control {0: collective pitch; 1: individual pitch control} (switch) [used only with Bladed Interface]
          0   Ptch_SetPnt  - Record  5: Below-rated pitch angle set-point (deg) [used only with Bladed Interface]
          0   Ptch_Min     - Record  6: Minimum pitch angle (deg) [used only with Bladed Interface]
          0   Ptch_Max     - Record  7: Maximum pitch angle (deg) [used only with Bladed Interface]
          0   PtchRate_Min - Record  8: Minimum pitch rate (most negative value allowed) (deg/s) [used only with Bladed Interface]
          0   PtchRate_Max - Record  9: Maximum pitch rate  (deg/s) [used only with Bladed Interface]
          0   Gain_OM      - Record 16: Optimal mode gain (Nm/(rad/s)^2) [used only with Bladed Interface]
          0   GenSpd_MinOM - Record 17: Minimum generator speed (rpm) [used only with Bladed Interface]
          0   GenSpd_MaxOM - Record 18: Optimal mode maximum speed (rpm) [used only with Bladed Interface]
          0   GenSpd_Dem   - Record 19: Demanded generator speed above rated (rpm) [used only with Bladed Interface]
          0   GenTrq_Dem   - Record 22: Demanded generator torque above rated (Nm) [used only with Bladed Interface]
          0   GenPwr_Dem   - Record 13: Demanded power (W) [used only with Bladed Interface]
---------------------- BLADED INTERFACE TORQUE-SPEED LOOK-UP TABLE -------------
          0   DLL_NumTrq   - Record 26: No. of points in torque-speed look-up table {0 = none and use the optimal mode parameters; nonzero = ignore the optimal mo
 GenSpd_TLU   GenTrq_TLU
 (rpm)          (Nm)
---------------------- OUTPUT --------------------------------------------------
True          SumPrint     - Print summary data to <RootName>.sum (flag) (currently unused)
          1   OutFile      - Switch to determine where output will be placed: {1: in module output file only; 2: in glue code output file only; 3: both} (currentl
True          TabDelim     - Use tab delimiters in text tabular output file? (flag) (currently unused)
"ES10.3E2"    OutFmt       - Format used for text tabular output (except time).  Resulting field should be 10 characters. (quoted string) (currently unused)
          0   TStart       - Time to begin tabular output (s) (currently unused)
              OutList      - The next line(s) contains a list of output parameters.  See OutListParameters.xlsx for a listing of available output channels, (-)
"GenPwr"                  - Electrical generator power and torque
"GenTq"                   - Electrical generator power and torque
END of input file (the word "END" must appear in the first 3 columns of this last OutList line)
---------------------------------------------------------------------------------------
```

14.6 ASE Validation - Comparison of Timeseries

The following plots contain the timeseries where the extreme FA bending moment (TwrBsMyt) in the comparison study of Jonkman [110] and in the own ASE load simulation occurred. The first plot contains the bending moments, the second plot shows the corresponding pitch angles of blade 1 (BldPitch1), and the third plot shows the main wind speed (Wind1Velx). Note that the values of these load sensors are raw without post processing, such as applied partial safety factors. The arithmetic mean and the standard deviation of TwrBsMyt in both timeseries were: $\bar{\mu} = 29.164$ MNm and $\sigma = 11.046$ MNm in the comparison study [110] and $\bar{\mu} = 29.501$ MNm and $\sigma = 11.438$ MNm from the own ASE simulation. Note that a dip in the pitch angle occurred simultaneous with the extreme load of the FA tower base bending moment, but no numerical instability was observed. The dip was caused by an abrupt drop in the wind speed, which forced the control region to change from 3 to 2. The fast pitch angle adjustment should improve the wind energy conversion efficiency, while higher loads were tolerated. The wind speed in the comparison study dropped at the interesting point of time to a slightly lower value than in the own analysis, wherefore a higher resulting TwrBsMyt was reasonable.

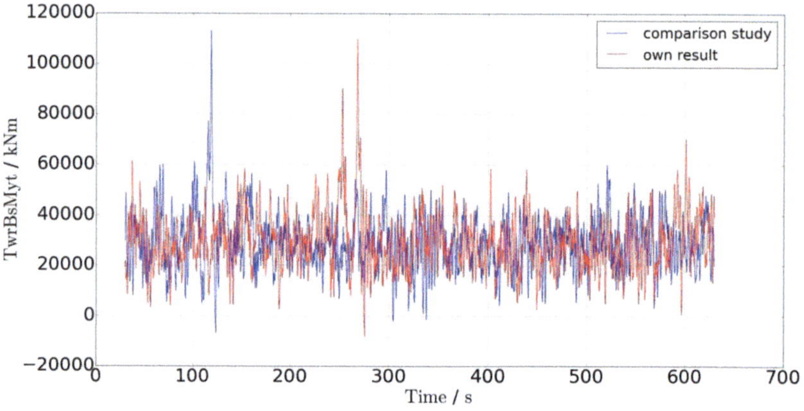

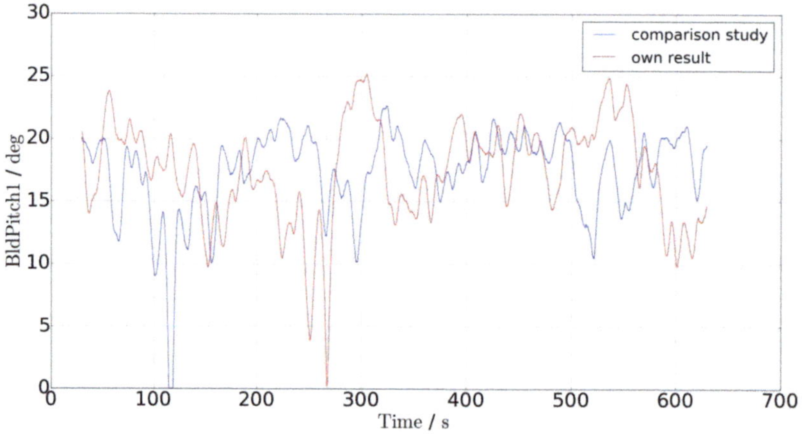

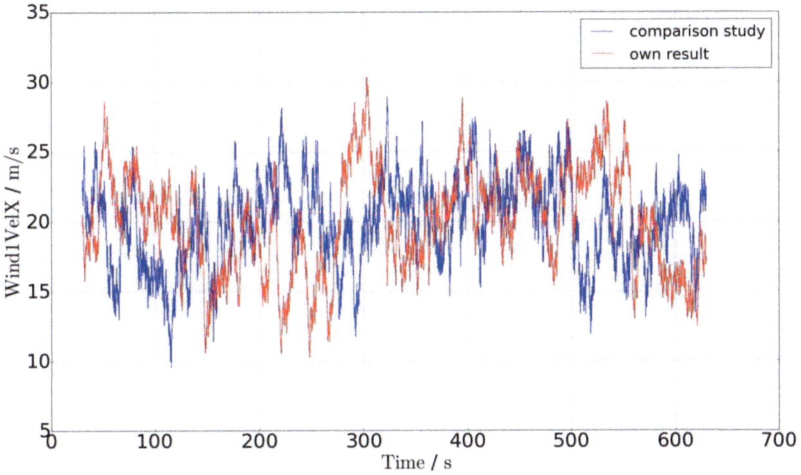

14.7 ASE Validation - FAST Tower Aerodynamics Input File

This is the new tower aerodynamics input file for calculation of the reference WT extreme loads.

AeroDyn Tower Input

```
NREL 5.0 MW baseline land-based aerodynamic tower CD input properties.
Used with AeroDyn 13.00.00 hidden tower influence feature.
  12        NTwrHt      - Number of tower input height stations listed (-)
  16        NTwrRe      - Number of tower Re values (-)
   1        NTwrCD      - Number of tower CD columns (-) Note: For current versions, this MUST be 1.
   0.0      Tower_Wake_Constant - Tower wake constant (-) {0.0: full potential flow, 0.1: Bak model}
---------------------- DISTRIBUTED TOWER PROPERTIES ----------------------------
TwrHtFr  TwrWid  NTwrCDCol
0.00000  6.000   1
0.09733  5.787   1
0.19467  5.574   1
0.29200  5.361   1
0.38933  5.148   1
0.48667  4.935   1
0.58400  4.722   1
0.68133  4.509   1
0.77867  4.296   1
0.87600  4.083   1
0.97333  3.870   1
1.00000  3.870   1
---------------------- Re v CD PROPERTIES --------------------------------------
TwrRe  TwrCD1  TwrCD2  TwrCD2  ...
 0.010   1.11
 0.020   1.20
 0.122   1.20
 0.200   1.17
 0.300   0.90
 0.400   0.54
 0.500   0.31
 1.000   0.38
 1.500   0.46
 2.000   0.53
 2.500   0.57
 3.000   0.61
 3.500   0.64
 4.000   0.67
 5.000   0.70
10.000   0.70
```

14.8 Rotor Blades Extreme Root Bending Moments Under Different Conditions and DLCs

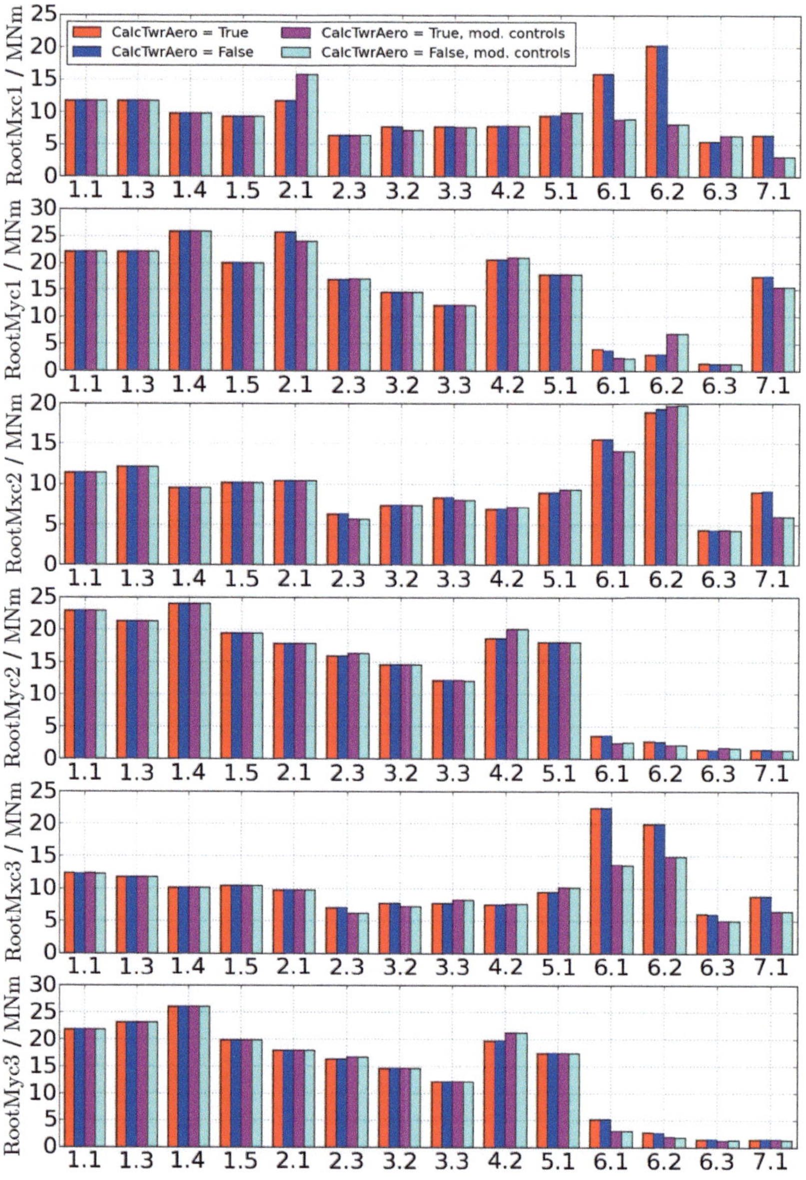

14.9 Extreme Loads from ASE Simulation of the Reference WT

Some extreme values of relevant sensors from the ASE simulation of the 5 MW baseline land-based NREL reference WT are summarized in the following tables. Thereby, the adjusted control settings for DLCs with shut-down event or extreme wind parking situations are applied. More detailed descriptions of these settings may be reviewed in subsection 3.5.3.

Wind Speed

Parameter	Type	File Name	Wind1VelX (m/s)	Wind1VelY (m/s)	Wind1VelZ (m/s)	Time (s)
Wind1VelX	Min	Dlc_21T_+08deg_Vout_S6.out	**-2.21E+04**	4.88E+03	4.42E+01	3.22E+03
Wind1VelX	Max	Dlc_21T_-08deg_Vout_S6.out	**2.15E+04**	4.88E+03	3.97E+01	3.31E+03
Wind1VelY	Min	Dlc_33_2.0_-_Vr+2.out	-3.97E+02	**-7.17E+02**	-3.47E+01	-2.89E+03
Wind1VelY	Max	Dlc_23_8.4_Vout.out	3.60E+03	**5.57E+03**	3.90E+01	5.31E+03
Wind1VelZ	Min	Dlc_33_0.0_-_Vr-2.out	-1.13E+03	-3.45E+02	**-5.33E+01**	-4.50E+03
Wind1VelZ	Max	Dlc_13T_-08deg_V24_S3.out	5.59E+03	4.75E+03	**4.74E+01**	4.52E+03

Tower Top Displacements

Parameter	Type	File Name	TTDspFA (m)	TTDspSS (m)	TTDspTwst (deg)	Wind1VelX (m/s)	Wind1VelY (m/s)	Time (s)
TTDspFA	Min	Dlc_23_6.3_Vout.out	**-4.49E-01**	-2.31E-02	0.00E+00	2.46E+01	0.00E+00	6.94E+01
TTDspFA	Max	Dlc_23_10.5_Vr+2.out	**7.30E-01**	-6.96E-02	0.00E+00	1.91E+01	-7.54E-04	6.64E+01
TTDspSS	Min	Dlc_21T_+00deg_Vout_S7.out	2.97E-01	**-3.60E-01**	0.00E+00	2.29E+01	1.78E+00	6.51E+01
TTDspSS	Max	Dlc_21T_+08deg_Vout_S6.out	-4.13E-01	**4.17E-01**	0.00E+00	2.80E+01	4.68E+00	6.68E+01
TTDspTwst	Min	Dlc_11T_-08deg_V04_S1.out	4.42E-02	-2.53E-02	**0.00E+00**	4.71E+00	-2.05E-01	3.00E+01
TTDspTwst	Max	Dlc_11T_-08deg_V04_S1.out	4.42E-02	-2.53E-02	**0.00E+00**	4.71E+00	-2.05E-01	3.00E+01

Rotor and Generator

Parameter	Type	File Name	RotTorq (kN·m)	GenTq (kN)	RotSpeed (rpm)	GenSpeed (rpm)	RotThrust (kN)	GenPwr (kN)	Wind1VelX (m/s)	Wind1VelY (m/s)	Time (s)
RotTorq	Min	Dlc_61T_-08deg_S5.out	**-2.65E+04**	-3.15E+02	1.01E-02	0.00E+00	3.31E+02	2.16E+02	6.46E+01	-9.75E+00	5.57E+02
RotTorq	Max	Dlc_62T_+160deg_S2.out	**2.53E+04**	1.91E+02	-1.14E-02	0.00E+00	-1.96E+02	-3.42E+02	-5.62E+01	1.95E+01	3.77E+02
GenTq	Min	Dlc_21T_+00deg_Vout_S7.out	5.66E+03	**-6.92E+02**	1.09E+01	1.06E+03	3.96E+02	3.12E+02	2.29E+01	1.78E+00	6.51E+01
GenTq	Max	Dlc_21T_+08deg_Vout_S6.out	5.25E+03	**9.93E+02**	1.09E+01	1.05E+03	-1.77E+01	-4.13E+02	2.71E+01	5.02E+00	6.67E+01
RotSpeed	Min	Dlc_33_4.0_-_Vr-2.out	3.51E+02	-5.00E+01	**-1.96E+00**	-1.89E+02	1.61E+02	1.92E+01	6.99E+00	6.47E+00	1.10E+02
RotSpeed	Max	Dlc_23_6.3_Vr+2.out	-3.78E+01	-7.22E+01	**1.79E+01**	1.75E+03	1.43E+02	9.74E+01	1.19E+01	0.00E+00	6.94E+01
GenSpeed	Min	Dlc_33_4.0_-_Vr-2.out	-2.33E+01	-7.09E+01	-1.95E+00	**-1.90E+02**	1.68E+02	3.61E+01	7.00E+00	6.28E+00	1.11E+02
GenSpeed	Max	Dlc_23_6.3_Vr+2.out	-3.78E+01	-7.22E+01	1.79E+01	**1.75E+03**	1.43E+02	9.74E+01	1.19E+01	0.00E+00	6.94E+01
RotThrust	Min	Dlc_23_6.3_Vout.out	-1.17E+03	-7.23E+01	1.47E+01	1.43E+03	**-7.60E+02**	-8.90E+02	2.43E+01	0.00E+00	6.95E+01
RotThrust	Max	Dlc_42_6.3_Vr.out	6.34E+03	-1.37E+01	1.35E+01	1.30E+03	**1.72E+03**	1.28E+03	1.72E+01	-1.07E-04	6.70E+01
GenPwr	Min	Dlc_23_6.3_Vout.out	-9.93E+02	-7.39E+01	1.48E+01	1.44E+03	-7.58E+02	**-8.93E+02**	2.46E+01	0.00E+00	6.94E+01
GenPwr	Max	Dlc_23_6.3_Vr+2.out	5.57E+03	-2.90E+01	1.28E+01	1.24E+03	1.35E+03	**1.31E+03**	1.90E+01	0.00E+00	6.63E+01

Blade1 Root

Parameter	Type	File Name	RootFxc1 (kN)	RootFyc1 (kN)	RootFzc1 (kN)	RootMxc1 (kN·m)	RootMyc1 (kN·m)	RootMzc1 (kN·m)	RootFMxy1 (kN)	RootMMxy1 (kNm)	Wind1VelX (m/s)	Wind1VelY (m/s)	Time (s)
RootFxc1	Min	Dlc_62T_+100deg_S4.out	**-2.58E+02**	-5.33E-01	-1.97E+02	8.09E+02	-6.54E+03	-1.65E+02	2.58E+02	6.59E+03	-1.01E+01	5.74E+01	2.76E+02
RootFxc1	Max	Dlc_21T_+08deg_Vout_S3.out	**7.06E+02**	-5.22E+02	6.00E+02	1.45E+04	2.38E+04	-2.00E+02	8.78E+02	2.79E+04	2.02E+01	2.65E+00	6.59E+01
RootFyc1	Min	Dlc_21T_+08deg_Vout_S6.out	6.51E+02	**-5.74E+02**	8.14E+02	1.56E+04	2.11E+04	-1.18E+02	8.68E+02	2.62E+04	4.65E+01	3.84E+00	6.58E+01
RootFyc1	Max	Dlc_61T_+08deg_S6.out	7.41E+00	**3.43E+02**	-2.39E+02	-8.33E+03	-1.05E+03	-1.17E+02	3.43E+02	8.40E+03	5.47E+01	9.62E+00	3.96E+02
RootFzc1	Min	Dlc_61T_+08deg_S5.out	3.41E+01	5.61E+01	**-2.49E+02**	2.45E+03	6.98E+01	-1.53E+02	6.56E+01	2.45E+03	5.87E+01	8.58E+00	1.90E+02
RootFzc1	Max	Dlc_42_6.3_Vout.out	2.91E+02	7.63E+01	**1.74E+03**	-3.34E+03	5.12E+03	-1.34E+02	3.01E+02	6.12E+03	3.41E+01	-2.25E-03	6.74E+01
RootMxc1	Min	Dlc_61T_-08deg_S3.out	7.23E+01	3.12E+02	-2.44E+02	**-8.90E+03**	1.36E+03	-8.37E+01	3.21E+02	9.00E+03	4.59E+01	-3.88E+00	6.35E+02
RootMxc1	Max	Dlc_21T_+08deg_Vout_S6.out	6.06E+02	-5.57E+02	8.20E+02	**1.58E+04**	2.00E+04	-1.60E+02	8.23E+02	2.55E+04	3.99E+01	3.57E+00	6.59E+01
RootMyc1	Min	Dlc_23_6.3_Vout.out	-2.29E+02	-1.96E+02	8.24E+02	3.16E+03	**-1.09E+04**	9.23E+01	3.01E+02	1.14E+04	2.42E+01	0.00E+00	7.00E+01
RootMyc1	Max	Dlc_14_-08deg_-_Vr+2.out	6.78E+02	-3.53E+02	4.94E+02	9.72E+03	**2.59E+04**	-2.55E+02	7.64E+02	2.76E+04	7.65E+00	2.49E+01	7.53E+01
RootMzc1	Min	Dlc_14_+00deg_-_Vr+2.out	-1.28E+02	-2.46E+00	7.11E+02	6.56E+01	-4.84E+03	**-2.73E+02**	1.28E+02	4.84E+03	1.78E+01	2.16E+01	6.86E+01
RootMzc1	Max	Dlc_23_6.3_Vr+2.out	4.55E+02	-5.94E+01	9.50E+02	5.07E+02	1.53E+04	**2.28E+02**	4.59E+02	1.53E+04	1.89E+01	0.00E+00	6.66E+01
RootFMxy1	Min	Dlc_62T_+180deg_S4.out	3.38E-02	1.30E-01	-1.93E+02	-1.74E+03	6.46E+02	-5.68E+01	**1.35E-01**	1.86E+03	-4.59E+01	7.76E-02	4.30E+02
RootFMxy1	Max	Dlc_21T_+08deg_Vout_S6.out	6.82E+02	-5.68E+02	8.06E+02	1.52E+04	2.17E+04	-7.85E+01	**8.88E+02**	2.65E+04	4.74E+01	4.25E+00	6.57E+01
RootMMxy1	Min	Dlc_62T_-80deg_S6.out	6.96E+01	-1.54E+01	-1.97E+02	-1.31E-01	-1.48E+00	-1.18E+02	7.13E+01	**1.49E+00**	8.79E+00	-5.34E+01	6.59E+02
RootMMxy1	Max	Dlc_21T_+08deg_Vout_S3.out	7.06E+02	-5.10E+02	5.89E+02	1.44E+04	2.40E+04	-2.01E+02	8.71E+02	**2.80E+04**	2.20E+01	2.80E+00	6.59E+01

Blade2 Root

Parameter	Type	File Name	RootFxc2 (kN)	RootFyc2 (kN)	RootFzc2 (kN)	RootMxc2 (kN·m)	RootMyc2 (kN·m)	RootMzc2 (kN·m)	RootFMxy2 (kN)	RootMMxy2 (kNm)	Wind1VelX (m/s)	Wind1VelY (m/s)	Time (s)
RootFxc2	Min	Dlc_23_6.3_Vout.out	**-2.99E+02**	2.68E+01	1.12E+03	-1.64E+03	-1.35E+04	1.22E+02	3.01E+02	1.36E+04	2.46E+01	0.00E+00	6.94E+01
RootFxc2	Max	Dlc_11T_+08deg_V14_S2.out	**6.47E+02**	-3.94E+02	8.04E+02	9.91E+03	2.29E+04	-1.06E+02	7.57E+02	2.50E+04	3.16E+01	1.68E+00	7.95E+01
RootFyc2	Min	Dlc_62T_+120deg_S5.out	-9.58E+01	**-6.92E+02**	4.24E+01	1.96E+04	-2.04E+03	-3.65E+02	6.99E+02	1.98E+04	-2.75E+01	5.72E+01	5.64E+02
RootFyc2	Max	Dlc_61T_-08deg_S3.out	5.42E+01	**4.31E+02**	7.35E+01	-1.40E+04	-2.24E+02	-1.57E+02	4.35E+02	1.40E+04	5.81E+01	-6.71E+00	6.50E+02
RootFzc2	Min	Dlc_21T_-08deg_Vout_S12.out	1.11E+01	-6.33E+01	**-2.37E+02**	3.23E+03	-1.32E+02	-1.30E+01	6.43E+01	3.23E+03	2.78E+01	-1.83E+00	5.82E+02
RootFzc2	Max	Dlc_42_6.3_Vr+2.out	1.26E+02	-6.10E+01	**1.71E+03**	1.38E+03	3.22E+03	-1.51E+02	1.40E+02	3.50E+03	1.29E+01	-1.50E-03	6.95E+01
RootMxc2	Min	Dlc_61T_-08deg_S3.out	6.53E+01	4.31E+02	7.53E+01	**-1.41E+04**	1.27E+01	-1.54E+02	4.36E+02	1.41E+04	5.68E+01	-6.92E+00	6.50E+02
RootMxc2	Max	Dlc_62T_+120deg_S5.out	-9.47E+01	-6.89E+02	4.01E+01	**1.97E+04**	-1.99E+03	-3.59E+02	6.95E+02	1.98E+04	-2.83E+01	5.70E+01	5.64E+02
RootMyc2	Min	Dlc_42_6.3_Vout.out	-2.86E+02	8.94E-03	1.24E+03	-9.96E+02	**-1.35E+04**	8.66E+01	2.86E+02	1.36E+04	2.39E+01	-1.66E-03	6.96E+01
RootMyc2	Max	Dlc_14_+08deg_+_Vr+2.out	6.23E+02	1.34E+02	6.50E+02	-6.62E+02	**2.40E+04**	1.40E+02	6.37E+02	2.40E+04	1.89E+01	-2.50E+01	7.63E+01
RootMzc2	Min	Dlc_62T_+160deg_S2.out	-1.14E+02	-6.14E+02	6.39E+01	1.61E+04	-1.85E+03	**-4.84E+02**	6.25E+02	1.62E+04	-6.01E+01	2.61E+01	2.88E+02
RootMzc2	Max	Dlc_23_6.3_Vr+2.out	4.57E+02	1.74E+02	7.34E+02	-4.03E+03	1.55E+04	**3.38E+02**	4.89E+02	1.60E+04	1.89E+01	0.00E+00	6.66E+01
RootFMxy2	Min	Dlc_62T_-140deg_S3.out	-1.82E-01	-1.61E-01	8.37E+01	-2.32E+03	4.20E+02	1.78E+02	**2.43E-01**	2.36E+03	-4.60E+01	-3.28E+01	5.03E+02
RootFMxy2	Max	Dlc_11T_+08deg_V14_S2.out	6.47E+02	-3.94E+02	8.04E+02	9.91E+03	2.29E+04	-1.06E+02	**7.57E+02**	2.50E+04	3.16E+01	1.68E+00	7.95E+01
RootMMxy2	Min	Dlc_21T_-08deg_Vout_S1.out	1.13E+01	4.66E+01	-2.14E+02	-2.16E+00	-1.44E-02	-1.00E+01	4.79E+01	**2.16E+00**	2.36E+01	8.45E-02	1.33E+02
RootMMxy2	Max	Dlc_11T_+08deg_V14_S2.out	6.42E+02	-3.99E+02	7.87E+02	1.01E+04	2.29E+04	-1.10E+02	7.56E+02	**2.50E+04**	4.24E+01	1.91E+00	7.95E+01

Blade3 Root

Parameter	Type	File Name	RootFxc3 (kN)	RootFyc3 (kN)	RootFzc3 (kN)	RootMxc3 (kN·m)	RootMyc3 (kN·m)	RootMzc3 (kN·m)	RootFMxy3 (kN)	RootMMxy3 (kNm)	Wind1VelX (m/s)	Wind1VelY (m/s)	Time (s)
RootFxc3	Min	Dlc_21T_-08deg_Vout_S11.out	**-2.79E+02**	-7.55E+01	7.88E+02	-9.27E+02	-1.26E+04	4.63E+01	2.89E+02	1.26E+04	6.00E+01	-4.30E+00	6.78E+01
RootFxc3	Max	Dlc_14_-08deg_-_Vr+2.out	**6.83E+02**	-3.21E+02	5.63E+02	8.91E+03	2.61E+04	-1.99E+02	7.55E+02	2.75E+04	7.32E+00	2.49E+01	7.69E+01
RootFyc3	Min	Dlc_11T_-08deg_V24_S3.out	4.25E+02	**-5.16E+02**	9.02E+02	1.23E+04	1.22E+04	-5.33E+01	6.68E+02	1.73E+04	2.25E+01	-3.63E+00	2.31E+02
RootFyc3	Max	Dlc_62T_-140deg_S5.out	3.89E+00	**5.37E+02**	6.76E+01	-1.49E+04	5.66E+02	3.36E+02	5.37E+02	1.49E+04	-4.94E+01	-3.82E+01	3.32E+02
RootFzc3	Min	Dlc_21T_+08deg_Vout_S3.out	2.19E-01	1.17E+02	**-2.40E+02**	-3.91E+03	-6.05E+02	-4.86E+01	1.17E+02	3.95E+03	2.71E+01	2.73E+00	3.84E+02
RootFzc3	Max	Dlc_42_6.3_Vout.out	4.30E+01	-9.10E+01	**1.82E+03**	4.78E+02	-2.36E+03	-1.87E+02	1.01E+02	2.41E+03	3.26E+01	-8.12E-03	6.80E+01
RootMxc3	Min	Dlc_62T_-140deg_S5.out	3.89E+00	5.37E+02	6.76E+01	**-1.49E+04**	5.66E+02	3.36E+02	5.37E+02	1.49E+04	-4.94E+01	-3.82E+01	3.32E+02
RootMxc3	Max	Dlc_61T_+08deg_S5.out	-6.76E-01	-2.75E+02	6.07E+01	**1.37E+04**	-1.76E+03	-4.86E+01	2.75E+02	1.38E+04	5.89E+01	1.22E+01	1.70E+02
RootMyc3	Min	Dlc_21T_-08deg_Vout_S11.out	-2.79E+02	-7.55E+01	7.88E+02	-9.27E+02	**-1.26E+04**	4.63E+01	2.89E+02	1.26E+04	6.00E+01	-4.30E+00	6.78E+01
RootMyc3	Max	Dlc_14_-08deg_-_Vr+2.out	6.78E+02	-3.45E+02	5.98E+02	9.62E+03	**2.61E+04**	-2.36E+02	7.61E+02	2.78E+04	5.95E+00	2.49E+01	7.70E+01
RootMzc3	Min	Dlc_62T_+160deg_S6.out	-4.36E+01	-1.31E+02	8.14E+01	5.12E+03	-6.70E+01	**-4.07E+02**	1.38E+02	5.12E+03	-6.18E+01	2.20E+01	3.88E+02
RootMzc3	Max	Dlc_62T_-160deg_S6.out	-1.18E+01	4.37E+02	7.13E+01	-1.32E+04	5.27E+02	**3.55E+02**	4.37E+02	1.32E+04	-6.31E+01	-1.91E+01	3.17E+02
RootFMxy3	Min	Dlc_62T_+60deg_S4.out	-1.91E-02	2.05E-02	9.02E+01	8.00E+02	-3.03E+01	-1.56E+01	**2.80E-02**	8.01E+02	2.34E+01	3.54E+01	4.26E+02
RootFMxy3	Max	Dlc_14_-08deg_-_Vr+2.out	6.71E+02	-3.74E+02	7.11E+02	1.01E+04	2.51E+04	-1.59E+02	**7.68E+02**	2.70E+04	9.83E+00	2.48E+01	7.72E+01
RootMMxy3	Min	Dlc_62T_+120deg_S1.out	-3.71E+00	4.05E+01	9.65E+01	-6.14E-01	1.05E-01	-1.06E+02	4.07E+01	**6.23E-01**	-2.43E+01	4.09E+01	5.00E+02
RootMMxy3	Max	Dlc_14_-08deg_-_Vr+2.out	6.75E+02	-3.54E+02	6.17E+02	9.89E+03	2.60E+04	-2.45E+02	7.62E+02	**2.78E+04**	5.95E+00	2.49E+01	7.70E+01

Tower Top

Parameter	Type	File Name	YawBrFxp (kN)	YawBrFyp (kN)	YawBrFzp (kN)	YawBrMxp (kN·m)	YawBrMyp (kN·m)	YawBrMzp (kN·m)	YawBrFMxyp (kN)	YawBrMMxyp (kNm)	Wind1VelX (m/s)	Wind1VelY (m/s)	Time (s)
YawBrFxp	Min	Dlc_42_8.4_Vout.out	**-9.94E+02**	-3.07E+01	-4.53E+03	2.91E+03	9.66E+02	-4.48E+02	9.94E+02	3.06E+03	2.68E+01	-1.94E-03	6.90E+01
YawBrFxp	Max	Dlc_42_6.3_Vr+2.out	**1.76E+03**	-3.08E+01	-4.76E+03	6.89E+03	2.32E+03	1.73E+02	1.76E+03	7.27E+03	1.92E+01	-9.58E-04	6.64E+01
YawBrFyp	Min	Dlc_21T_+00deg_Vout_S7.out	4.22E+02	**-6.91E+02**	-4.84E+03	8.28E+03	1.86E+04	1.46E+04	8.10E+02	2.04E+04	2.29E+01	1.78E+00	6.51E+01
YawBrFyp	Max	Dlc_21T_+08deg_Vout_S6.out	-5.58E+02	**9.94E+02**	-4.57E+03	2.94E+03	-2.74E+04	-4.54E+03	1.14E+03	2.75E+04	2.71E+01	5.02E+00	6.67E+01
YawBrFzp	Min	Dlc_11T_+08deg_V24_S5.out	2.97E+02	-1.16E+02	**-5.40E+03**	6.53E+03	-2.33E+03	5.75E+03	3.19E+02	6.94E+03	7.35E+01	3.43E+00	3.56E+02
YawBrFzp	Max	Dlc_62T_+00deg_S6.out	1.85E+01	2.57E+02	**-3.28E+03**	-4.91E+03	2.37E+03	-2.72E+02	2.57E+02	5.45E+03	5.64E+01	1.91E+00	5.48E+02
YawBrMxp	Min	Dlc_61T_-08deg_S5.out	3.12E+02	-2.78E+02	-4.35E+03	**-2.58E+04**	2.01E+03	1.74E+03	4.18E+02	2.59E+04	6.47E+01	-1.06E+01	5.57E+02
YawBrMxp	Max	Dlc_62T_+160deg_S2.out	-3.73E+02	1.71E+02	-3.82E+03	**2.47E+04**	-3.22E+03	-4.04E+03	4.10E+02	2.49E+04	-5.63E+01	1.97E+01	3.77E+02
YawBrMyp	Min	Dlc_21T_+08deg_Vout_S6.out	-2.72E+02	7.20E+02	-4.79E+03	4.36E+03	**-2.98E+04**	7.34E+03	7.70E+02	3.02E+04	2.79E+01	4.42E+00	6.63E+01
YawBrMyp	Max	Dlc_21T_-08deg_Vout_S6.out	4.88E+02	-3.44E+02	-4.58E+03	4.48E+03	**2.90E+04**	-4.45E+03	5.97E+02	2.93E+04	2.33E+01	-2.21E+00	6.47E+01
YawBrMzp	Min	Dlc_21T_-08deg_Vout_S11.out	2.83E+01	1.11E+02	-4.27E+03	2.03E+03	-2.98E+03	**-3.03E+04**	1.15E+02	3.60E+03	6.00E+01	-4.30E+00	6.78E+01
YawBrMzp	Max	Dlc_21T_+08deg_Vout_S3.out	5.20E+02	-2.45E+01	-4.92E+03	7.79E+03	2.89E+03	**2.87E+04**	5.21E+02	8.31E+03	-4.01E+01	3.28E+00	6.61E+01
YawBrFMxyp	Min	Dlc_51T_+00deg_Vr-2_S1.out	-3.27E-02	3.25E-02	-4.62E+03	-4.08E+02	-1.86E+03	-3.66E+00	**4.61E-02**	1.90E+03	1.13E+01	4.50E-01	1.24E+02
YawBrFMxyp	Max	Dlc_42_6.3_Vr+2.out	1.76E+03	-3.08E+01	-4.76E+03	6.89E+03	2.32E+03	1.73E+02	**1.76E+03**	7.27E+03	1.92E+01	-9.58E-04	6.64E+01
YawBrMMxyp	Min	Dlc_21T_+08deg_Vr_S2.out	7.29E+01	-1.01E+01	-4.62E+03	-3.32E-02	-1.70E+00	3.67E+02	7.36E+01	**1.70E+00**	1.42E+01	2.19E+00	1.45E+02
YawBrMMxyp	Max	Dlc_21T_+08deg_Vout_S6.out	-2.72E+02	7.20E+02	-4.79E+03	4.36E+03	-2.98E+04	7.34E+03	7.70E+02	**3.02E+04**	2.79E+01	4.42E+00	6.63E+01

Tower Base

Parameter	Type	File Name	TwrBsFxt (kN)	TwrBsFyt (kN)	TwrBsFzt (kN)	TwrBsMxt (kN·m)	TwrBsMyt (kN·m)	TwrBsMzt (kN·m)	TwrBsFMxyt (kN)	TwrBsMMxyt (kNm)	Wind1VelX (m/s)	Wind1VelY (m/s)	Time (s)
TwrBsFxt	Min	Dlc_62T_+160deg_S2.out	**-1.25E+03**	4.78E+02	-7.59E+03	-1.15E+04	-8.12E+04	-3.86E+03	1.34E+03	8.20E+04	-6.04E+01	2.42E+01	6.89E+02
TwrBsFxt	Max	Dlc_42_6.3_Vr+2.out	**1.85E+03**	-3.08E+01	-9.38E+03	9.82E+03	1.64E+05	1.25E+02	1.85E+03	1.64E+05	1.93E+01	-9.56E-04	6.64E+01
TwrBsFyt	Min	Dlc_62T_-120deg_S2.out	-3.97E+02	**-1.17E+03**	-7.62E+03	6.11E+04	-1.89E+04	3.18E+03	1.23E+03	6.39E+04	-3.21E+01	-5.83E+01	6.82E+02
TwrBsFyt	Max	Dlc_62T_+60deg_S6.out	1.60E+02	**1.29E+03**	-7.47E+03	-7.05E+04	-1.69E+03	-3.61E+03	1.30E+03	7.06E+04	3.07E+01	5.31E+01	1.81E+02
TwrBsFzt	Min	Dlc_11T_+08deg_V24_S5.out	4.44E+02	-1.03E+02	**-1.05E+04**	1.65E+04	3.09E+04	5.75E+03	4.56E+02	3.51E+04	7.35E+01	3.43E+00	3.56E+02
TwrBsFzt	Max	Dlc_62T_+00deg_S6.out	6.41E+02	3.07E+02	**-7.03E+03**	-3.07E+04	3.24E+04	-2.76E+02	7.10E+02	4.47E+04	5.64E+01	1.91E+00	5.48E+02
TwrBsMxt	Min	Dlc_21T_+08deg_Vout_S6.out	-4.86E+02	1.16E+03	-9.17E+03	**-9.52E+04**	-7.51E+04	-6.28E+03	1.25E+03	1.21E+05	2.80E+01	4.68E+00	6.68E+01
TwrBsMxt	Max	Dlc_33_6.0_+_Vout.out	5.71E+02	-7.95E+02	-9.28E+03	**7.38E+04**	4.95E+04	4.22E+02	9.79E+02	8.88E+04	2.09E+01	-1.36E+01	1.27E+02
TwrBsMyt	Min	Dlc_42_8.4_Vout.out	-9.41E+02	-2.12E+01	-9.13E+03	4.86E+03	**-8.59E+04**	-5.15E+02	9.42E+02	8.61E+04	2.66E+01	-2.64E-03	6.91E+01
TwrBsMyt	Max	Dlc_42_6.3_Vr+2.out	1.85E+03	-3.16E+01	-9.38E+03	9.80E+03	**1.64E+05**	1.90E+02	1.85E+03	1.64E+05	1.92E+01	-9.58E-04	6.64E+01
TwrBsMzt	Min	Dlc_21T_-08deg_Vout_S11.out	7.65E+01	7.14E+00	-8.87E+03	-3.34E+03	9.37E+02	**-3.03E+04**	7.68E+01	3.47E+03	6.00E+01	-4.30E+00	6.78E+01
TwrBsMzt	Max	Dlc_21T_+08deg_Vout_S3.out	6.90E+02	-4.56E+01	-9.53E+03	1.12E+04	5.84E+04	**2.87E+04**	6.92E+02	5.94E+04	-4.01E+01	3.28E+00	6.61E+01
TwrBsFMxyt	Min	Dlc_51T_+08deg_Vr-2_S6.out	-7.31E-02	-9.54E-02	-9.23E+03	-4.14E+02	-2.46E+03	-2.81E+01	**1.20E-01**	2.49E+03	7.53E+00	1.46E+00	1.91E+02
TwrBsFMxyt	Max	Dlc_42_8.4_Vr+2.out	1.85E+03	-3.16E+01	-9.38E+03	9.83E+03	1.64E+05	9.26E+01	**1.85E+03**	1.64E+05	1.93E+01	-1.18E-03	6.64E+01
TwrBsMMxyt	Min	Dlc_21T_+00deg_Vr_S12.out	3.18E+01	-8.27E-01	-9.23E+03	1.01E+00	-1.37E+00	2.58E+01	3.18E+01	**1.70E+00**	9.82E+00	-4.02E-01	5.98E+02
TwrBsMMxyt	Max	Dlc_42_8.4_Vr+2.out	1.85E+03	-3.16E+01	-9.38E+03	9.83E+03	1.64E+05	9.26E+01	1.85E+03	**1.64E+05**	1.93E+01	-1.18E-03	6.64E+01

14.10 Expertise Survey - Yaw Bearing at WT Tower Base

S·K·I
Schaumann • Keindorf • Ingenieure

SKI Ingenieurges. mbH
Mengendamm 12
30177 Hannover
Tel. +49-511-261847-0
Fax +49-511-261847-29
www.ski-consult.de

Gutachtliche Stellungnahme

Auftraggeber:	**Wind Energy Technology Institute** **Nordstraße 2** **24937 Flensburg**
Anlass:	**Verlegung des Azimutlagers einer Windenergieanlage von der Gondel zum Turmfuß**
Auftragnehmer:	**SKI Ingenieurges. mbH** **Mengendamm 12** **30177 Hannover**
Gutachten:	**2015-09-GTA-01**
Umfang:	**8 DIN A4 Seiten**

Hannover, 28.01.2015	00	C. Keindorf
Ort, Datum	Revision	Prof. Dr.-Ing. C. Keindorf

Inhaltsverzeichnis

1. Einleitung

Bei Windenergieanlagen mit horizontaler Rotorachse wird die Ausrichtung der Rotorblätter in den Wind durch ein drehbares Lager, welches direkt unter der Gondel liegt, realisiert. Untersuchungen haben gezeigt, dass es sowohl wirtschaftlich als auch umwelttechnisch interessant ist, das Azimutlager vom Turmkopf zum Turmfuß zu verlagern, so dass der komplette Turm mit gedreht wird. Ein drehbarer Turm hätte die Vorteile, dass dieser geometrisch bezüglich Luftwiderstand und Tragfähigkeit optimaler als bisherige rotationssymmetrische Türme gestaltet werden könnte. Die Ausrichtung auf die aktuelle Windrichtung bei einem strömungsoptimierten Querschnitt hätte zur Folge, dass die Windlasten, die auf den Turm einwirken, reduziert werden können. Durch ein an die Hauptbelastungsrichtung angepasstes Flächenträgheitsmoment des Turmquerschnittes könnte Material und somit Kosten eingespart werden. Darüber hinaus könnte eine Geometrie realisiert werden, die den Luftwiderstand und die turbulenten Wirbelanregungen auf den Turm reduzieren.

2. Gutachtliche Stellungnahme

Durch die Verlegung des Azimutlagers von der Gondel zum Turmfuß verändern sich die Belastungen auf das Lager um ein hohes Maß. Das Lager muss neben dem zusätzlichen Gewicht des Turmes wesentlich größere Momente durch den verlängerten Hebelarm zwischen Rotor und Turmfuß aufnehmen. Die Umgestaltung des Turmes begünstigt jedoch eine im Turmfuß gelegene Lagerung in Bezug auf die Lasten. Das Gutachten ist eine methodische Betrachtung eines Azimutlager Konzeptes, um die Machbarkeit zu prüfen. Die Machbarkeit der Lagerverlegung an den Turmfuß ist gewährleistet, sobald ein Konzept gefunden wird, welches ausreichendes Potenzial für die wirtschaftliche Umsetzung aufweist.

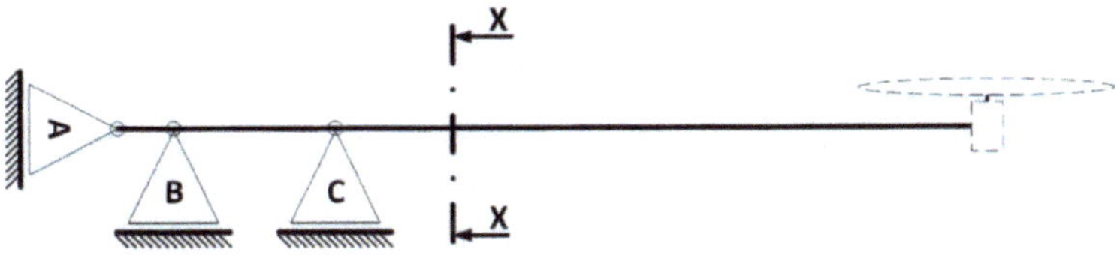

Abb. 2.1: Mögliches Lagergesamtkonzept

Die Abb. 2.1 zeigt ein mögliches Lagerkonzept, welches einen großen Vorteil gegenüber den bisher betrachteten Lösungen aus den vorangegangenen Arbeiten bietet. In [3.1] wurden bislang ein Gleitlager und ein zweireihiges Vierpunktlager am Turmfuß beschrieben. Die Nachteile dieser Ansätze ist das hohe Kippmoment, welches aufgenommen werden muss. Aufgrund des größeren Abstandes zwischen den Lagern B und C wird das Kippmoment aufgeteilt. Es können somit relativ klein dimensionierte Bauteile verwendet werden, wodurch deutliche Einsparungen im Vergleich zu den vorangegangenen Konzepten zu erwarten sind.

Die Separierung zwischen Lagern für axiale und radiale Kräfte ermöglicht zusätzlich zielgerichtetes Konstruieren ohne Kompromisse in Bezug auf die Belastungsrichtung. Ein weiterer Ansatz wäre das Lager B zu entfernen und Lager A als Festlager auszulegen. Dies würde den Vorteil der Separierung der Kräfte eliminieren – könnte jedoch durch die Einsparung eines Lagers noch kostengünstiger sein. Es lassen sich in jedem Fall alle Verschleißteile ohne Demontage des Turmes austauschen und warten. Das axiale Lager A kann durch einfaches Anheben des Turmes mittels Hydraulikstempeln ausgetauscht werden.

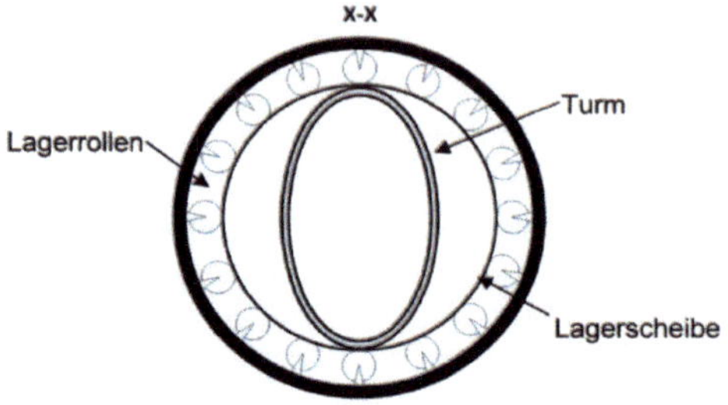

Abb. 2.2: Schematische Schnittdarstellung X-X eines radialen Lagerungsansatzes für Lager B und C

Die vorgespannten Lagerrollen der radialen Lager B und C können nach dem Entwurf in Abb. 2.2 einzeln entspannt und ausgetauscht werden ohne den Turm demontieren zu müssen. Turmseitig dient eine runde Lagerscheibe dazu, die Kräfte vom profilierten Turmquerschnitt in die Lagerrollen einzuleiten. Durch Optimierung der Topologie können sämtliche Bauteile noch materialsparender ausgelegt werden. Die Lager B und C können sowohl überirdisch als auch unterirdisch vorgesehen werden und somit mögliche Anforderungen an die Optik der Anlage im Landschaftsbild berücksichtigen.

3. Zusammenfassung

Zusammenfassend gibt es mit den begutachteten Lagerkonzepten mindestens zwei Lösungen in der die Probleme der

- hohen Momente am Turmfuß
- hohen Kosten einer Lagerung am Turmfuß

gelöst werden können. Es ist durchaus denkbar, dass eine Optimierung dieses Konzeptes oder weitere Konzepte zu einer noch besseren Wirtschaftlichkeit führen können. Die Verlegung des Azimutlagers zum Turmfuß bietet sehr hohes Potential für eine ressourcenschonende und somit umweltschützende Alternative zu konventionellen Bauweisen von Türmen für Windenergieanlagen.

Das Vorhaben kann dazu führen, dass die Bauweise zukünftiger Windenergieanlagen neu überdacht wird. Die Machbarkeit einer Verlegung des Azimutlagers in den Turmfuß ist gewährleistet. Da durch das Forschungsvorhaben ein innovatives, wirtschaftliches und umweltschonenderes Potential freigesetzt werden würde, empfehle ich das Promotionsstipendium für das Thema "*Entwicklung eines energieeffizienten und nachhaltigen Windenergieanlagenturm-Konzeptes mit Azimutlager am Turmfuß (Profilierter Windenergieanlagenturm)*" zu bewilligen.

4. Literatur

4.1 Normen und Regelwerke

[1.1] DIN EN 61400-1: Windenergieanlagen – Teil 1: Auslegungsanforderungen, Deutsches Institut für Normung, Beuth Verlag, 2001.

[1.2] DIN EN 61400-2: Windenergieanlagen – Teil 2: Sicherheit kleiner Windenergieanlagen, Deutsche Fassung EN 61400-2:2006.

4.2 Richtlinien

[2.1] Deutsches Institut für Bautechnik: Bauregelliste A, B und C, Mitteilungen DIBt 33, Sonderheft Nr. 26, 2002.

[2.2] DIBt: Richtlinie für Windenergieanlagen - Einwirkungen und Standsicherheitsnachweise für Turm und Gründung, DIBt - Schriften des Deutschen Instituts für Bautechnik, Reihe B, Heft 8, Berlin, Fassung 2012.

4.3 Projektunterlagen

[3.1] Beye, T.: Bachelor Thesis: Konzeptstudie über die Verlegung des Azimutlagers an den Turmfuß einer Windenergieanlage. Fachhochschule Flensburg, 05.2012.

4.4 Veröffentlichungen zum Thema

[4.1] Peterson, C.: Dynamik der Baukonstruktionen, Braunschweig/Wiesbaden, Vieweg & Sohn Verlag, 1996.

5. Verzeichnisse

5.1 Abbildungsverzeichnis

14.11 Transition between Shells and Plates

Local stability failures are based on the stresses inside a shell or a plate and its boundary conditions where plates are shells with radius $r = \infty$. In general, plates are treated within the plate theory and shells are treated within the membrane theory. Different profiles, such as plate like, ideal shell like and intermediates are expectable for rotatable tubular WT towers. Therefore, following statements should lead to extract the most relevant influences in terms of stability for arbitrary plates and shells, respectively. Shells with an ideal shape and without imperfections such as small local buckles and residual stresses sustain the membrane state of stress if deformations are not hindered [149]. Local shell buckling depends on the relation r/t, with r as the shell radius. This is visible through the statements within the Eurocode 3 DIN EN 1993-1-6:2010-12 [44], which regulates steel shell designs and provides r/t limits according to meridian buckling

$$\frac{r}{t} \leq 0.03 \frac{E}{f_{\text{yk}}}, \tag{14.1}$$

circumferential buckling

$$\frac{r}{t} \leq 0.21 \sqrt{\frac{E}{f_{\text{yk}}}} \tag{14.2}$$

and shear buckling

$$\frac{r}{t} \leq 0.16 \left(\frac{E}{f_{\text{yk}}} \right)^{0.67} \tag{14.3}$$

for unstiffed cylinder shells with constant wall thickness. If these limits are respected, corresponding buckling checks need not to be done. The general meaning of this is that local shell buckling resistance increases with decreasing r/t. This relation alone is insufficient to represent plate like profiles and to compare them with shell-like profiles, because $r/t = \infty$ for plates. But plates have the ability to sustain normal stresses, which is not suggested by the r/t ratio. However, to acquire this problem, explanations from Wiedemann [195] are used. Plate buckling can be described according to the expansion of the fourth order bending line theory differential equation for beams with constant bending stiffness under axial force N

$$E\, I_{\text{B}} \frac{\partial^4 w}{\partial x^4} + N \frac{\partial w}{\partial x^2} = q(x), \tag{14.4}$$

where $q(x)$ equals to a line load. The expansion of equation 14.4 for plates considers pressure loads N_x, N_y and shear loads $N_{xy} = N_{yx}$ at the borders according to Petersen [149].

$$\frac{E\, I_{\text{B}}}{1 - \nu^2} \left(\frac{\partial^4 w}{\partial x^4} + 2 \frac{\partial^4 w}{\partial x^2\, \partial y^2} + \frac{\partial^4 w}{\partial y^4} \right) + N_x \frac{\partial^2 w}{\partial x^2} + 2\, N_{xy} \frac{\partial^2 w}{\partial x\, \partial y} + N_y \frac{\partial^2 w}{\partial y^2} = q(x, y). \tag{14.5}$$

Practical civil engineering checks without the over critical nonlinear behaviour of a plate are based on equation 14.5. Solving this equation for specific load cases leads to the critical plate buckling stress

$$\sigma_{\text{Ki}} = k_\sigma\, E \left(\frac{t}{b} \right)^2 \tag{14.6}$$

with the buckling value k_σ and the geometric influence of plate wall thickness t divided by its width b squared. Buckling value k_σ contains the influence of the outer plate dimensions, such as its width b and length a. Plates under normal stresses $\sigma_{\mathrm{N}} \geq \sigma_{\mathrm{Ki}}$ will experience a certain amount of buckling half waves m in longitudinal and n in lateral direction dependent on the aspect ratio a/b. Kinds of plate support do have also influence, but are assumed as monovalent at each edge. For practical applications m and n should be selected according to the minimum of k_σ within

$$k_\sigma = \frac{\pi^2}{12\left(1-\nu^2\right)} n^2 \left(\frac{n\,a}{m\,b} + \frac{m\,b}{n\,a}\right)^2, \tag{14.7}$$

because this leads to the first and therefore most relevant failure mode. Incidentally, σ_{Ki} equals to the bifurcation stress until a plate starts to buckle and can also be found in the Eurocode 3 norm DIN EN 1993-1-5 [43] where term $\pi^2/\left[12\left(1-\nu^2\right)\right]$ is included in equation 14.6 instead of in the definition of k_σ in Equation 14.7. Furthermore, occurring shear stresses have to be mentioned similar to Equation 14.6 within

$$\tau_{\mathrm{Ki}} = k_\tau\, E \left(\frac{t}{b}\right)^2. \tag{14.8}$$

Previous remarks are only valid for plates with monovalent supported borders and constant stress distributions along the edge. Therefore, other configurations will lead to different critical buckling stresses, such as shown in Figure 14.1 where different plates with varying stress distributions and supports are compared.

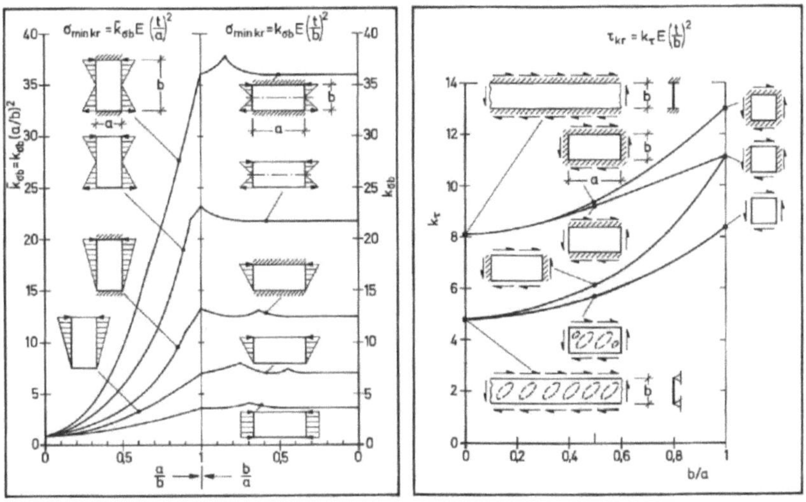

Figure 14.1: Buckling value dependencies for different aspect ratios a/b and edge supports [195]

Buckling value k_σ got the little additional index $_\mathrm{b}$ to mark it as bending load distribution dependent buckling value. The monovalent supported plate under uniform distributed in-plane compression loads is the most critical situation except for plates with free edges. Any kind of increasing border support in terms of divalent or trivalent borders will increase the critical buckling stresses. To compare shell and plate like behaviours Wiedemann [195] proposes a curvature parameter Ω_{CP} based on Timoshenko and Gere [186] to describe the proportion between membrane and bending support of a shell. It is build out of the plate bending stiffness $B = E\,t^3/\left[12\left(1-\nu^2\right)\right]$ and the membrane stiffness $D = E\,t$ according to

$$\Omega_{\text{CP}} \equiv \frac{D\,b^4}{B\,r^2} = 12\left(1-\nu^2\right)\left(\frac{b^2}{r\,t}\right)^2 . \tag{14.9}$$

Small lateral curvatures increase the buckling values compared to plates ($\Omega_{\text{CP}} = 0$) and lead to a curvature parameter dependent buckling value

$$k_\sigma\left(\Omega_{\text{CP}}\right) = \frac{\pi^2\,n^2}{12\left(1-\nu^2\right)}\left(\frac{n\,a}{m\,b}\right)^2\left\{\left[\left(\frac{m\,b}{n\,a}\right)^2+1\right]^2+\frac{\Omega_{\text{CP}}}{\pi^4\,n^4}\left[\left(\frac{n\,a}{m\,b}\right)^2+1\right]^{-2}\right\} \tag{14.10}$$

Low curvatures result in only one ($n = 1$) buckle over b and in pressure direction in a half wave length $a/m = b$. This means according to Wiedemann [195] for $a/b > 1$:

$$k_\sigma \approx 3.6\left(1+\frac{\Omega_{\text{CP}}}{16\,\pi^4}\right) \tag{14.11}$$

Relation 14.11 is valid up to $\sqrt{\Omega_{\text{CP}}} < 40$. Critical buckling stresses for plates with greater Ω_{CP} becomes proportional to $\sqrt{\Omega_{\text{CP}}}$ and will be calculated according to

$$\sigma_{\text{Ki}} = k_\sigma\,E\left(\frac{t}{b}\right)^2 = 0.61\,E\frac{t}{r} \tag{14.12}$$

because of the increasing amount of buckles over b regarding great $b^2/(r\,t)$ ratios. Therefore, σ_{Ki} becomes independent from b and leads to equation 14.12, which is valid up to closed cylindrical shells. Figure 14.2 shows the curvature parameter dependent buckling value for monovalent supported plates. The intersection of the blue and red dashed lines marks the transition from curved plate to cylindrical shell buckling.

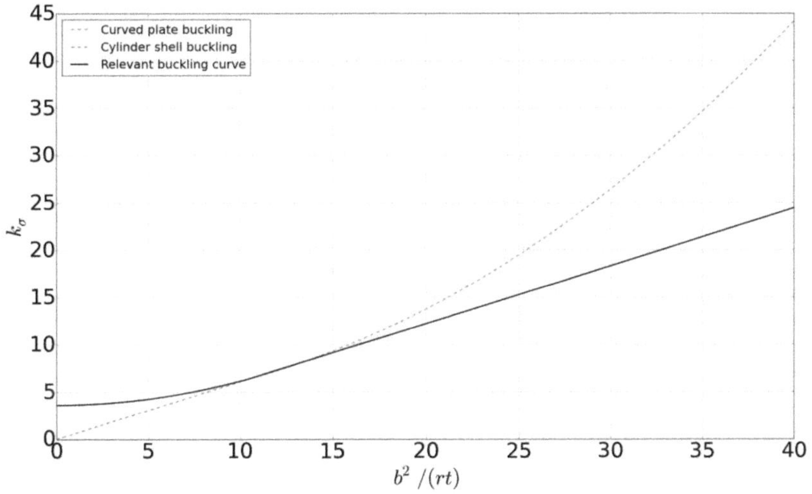

Figure 14.2: Buckling value dependency according to different plate curvatures for plates with monovalent supported borders

The intersection point can be determined by equating Equation 14.11 and 14.12 and leads to the quadratic equation

$$\begin{aligned}
&\left(\frac{b^2}{r\,t}\right)^2 - \frac{0.61 \cdot 4\,\pi^4}{3.6 \cdot 3\left(1-\nu^2\right)}\frac{b^2}{r\,t} + \frac{4\,\pi^4}{3\left(1-\nu^2\right)} = 0\\
\Rightarrow &\frac{b^2}{r\,t}_{1,2} = \frac{24.184}{2} \pm \sqrt{\left(\frac{24.184}{2}\right)^2 - 142.724}\\
&\frac{b^2}{r\,t}_1 = 10.223\\
&\frac{b^2}{r\,t}_2 = 13.961
\end{aligned} \tag{14.13}$$

To assume the most critical k_σ curve, the second intersection point with $b^2\ /\ (r\ t) = 13.961$ was chosen to mark the limit for Equation 14.11 to be valid. Furthermore, curvature parameter dependent shear buckling values for $a/b = \infty$ equal to

$$k_\tau \approx 4.8\sqrt[4]{1 + \frac{\Omega_{\mathrm{CP}}}{770}} \tag{14.14}$$

and will increase for $1 < a/b < \infty$. Applying WT relevant dimensions for the profile width $b = const. = 6\ m$ led to radius dependent critical normal stresses, such as shown in Figure 14.3 for different wall thicknesses t. The lower radius limit was set to the half of b so that the minimum allowable curvature was given through a half cylinder of diameter b. Such a supported half cylinder can likewise be seen as a full cylinder without supports. Figure 14.3 shows that small radii and great wall thickness led to higher critical normal buckling stresses. Furthermore, yield strengths of f_{yk} = 275 MPa, f_{yk} = 355 MPa and f_{yk} = 460 MPa for common WT steel types were introduced. They show that it was possible to reach the yield limit before the plate starts to buckle for small radii and great wall thickness likewise.

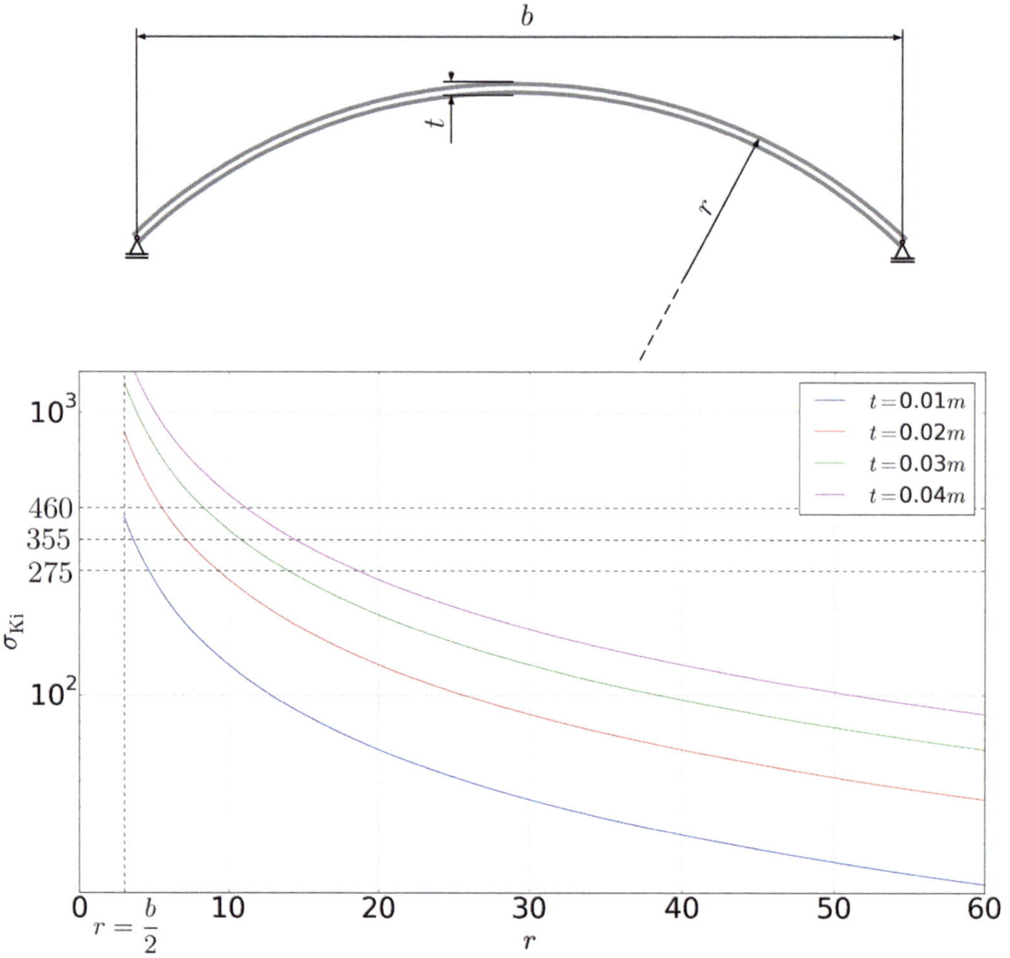

Figure 14.3: Critical normal buckling stresses for large WT tower dimensions depending on curvature radius r and wall thickness t for constant plate width $b = 6\ m$

The topic of local buckling becomes more relevant for LDD shell towers than for conventional circular ones. Due to decreasing curvature of steel sheets at elongated tower flanks, lower local buckling resistance results.

14.12 Conversation with 5 MW NREL Baseline WT Developer

This conversation[1] with one of the NREL Baseline WT developers shows that the wall thickness of the tubular reference tower was increased to avoid resonance within the operational range. Since the cut-in rotational speed was reduced throughout the development of the WT, the reason for increasing the wall thickness was not more given. Nevertheless, a corresponding new adjustment of the tower wall thickness was never been carried out, because no relevant resonance problems occurred with the old tower setting.

Re: Campbell diagram for 5MW turbine rotor

by Achim.Struve » Wed Jun 13, 2018 6:04 am

Achim.Struve
Posts: 12
Joined: Thu Aug 06, 2015 2:10 am
Organization: Wind Energy Technology Institute of Fh Flensburg
Location: Germany - Schleswig-Holstein

Dear Jason,

I just reviewed your derived Campbell diagram for the land-based 5 MW baseline wind turbine. It was in your post from Thu Sep 22, 2016 8:24 am.

Furthermore I reviewed the tower section 6 in Jonkman, J., Butterfield, S., Musial, W., and Scott, G: Definition of a 5-MW Reference Wind Turbine for Offshore System Development. February 2009.

In this section 6 on page 15 in the second paragraph, one writes:

> We based the distributed properties of the land-based tower for the NREL 5-MW baseline wind turbine on the base diameter (6 m) and thickness (0.027 m), top diameter (3.87 m) and thickness (0.019 m), and effective mechanical steel properties of the tower used in the DOWEC study (as given in Table 9 on page 31 of Ref. [14]) ... Because the REpower 5M machine had a larger tower-top mass than the DOWEC wind turbine, we scaled up the thickness of the tower relative to the values given earlier in this paragraph to strengthen the tower. We chose an increase of 30% to ensure that the first fore-aft and side-to-side tower frequencies were placed between the one- and three-per-rev frequencies throughout the operational range of the wind turbine in a Campbell diagram....

This sounds like the wall thickness of the baseline tower is driven by its first two eigenfrequencies. But when I go back to your Campbell diagram I see, that these tower eigenfrequencies are very close to the three-per-rev frequencie at cut-in rotational speed. In fact, neglecting the proposed wall thickness increasement of 30 % would decrease the tower eigenfrequencies to about 0.282 Hz and a larger distance to the exciting three-per-rev frequency can be obtained. Note, that it would still have enough distance to the one-per-rev excitation frequency of 12.1 1/min / 60 s/min * 1.05 = 0.21 Hz.

What do you think about this?

Best regards,
Achim

Re: Campbell diagram for 5MW turbine rotor

by Jason.Jonkman » Wed Jun 13, 2018 8:52 am

Jason.Jonkman
Posts: 3431
Joined: Thu Nov 03, 2005 4:38 pm
Location: Boulder, CO
Contact:

Dear Achim,

What you say sounds reasonable.

If I recall correctly (although this was over 10 years ago now), I think the cut-in speed dropped a bit over the course of the NREL 5-MW development and perhaps the tower thickness was set before this and not later updated.

I agree that the final first tower fore-aft and side-to-side mode frequencies are bit close to the 3P frequency at the cut-in rotor speed. That said, I don't recall finding a problem with resonance of the tower from 3P excitation.

Best regards,

Jason Jonkman, Ph.D.
Senior Engineer | National Wind Technology Center (NWTC)

National Renewable Energy Laboratory (NREL)
15013 Denver West Parkway | Golden, CO 80401
+1 (303) 384 - 7026 | Fax: +1 (303) 384 - 6901
nwtc.nrel.gov

[1] Internet forum of NREL's National Wind Technology Center: `https://wind.nrel.gov/forum/wind/viewtopic.php?f=3&t=1056&p=11358#p11358`; Accessed 14-June-2018

14.13 Calculating Tower Natural Frequencies by Lagrange's Equation of Motion

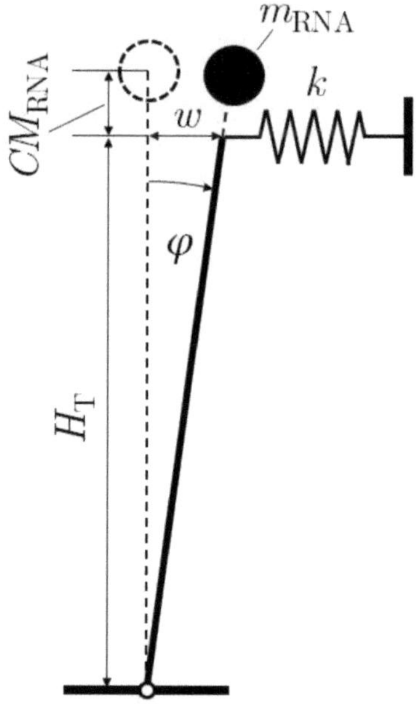

Figure 14.4: Lagrangian tower model

Calculation of tower natural frequencies may be carried out by means of Lagrange's equations of motion [8]:

$$\frac{\mathrm{d}}{\mathrm{d}t}\left(\frac{\partial L}{\partial \dot{\underline{q}}}\right) - \frac{\partial L}{\partial \underline{q}} = \underline{Q}^{nc} \tag{14.15}$$

where the Lagrangian L is a function of the systems generalized coordinates, their time derivatives, and time, thus $L = L\left(\underline{q}, \dot{\underline{q}}, t\right)$. In this consideration, damping and friction is neglected, wherefore the non-conservative term $\underline{Q}^{nc} = 0$ vanishes. The constant Lagrangian is defined as

$$L = T - V = \text{const.} \tag{14.16}$$

where T and V denote the kinetic and potential energy of the system. For the system in Figure 14.4, they may be written as

$$\begin{aligned} T &= \frac{1}{2} J_0\, \dot{\varphi}^2 \\ V &= \frac{1}{2} k\, w^2 \end{aligned} \tag{14.17}$$

where J_0 is the systems moment of inertia around the artificial rotation point at the bottom of the tower. It is calculated by Equation 14.18.

$$\begin{aligned} J_0 &= \int_V \underline{r}_\perp^2 \, \rho\left(\underline{r}\right) \, \mathrm{d}V \\ &= \rho_{\mathrm{steel}} \int_0^{H_\mathrm{T}} z^2 \, A(z) \, \mathrm{d}z + m_{\mathrm{RNA}} \left(H_\mathrm{T} + CM_{\mathrm{RNA}}\right)^2 \end{aligned} \tag{14.18}$$

In Equation 14.18 $\underline{r}_\perp$ is the perpendicular position vector of volume increments $\mathrm{d}V$ to the considered rotation axis and $\rho\left(\underline{r}\right)$ is its material density. For the given configuration, one may integrate along the height of the tower H_T with height coordinate z and a corresponding tower cross sectional area $A(z)$. The RNA's moment of inertia is included by adding the RNA mass with its squared vertical distance to the ground. Here CM_{RNA} denotes the vertical RNA center of mass measured from the tower top.

The bending stiffness of the tower is represented by a virtual spring at the tower top. Its spring constant k is calculated from Equation 14.19.

$$k = \frac{F}{w} = \frac{3\, E\, I_{\mathrm{rep}}}{H_\mathrm{T}^3} \tag{14.19}$$

where F is a unit force, w is the horizontal tower top deflection, E is the elasticity modulus, and I_{rep} is a representative second area moment of inertia for the whole tower. Since the stiffness at the bottom has the most influence to the tower top deflection and the stiffness at the top has no such influence, the linear distance weighted second are moment of inertia in Equation 14.20 is proposed as being representative for the tower. Note that the height coordinate runs this time from the top to the bottom.

$$I_{\mathrm{rep}} = \frac{\int_0^{H_\mathrm{T}} I(z)\, z \, \mathrm{d}z}{\int_0^{H_\mathrm{T}} z \, \mathrm{d}z} \tag{14.20}$$

In the present case, the towers rotational deflection angle φ is the only required generalized coordinate. Note that the tower itself is considered as being rigid. The horizontal tower top deflection in the potential energy in Equation 14.17 becomes $w = \sin(\varphi)\, H_\mathrm{T} \approx \varphi\, H_\mathrm{T}$ by means of the small angle assumption. Thus, using Equations 14.16 and 14.17 in 14.15 delivers the Equation of motion 14.21.

$$0 = \ddot{\varphi} + \underbrace{\frac{k\, H_\mathrm{T}^2}{J_0}}_{\omega^2} \varphi \tag{14.21}$$

The first tower eigenfrequency follows consequently as

$$f = \frac{\omega}{2\,\pi} = \frac{\sqrt{\frac{k\, H_\mathrm{T}}{J_0}}}{2\,\pi} \tag{14.22}$$

The numerical integration of Equation 14.20 avoids the development of fourth-order polynomials and is carried out in a python script, such as shown in Figure 14.5.

```python
from __future__ import division
import numpy as np

### parameter set for eigenfrequency calculation ###
H_T = 87.6 # tower height / m
CM_RNA = 1.75 # vertical distance from tower top to RNA center of mass / m
m_RNA = 3.5e5 # RNA mass / kg
t_T = 0.019 * 1.3 # tower top wall thickness / m
t_B = 0.027 * 1.3 # tower bottom wall thickness / m
R_T = 3.87 / 2.0 # tower top outer diameter / m
R_B = 6 / 2.0 # tower bottom outer diameter / m
rho_steel = 8500 # assumed tower material density (includes bolts, paintings etc.) / kg/m^3
E = 210000 * 1e6 # elasticity modulus / N/m^2


### tower moment of inertia and linear distance weighting (numerical integration) ###
Irep = 0 # initial representative second area moment of inertia / m^4
J_T = 0 # initial tower moment of inertia / kg*m^2
z_a = np.linspace(0,H_T,1000) # hight coordinate / m
dz = z_a[1] - z_a[0]
for zi in range(len(z_a)):
    z = z_a[zi] # actual hight / m
    t = t_T + (t_B - t_T) / np.max(z_a) * z # actual wall thickness / m
    R = R_T + (R_B - R_T) / np.max(z_a) * z # actual outer tower radius / m
    r = R - t # actual inner tower radius / m
    A = (R**2 - r**2) * np.pi # actual tower cross sectional area / m^2
    J_T += (H_T-z)**2 * A * rho_steel * dz
    I = np.pi / 4.0 * (R**4 - r**4) # actual second are moment of inertia / m^4
    Irep += I * z * dz

Irep = Irep / np.sum(z_a * dz)
J0 = J_T + m_RNA * (H_T + CM_RNA)**2 # systems moment of inertia w.r.t. tower base / kg*m^2

k = 3 * E * Irep / H_T**3 # tower stiffness representing spring constant / N/m

f = np.sqrt(k * H_T**2 / J0) / (2 * np.pi) # towers first eigenfrequency / Hz

print "Frequency (linear distance weighted second area moment of inertia): \n",round(f,4)," Hz"
```

Figure 14.5: Tower eigenfrequency calculation within Python

Usage of the NREL reference WT parameters in the Python code leads to an eigenfrequency of 0.313 Hz. It is very close to the linearized model SS tower frequency of 0.312 Hz, such as listed in the reference WT documentation [109, p.30]. Thereby, this approach to calculate the first tower bending eigenfrequency can be seen as validated.

14.14 Polynomial Coefficients for Tower Mass to Tower Height Relation

The following Table 14.1 contains the polynomial coefficients to describe tower masses as cubic function of the tower height

$$\text{mass}_{\text{T},\text{C}_i} = a_3\, H_\text{T}^3 + a_2\, H_\text{T}^2 + a_1\, H_\text{T} + a_0 \tag{14.23}$$

for different cases according to Table 5.4. Thereby, the coefficients were optimized by means of a least squares approach, implemented in the python scipy package under scipy.optimize.curve_fit [104]. Note that it was added a point $\{0,0\}$ to the data of the first diagram in Figure 5.7 for the cubic polynomial fits.

i	a_3	a_2	a_1	a_0
C1	2.51e−5	1.72e−2	7.99e−1	5.49
C2	1.84e−5	3.02e−2	8.17e−1	-6.62e−2
C3	1.25e−5	2.47e−2	1.70e−1	7.19
C3,wl	9.35e−6	2.66e−2	4.96e−2	7.35

Table 14.1: Cubic polynomial coefficients for tower mass description of different cases as function of WTs tower height

14.15 Offer for Lockbolts

Offer mail from KVT-Fastening GmbH from the 03. January 2018 for 10,000 Bobtail lockbolts. The request was differentiated with respect to two different bolt diameters 22.2 mm and 25.4 mm, respectively. Thereby, a clamped length between 19.1 mm and 28.4 mm is has been assumed for both zinc galvanized lockbolts.

Sehr geehrter Herr Struve,

wir danken für Ihre Anfrage und unterbreiten nachstehend unser – freibleibendes - Angebot:

Bezeichnung:	**Art.Nr.:**	**Auftragsmenge:**	**PREIS netto / 100:**
BTR-BR28-12GA	-	**10.000 Stück MM**	**EUR 423,05**
BTR-BR32-12GA	-	**10.000 Stück MM**	**EUR 606,00**

LIEFERZEIT: ca. 18-20 Wochen nach Auftragseingang unter Vorbehalt.

14.16 Python Script to Solve for the Constant Lattice Tower Bracing Angle

The following python function calculates the vertical joint heights h_i and the constant lattice tower bracing angle α_B for a given tower bottom and tower top width and tower height H_T.

```
from __future__ import division
import numpy as np
from sympy import nsolve, var

# function to solve for lattice tower joint heights h_i
# with arbitrary number of bracing segments Nseg,
# tower bottom and top width wB and wT, and
# tower Heights HT
def solve_hights(Nseg, HT, wB, wT):

    k = (wT-wB) / HT                        # tower slope
    tan_beta = (1/2.0 * (wB-wT)) / HT     # tower taper angle

    h_vars = var('h:'+str(Nseg))            # height variable names
    c = var('c')                            # c variable name

    # write variable names to one array
    variables = ['' for x in range(Nseg+1)]
    variables[:Nseg] = h_vars
    variables[Nseg] = c

    # initialize the Nseg+1 functions
    functions = [0 for x in range(Nseg+1)]

    # write all functions to the function array.
    # Note, that they are all rearranged to be equal to zero
    for i in range(Nseg):
        if i == 0:
            functions[i] = wB * variables[-1] - variables[0]

        if i > 0:
            functions[i] = wB * variables[-1] - variables[i]
            for j in range(i):
                functions[i] = functions[i] + k * variables[-1] * variables[j]

    for i in range(Nseg):
        functions[-1] = functions[-1] + variables[i]

    functions[-1] = functions[-1] - HT

    # initialize the height results
    result_estimation = [HT/Nseg for x in range(Nseg+1)]

    # solve the system of equations
    result = nsolve(functions, variables, result_estimation)

    # write results to new variables
    h = result[:Nseg]
    c = float(result[Nseg])

    # calculate the constant bracing angle alpha_B
    alpha_B = np.arctan(c/(1-c * tan_beta))
    return h, alpha_B
```

14.17 Derivation of X-Bracing Member Intersection Point for Arbitrary Lattice Tower Shapes

This is a derivation for the x-bracing member intersection point where the shortest distance between two bracing member lines

$$\begin{aligned} \mathcal{L}_1 &= \underline{x}_{\mathrm{ll}} + \bar{n}_1\, \lambda_1 \\ \mathcal{L}_2 &= \underline{x}_{\mathrm{lr}} + \bar{n}_2\, \lambda_2 \end{aligned} \tag{14.24}$$

is calculated. In Equation 14.24, $\underline{x}_{\mathrm{ll}}$ and $\underline{x}_{\mathrm{lr}}$ represent position vectors of the joints where the bracing members are attached to the legs. Here, one lattice tower wall is considered from a front view where one joint is on the left and the other joint is on the right side. $\bar{n}_1$ and $\bar{n}_2$ are unit vectors, which point along line one ($\mathcal{L}_1$) and two ($\mathcal{L}_2$) and λ_1 and λ_2 are scalars, which are used to describe a certain point on that lines, respectively. The distance between the two bracing member representing lines is described as

$$d = ||\mathcal{L}_2 - \mathcal{L}_1|| = ||\underline{d}|| = g(\lambda_1, \lambda_2) \tag{14.25}$$

for arbitrary values λ_1 and λ_2. Definition of function

$$f(\lambda_1, \lambda_2) = \underline{d}^T \underline{d} \tag{14.26}$$

is used in the optimization problem to find the minimum distance

$$d_{\min} = \min\left[||\underline{d}||\right] = \min\left[\sqrt{f(\lambda_1, \lambda_2)}\right] = \min\left[\sqrt{d_1^2 + d_2^2 + d_3^2}\right] \tag{14.27}$$

A minimum distance is found if

$$\frac{\partial}{\partial \lambda_1} g(\lambda_1, \lambda_2) \stackrel{!}{=} \frac{\partial}{\partial \lambda_2} g(\lambda_1, \lambda_2) \stackrel{!}{=} 0 \tag{14.28}$$

and

$$\frac{\partial^2}{\partial \lambda_1 \lambda_1} g(\lambda_1, \lambda_2) \frac{\partial^2}{\partial \lambda_2 \lambda_2} g(\lambda_1, \lambda_2) - \frac{\partial^2}{\partial \lambda_1 \lambda_2} g(\lambda_1, \lambda_2)^2 \stackrel{!}{>} 0 \tag{14.29}$$

In Equation 14.27, the three elements in $\underline{d}$ are

$$\begin{aligned} d_1 &= x_{\text{lr},1} + n_{2,1}\,\lambda_2 - x_{\text{ll},1} - n_{1,1}\,\lambda_1 \\ d_2 &= x_{\text{lr},2} + n_{2,2}\,\lambda_2 - x_{\text{ll},2} - n_{1,2}\,\lambda_1 \\ d_3 &= x_{\text{lr},3} + n_{2,3}\,\lambda_2 - x_{\text{ll},3} - n_{1,3}\,\lambda_1 \end{aligned} \tag{14.30}$$

and the general derivative of d with respect to λ_i is

$$\begin{aligned} \frac{\partial}{\partial \lambda_i} d &= \frac{\partial}{\partial \lambda_i}\left(d_1^2 + d_2^2 + d_3^2\right) \frac{\partial}{\partial \lambda_i}\sqrt{f(\lambda_1, \lambda_2)} \\ &= \left(d_1 \frac{\partial}{\partial \lambda_i} d_1 + d_2 \frac{\partial}{\partial \lambda_i} d_2 + d_3 \frac{\partial}{\partial \lambda_i} d_3\right) \frac{1}{\sqrt{f(\lambda_1, \lambda_2)}} \\ &= \frac{\underline{d}^T \overline{n}_i}{\sqrt{f(\lambda_1, \lambda_2)}} \end{aligned} \tag{14.31}$$

From Equation 14.31, a second derivative of d with respect to λ_j becomes

$$\frac{\partial^2}{\partial \lambda_i \lambda_j} d = \frac{\overline{n}_j^T \, \overline{n}_i}{\sqrt{f(\lambda_1, \lambda_2)}} - \frac{\underline{d}^T \, \overline{n}_i}{2\sqrt{(f(\lambda_1, \lambda_2))^3}} \underline{d}^T \, \overline{n}_j \tag{14.32}$$

Since Equation 14.28 requires $\frac{\partial}{\partial \lambda_i} g(\lambda_1, \lambda_2) \stackrel{!}{=} 0$, one can identify

$$\underline{d}^T \,\overline{n}_i \stackrel{!}{=} 0 \tag{14.33}$$

as condition for orthogonality between the distance vector $\underline{d}$ and the lines unit vectors $\overline{n}_i$. This condition results into

$$\begin{aligned} 0 &\stackrel{!}{=} d_1\, n_{i,1} + d_2\, n_{i,2} + d_3\, n_{i,3} \\ &\stackrel{!}{=} \sum_{k=1}^{3} x_{\mathrm{lr},k}\, n_{i,k} + n_{2,k}\, \lambda_2\, n_{i,k} - x_{\mathrm{ll},k}\, n_{i,k} - n_{1,k}\, \lambda_1\, n_{i,k} \end{aligned} \tag{14.34}$$

and delivers

$$\lambda_1 = \frac{\sum_{k=1}^{3} x_{\mathrm{lr},k}\, n_{i,k} - x_{\mathrm{ll},k}\, n_{i,k}}{\sum_{k=1}^{3} n_{1,k}\, n_{i,k}} + \frac{\sum_{k=1}^{3} n_{2,k}\, n_{i,k}}{\sum_{k=1}^{3} n_{1,k}\, n_{i,k}} \lambda_2 \tag{14.35}$$

Using Equation 14.35 in 14.34 where index i becomes j and $i \neq j$ results to

$$\lambda_2 = \frac{\underline{x}_{\mathrm{lr}}^T\, \overline{n}_j - \underline{x}_{\mathrm{ll}}^T\, \overline{n}_j - \overline{n}_1^T\, \overline{n}_j \left(\frac{\underline{x}_{\mathrm{lr}}^T\, \overline{n}_i - \underline{x}_{\mathrm{ll}}^T\, \overline{n}_i}{\overline{n}_1^T\, \overline{n}_i} \right)}{\overline{n}_1^T\, \overline{n}_j \left(\frac{\overline{n}_2^T\, \overline{n}_i}{\overline{n}_1^T\, \overline{n}_i} \right) - \overline{n}_2^T\, \overline{n}_j} \tag{14.36}$$

and with Equation 14.36 in the original Equation 14.35 one arrives at

$$\lambda_1 = \frac{\underline{x}_{\mathrm{lr}}^T\, \overline{n}_i - \underline{x}_{\mathrm{ll}}^T\, \overline{n}_i + \overline{n}_2^T\, \overline{n}_i\, \lambda_2}{\overline{n}_1^T\, \overline{n}_i} \tag{14.37}$$

With λ_1 and λ_2 one can calculate the closest points on both lines from Equation 14.24, which are denoted as $\mathcal{L}_{1\perp}$ and $\mathcal{L}_{2\perp}$. Thereby, one can calculate the distance vector between both points as

$$\underline{d}_{\mathrm{min}} = \mathcal{L}_{2\perp} - \mathcal{L}_{1\perp} \tag{14.38}$$

and its unit vector follows consequently as

$$\overline{d}_{\mathrm{min}} \begin{cases} \dfrac{\underline{d}_{\mathrm{min}}}{||\underline{d}_{\mathrm{min}}||} & \text{for} \quad ||\underline{d}_{\mathrm{min}}|| > 0 \\ 0 & \text{for} \quad ||\underline{d}_{\mathrm{min}}|| = 0 \end{cases} \tag{14.39}$$

The searched x-bracing intersection point is finally calculated as

$$\underline{P}_{\mathrm{x}} = \mathcal{L}_{1\perp} + \overline{d}_{\mathrm{min}} \frac{||\underline{d}_{\mathrm{min}}||}{2} \tag{14.40}$$

Note that in plane lattice walls lead to $d_{\min} = 0$ and both points $\mathcal{L}_{1\perp}$ and $\mathcal{L}_{2\perp}$ are at the same position, which is the searched bracing intersection point $\underline{P}_{\mathrm{x}}$.

14.18 SubDyn Rotatable Inclined Lattice Tower Input File

```
----------- SubDyn v1.03.xx MultiMember Support Structure Input File ------------
Input File for the 87.6 m high rotatable inclined lattice tower and the adjusted SubDyn Version
-------------------------- SIMULATION CONTROL ---------------------------------
True                Echo        - Echo input data to "<rootname>.SD.ech" (flag)
0.0125              SDdeltaT    - Local Integration Step. If "default", the glue-code integration step will be used.
              4     IntMethod   - Integration Method [1/2/3/4 = RK4/AB4/ABM4/AM2].
True                SttcSolve   - Solve dynamics about static equilibrium point
-------------------- FEA and CRAIG-BAMPTON PARAMETERS---------------------------
              1     FEMMod      - FEM switch: element model in the FEM. [1= anisotropic straigth tapered Timoshenko beam el
              1     NDiv        - Number of sub-elements per member
True                CBMod       - [T/F] If True perform C-B reduction, else full FEM dofs will be retained. If True, select
              7     Nmodes      - Number of internal modes to retain (ignored if CBMod=False). If Nmodes=0 --> Guyan Reduct
              1    JDampings   - Damping Ratios for each retained mode (% of critical) If Nmodes>0, list Nmodes structural
---- STRUCTURE JOINTS: joints connect structure members (~Hydrodyn Input File)---
               172   NJoints     - Number of joints (-)
JointID          JointXss               JointYss              JointZss  [Coordinates of Member joints in SS-Coordinate S
 (-)               (m)                    (m)                   (m)
  1              -3.4         -3.4          0.0
  2              3.4         -3.4          0.0
  3              3.4         3.4          0.0
  4              -3.4         3.4          0.0
  5              -4.0479            -3.3305           5.2914
  6              2.6132            -3.3305           5.2914
  7              2.6132            3.3305           5.2914
  8              -4.0479            3.3305           5.2914
  9              -4.6826            -3.2625           10.4746
  10              1.8424            -3.2625           10.4746
  11              1.8424            3.2625           10.4746
  12              -4.6826            3.2625           10.4746
  13              -5.3043            -3.1958           15.552
  14              1.0874            -3.1958           15.552
  15              1.0874            3.1958           15.552
                                                 ⋮
  162              -8.2781            -0.0            78.5327
  163              -10.6471            2.369            78.5327
  164              -13.0162            0.0            78.5327
  165              -11.147            -2.3206            82.2195
  166              -8.8263            -0.0            82.2195
  167              -11.147            2.3206            82.2195
  168              -13.4676            0.0            82.2195
  169              -11.6366            -2.2732            85.8311
  170              -9.3634            0.0            85.8311
  171              -11.6366            2.2732            85.8311
  172              -13.9098            0.0            85.8311
------------------- BASE REACTION JOINTS: 1/0 for Locked/Free DOF @ each Reaction Node ---------------------
              4   NReact      - Number of Joints with reaction forces; be sure to remove all rigid motion DOFs of the str
RJointID   RctTDXss    RctTDYss    RctTDZss    RctRDXss    RctRDYss    RctRDZss     [Global Coordinate System]
 (-)        (flag)      (flag)      (flag)      (flag)      (flag)      (flag)
  1          1           1           1           1           1           1
  2          1           1           1           1           1           1
  3          1           1           1           1           1           1
  4          1           1           1           1           1           1
------- INTERFACE JOINTS: 1/0 for Locked (to the TP)/Free DOF @each Interface Joint (only Locked-to-TP implemented thus
              4   NInterf     - Number of interface joints locked to the Transition Piece (TP):  be sure to remove all ri
IJointID   ItfTDXss    ItfTDYss    ItfTDZss    ItfRDXss    ItfRDYss    ItfRDZss     [Global Coordinate System]
 (-)        (flag)      (flag)      (flag)      (flag)      (flag)      (flag)
  81         1           1           1           1           1           1
  82         1           1           1           1           1           1
  83         1           1           1           1           1           1
  84         1           1           1           1           1           1
----------------------------------- MEMBERS -------------------------------------
              408   NMembers    - Number of frame members
MemberID   MJointID1   MJointID2   MPropSetID1   MPropSetID2   OriTpe   RotAngle   PointAXss   PointAYss   PointAZss
 (-)         (-)         (-)          (-)           (-)          (-)      (deg)       (m)         (m)        (m)4
  1          1           5            1             1            2        0.0        3.4        -3.4        0.0
  2          2           6            2             2            2        0.0        3.4         3.4        0.0
  3          3           7            2             2            2        0.0       -3.4         3.4        0.0
  4          4           8            1             1            2        0.0       -3.4        -3.4        0.0
  5          5           9            3             3            2        0.0        2.6132     -3.3305      5.2914
  6          6           10           4             4            2        0.0        2.6132      3.3305      5.2914
  7          7           11           4             4            2        0.0       -4.0479      3.3305      5.2914
  8          8           12           3             3            2        0.0       -4.0479     -3.3305      5.2914
  9          9           13           5             5            2        0.0        1.8424     -3.2625      10.4746
  10         10          14           6             6            2        0.0        1.8424      3.2625      10.4746
  11         11          15           6             6            2        0.0       -4.6826      3.2625      10.4746
                                                 ⋮
```

⋮

```
    396        169         81          60           60          2      0.0      -11.6366     -3.2641     85.8441
    397         78        170          60           60          2      0.0       -8.3743      0.0        85.9782
    398        170         83          60           60          2      0.0       -8.3743      0.0        85.9782
    399         79        170          60           60          2      0.0       -8.3743      0.0        85.9782
    400        170         82          60           60          2      0.0       -8.3743      0.0        85.9782
    401         79        171          60           60          2      0.0      -11.6366      3.2641     85.8441
    402        171         84          60           60          2      0.0      -11.6366      3.2641     85.8441
    403         80        171          60           60          2      0.0      -11.6366      3.2641     85.8441
    404        171         83          60           60          2      0.0      -11.6366      3.2641     85.8441
    405         80        172          60           60          2      0.0      -14.9024      0.0        85.7096
    406        172         81          60           60          2      0.0      -14.9024      0.0        85.7096
    407         77        172          60           60          2      0.0      -14.9024      0.0        85.7096
    408        172         84          60           60          2      0.0      -14.9024      0.0        85.7096
------------------ MEMBER X-SECTION PROPERTY data 1/3 [isotropic material: use this table for circular-tubular elements]
              0   NPropSets   - Number of structurally unique x-sections (i.e. how many groups of X-sectional properties a
PropSetID      YoungE           ShearG           MatDens            XsecD          XsecT
  (-)          (N/m2)           (N/m2)           (kg/m3)             (m)            (m)
------------------ MEMBER X-SECTION PROPERTY data 2/3 [use this table if any section other than circular or anisotropic m
              0   NXPropSets  - Number of structurally unique arbitrary cross sections defined by engineering constants (i
PropSetID      YoungE           ShearG           MatDens            XsecA          XsecAx              XsecAy              XsecAxy
  (-)          (N/m2)           (N/m2)           (kg/m3)            (m2)            (-)                 (-)                 (-)
------------------ MEMBER X-SECTION PROPERTY data 3/3 [use this table if any section other than circular or anisotropic m
              61   NXFSMPropSets  - Number of structurally unique arbitrary cross sections defined by full cross sectional
PropSetID - formatting: review the user manual - Give all values in SI-units
  (-)
1
2002240098.52   440002804.747   0   0   0   183629.111463
440002804.747   2002240098.52   0   0   0   -183629.111464
                0               0          14910110719.3   0.000503301620483  -0.000577449798584          0
                0               0       1.71661376953e-05      1730262806.17      -447667169.189          0
                0               0      -6.62803649902e-05     -447667169.189       1730262806.17          0
183629.111464   -183629.111464   0   0   0   840069.330429

     603.504481497              0                   0                  0                   0              0
                0   603.504481497                   0                  0                   0              0
                0               0       603.504481497                  0                   0              0
                0               0                   0      70.0302087943      -18.1176833742              0
                0               0                   0     -18.1176833742       70.0302087943              0
                0               0                   0                  0                   0  140.060417589
2
2002240098.52   440002804.747   0   0   0   183629.111463
440002804.747   2002240098.52   0   0   0   -183629.111464
                0               0          14910110719.3   0.000503301620483  -0.000577449798584          0
                0               0       1.71661376953e-05      1730262806.17      -447667169.189          0
                0               0      -6.62803649902e-05     -447667169.189       1730262806.17          0
183629.111464   -183629.111464   0   0   0   840069.330429

     603.504481497              0                   0                  0                   0              0
                0   603.504481497                   0                  0                   0              0
                0               0       603.504481497                  0                   0              0
                0               0                   0      70.0302087943      -18.1176833742              0
                0               0                   0     -18.1176833742       70.0302087943              0
                0               0                   0                  0                   0  140.060417589
3
1084643528.95   242512567.547   0   0   0   85142.3752916
242512567.547   1084643528.95   0   0   0   -85142.3752916
                0               0          8524927061.65  -7.89165496826e-05  -0.000153303146362          0
                0               0      -2.86102294922e-06      530133320.487      -140955755.335          0
                0               0       1.10864639282e-05     -140955755.335       530133320.487          0
85142.3752916   -85142.3752916   0   0   0   239014.400826

     345.056571543              0                   0                  0                   0              0
                0   345.056571543                   0                  0                   0              0
                0               0       345.056571543                  0                   0              0
                0               0                   0      21.4565098056      -5.70469435972              0
                0               0                   0     -5.70469435972       21.4565098056              0
                0               0                   0                  0                   0  42.9130196113
4
1612629846.43   356097588.639   0   0   0   140409.221731
356097588.639   1612629846.43   0   0   0   -140409.221732
                0               0          12222601766.1  -0.000216960906982   0.000446796417236          0
                0               0      -2.76565551758e-05      1146041602.54      -299028245.808          0
                0               0       0.000105619430542     -299028245.808       1146041602.54          0
140409.221731   -140409.221732   0   0   0   543485.968871

     494.724357198              0                   0                  0                   0              0
```

⋮

⋮

```

              87.448                0                0                  0                  0                0
                   0           87.448                0                  0                  0                0
                   0                0           87.448                  0                  0                0
                   0                0                0     0.847941889333  -8.79643715724e-19                0
                   0                0                0 -8.79643715724e-19      1.73158699733                0
                   0                0                0                  0                  0    2.57952888667
59
246519402.735   -0.000309286370557   0   0   0   -8.78719939905e-09
-0.000326400648572   162065432.152   0   0   0   1.02952082696e-08
                   0                0       2195760000  2.0978835765e-05  -2.10566283338e-06               0
                   0                0 -8.47032947254e-22   21989010.5509   2.05698217157e-06               0
                   0                0 -1.96511643763e-19 3.04899296872e-06      45322096.276               0
5.67324881315e-09   9.87120604153e-09   0   0   0   10318.1340334

              88.876                0                0                  0                  0                0
                   0           88.876                0                  0                  0                0
                   0                0           88.876                  0                  0                0
                   0                0                0     0.889971737333   5.89916096115e-18                0
                   0                0                0  5.89916096115e-18      1.83435682533                0
                   0                0                0                  0                  0    2.72432856267
60
253387370.629   -0.000101372294714   0   0   0   -2.5510042235e-08
-0.000125216550635   164601458.842   0   0   0   8.60219132628e-09
                   0                0       2231040000  1.18392920015e-05  -1.24132785714e-05               0
                   0                0                0     23063751.8092  -6.95268736085e-07               0
                   0                0 -2.90193487729e-18 3.92628083806e-07      47959963.7331               0
-1.58450274611e-08   1.6270171189e-08   0   0   0   10957.0032891

              90.304                0                0                  0                  0                0
                   0           90.304                0                  0                  0                0
                   0                0           90.304                  0                  0                0
                   0                0                0     0.933470385333  -3.68671090292e-17                0
                   0                0                0 -3.68671090292e-17      1.94112124533                0
                   0                0                0                  0                  0    2.87459163067
61
  2.00224009852e+14   4.40002804747e+13   0.0   0.0   0.0   18362911146.3
  4.40002804747e+13   2.00224009852e+14   0.0   0.0   0.0   -18362911146.4
  0.0   0.0   1.49101107193e+15   80.3301620483   -87.7449798584   0.0
  0.0   0.0   1.71661376953   1.73026280617e+14   -4.47667169189e+13   0.0
  0.0   0.0   -6.62803649902   -4.47667169189e+13   1.73026280617e+14   0.0
  18362911146.4   -18362911146.4   0.0   0.0   0.0   84006933042.9

  0.00603504481497   0.0   0.0   0.0   0.0   0.0
  0.0   0.00603504481497   0.0   0.0   0.0   0.0
  0.0   0.0   0.00603504481497   0.0   0.0   0.0
  0.0   0.0   0.0   0.00070030208794   -0.00018117683374   0.0
  0.0   0.0   0.0   -0.00018117683374   0.00070030208794   0.0
  0.0   0.0   0.0   0.0   0.0   0.00140060417589
------------------------ JOINT ADDITIONAL CONCENTRATED MASSES--------------------------
             0   NCmass      - Number of joints with concentrated masses; Global Coordinate System
CMJointID       JMass            JMXX             JMYY             JMZZ
  (-)            (kg)          (kg*m^2)         (kg*m^2)         (kg*m^2)
---------------------------- OUTPUT: SUMMARY & OUTFILE ------------------------------
False            SSSum       - Output a Summary File (flag).It contains: matrices K,M  and C-B reduced M_BB, M-BM, K_BB,
True             OutCOSM     - Output cosine matrices with the selected output member forces (flag)
True             OutAll      - [T/F] Output all members' end forces
             2   OutSwtch    - [1/2/3] Output requested channels to: 1=<rootname>.SD.out;  2=<rootname>.out (generated b
True             TabDelim    - Generate a tab-delimited output in the <rootname>.SD.out file
             1   OutDec      - Decimation of output in the <rootname>.SD.out file
"ES11.4e2"       OutFmt      - Output format for numerical results in the <rootname>.SD.out file
"All"            OutSFmt     - Output format for header strings in the <rootname>.SD.out file
------------------------- MEMBER OUTPUT LIST ------------------------------------------
             0   NMOutputs   - Number of members whose forces/displacements/velocities/accelerations will be output (-)
MemberID   NOutCnt    NodeCnt [NOutCnt=how many nodes to get output for [< 10]; NodeCnt are local ordinal numbers from t
  (-)        (-)        (-)
------------------------- SSOutList: The next line(s) contains a list of output parameters that will be output in <rootn
"ReactFXss, ReactFYss, ReactFZss, ReactMXss, ReactMYss, ReactMZss"    -Base reactions (forces onto SS structure)
"IntfFXss,  IntfFYss,  IntfFZss,  IntfMXss, IntfMYss, IntfMZss"       -Interface reactions (forces onto SS structure)
END of output channels and end of file. (the word "END" must appear in the first 3 columns of this line)
```

⋮

14.19 Modelling in RFEM

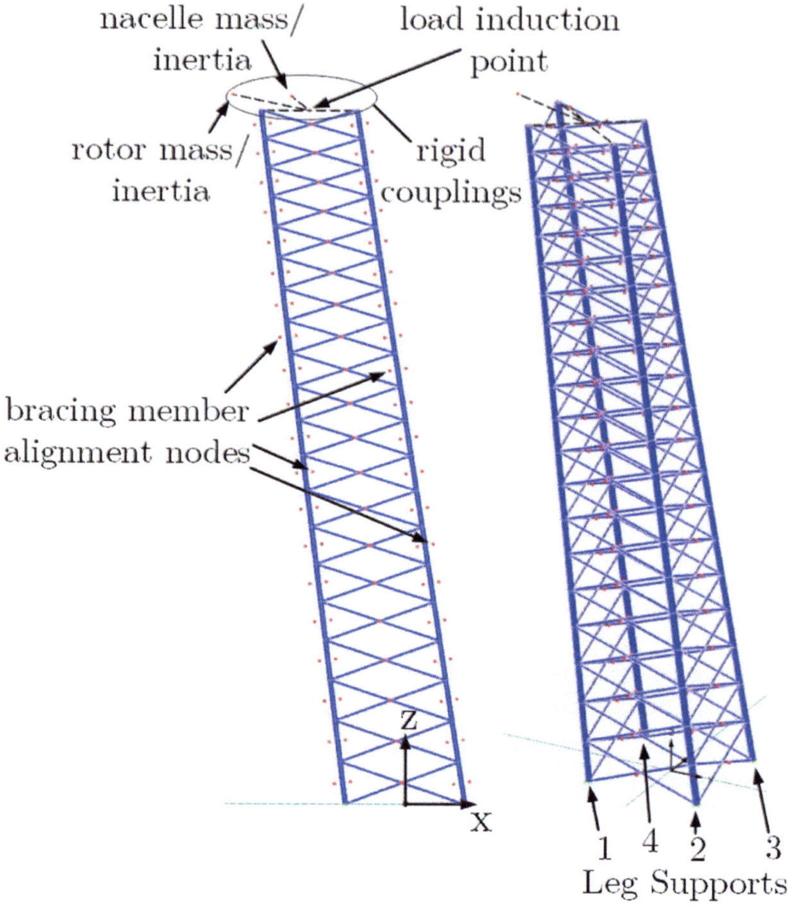

Figure 14.6: RFEM tower model

The commercial software package RFEM 5 from Dlubal [57] provides a state-of-the-art steel construction modelling environment. It can be used together with Dlubal's cross sectional analysis tool DUENQ 8 [58] to model the rotatable inclined lattice tower for extreme loads and eigenfrequency analyses. The later requires a stand alone add-on, named RF-DYNAM [54]. Strength analyses of the lattice members are performed with the RF-STAHL [59] module and buckling analyses are performed with RF-STABIL [55]. Another add-on, named RF-IMP [56], can be used to generate geometrical imperfections from a buckling or modal eigenmode. Since this work focused on the material saving potential of the tower, rather than on the tower base yaw bearing, all four legs were assumed to be rigidly connected to the ground, such as shown in Figure 14.6. Therefore, the modelled tower had slightly higher eigenfrequencies than in the final design. This higher stiffness had to be considered in the eigenfrequency analysis in which two rigidly coupled nodes at the top of the tower contain the rotor and nacelle masses and inertias. The machine foundation at the top of the tower was assumed to be very stiff, wherefore the four legs were coupled rigidly over one node at the top. This coupling node served simultaneously as load induction point and was located with the same distance to the front tower edge as for the conventional reference tower. Furthermore, additional bracing member alignment nodes served to find their correct alignments in space. A cross section similar to the leg in Figure 7.4 did not exist in the library, wherefore it was modelled in Dlubal's cross sectional analysis software DUENQ, such as shown in Figure 14.7. On the left side in Figure 14.7 is a detailed and on the right side a simplified model of the leg cross section. The detailed model accounted for folding radii of two times the material thickness and geometrical incompletenesses at the corners of two straight cross sectional parts. The simplified approach had the advantage, that it was modelled more quickly and saved time in terms of parameter studies.

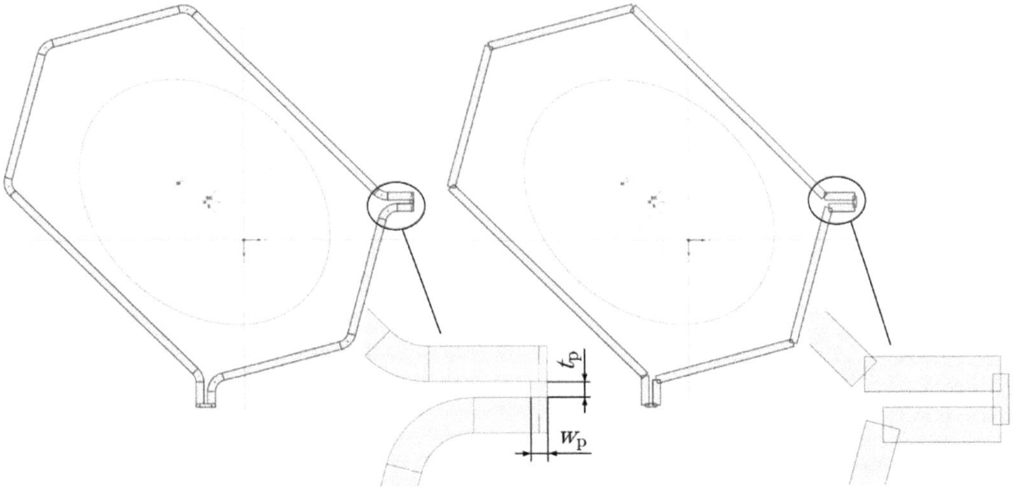

Figure 14.7: The detailed (left) and simplified (right) DUENQ leg cross sectional model

A comparison between both variants revealed that important properties, such as the cross sectional area, the second area moment of inertia and the torsional moment of inertia changed in magnitudes of < 1 %. Therefore, the simplified approach was seen as being appropriate to be used in the following analyses. t_p is representative for the junction plates thickness. Moreover, the dimension of w_p could be used to distribute the discrete filler plates properties as continuum over the members length.

Double hat profiles are likewise not provided by Dlubal's cross sectional library and they had to be modelled in DUENQ, such as shown in Figure 14.8. Due to their extruding manufacturing process, no big radii were expected and a simplified model was sufficient, such as for the legs. As mentioned earlier the leg and the bracing members consisted of two cross sectional parts, which were bolted together over their length. The interface between DUENQ and RFEM allows to use the results of the cross sectional analyses directly in RFEM.

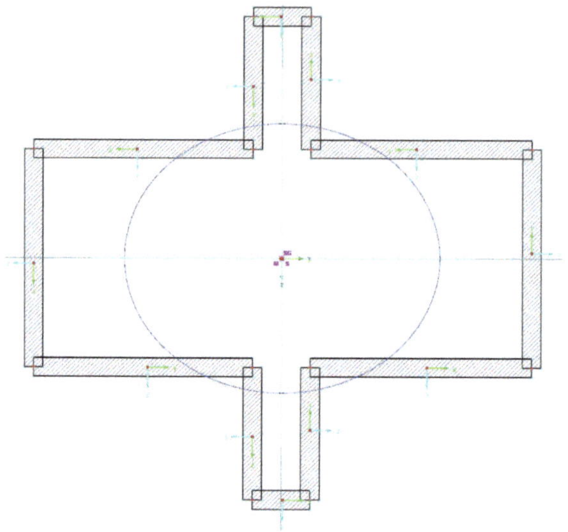

Figure 14.8: DUENQ double hat cross sectional model

Note that the modelled bracing members were slightly longer than in reality, because they were attached at junction plates, which were bolted to the leg flap, such as shown in Figure 7.6. This structural detail was not accounted for to keep the model simple and comparable to the parallel model in SubDyn, such as described in Chapter 8. However, this simplification is on the safe side. Due to the members elongation, its buckling resistance was reduced together with the torsional tower stiffness. Moreover, the additional bracing member material could be seen as compensation for the material of the lacking junction plates in the model.

14.20 Tower Inclination Sensitivity to Yield Strength Utilizations

The following figure shows the member yield strength utilization within the 10. segment of a RILT as function of its inclination angle α_{TI}. It was analysed within RFEM with extreme tower head loads, extracted from the ASE load simulation of the conventional tower of the 5 MW reference WT. All important modelling aspects are summarized in Appendix 14.19. The global RILT parameters in this analysis are H_{T} = 87.6 m, $a_{\mathrm{T}} = b_{\mathrm{T}}$ = 4.4 m, and $a_{\mathrm{B}} = b_{\mathrm{B}}$ = 6.7 m. An analysis in RFEM had the advantage of less required simulation time compared to a full ASE simulation for each tower inclination angle.

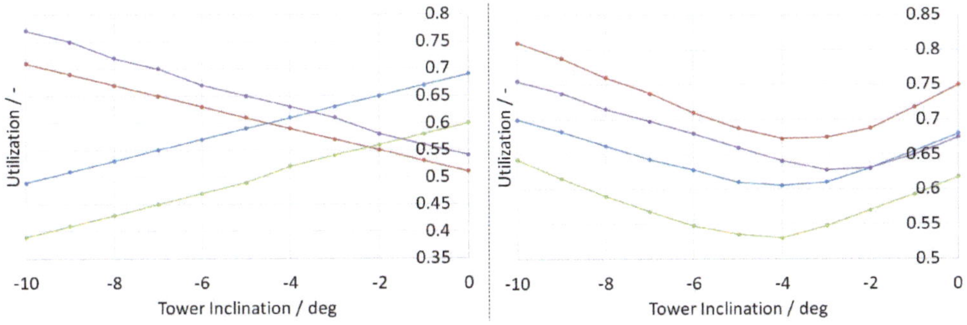

14.21 Rotatable Inclined Lattice Tower Reynold's Numbers

The following figure shows the Reynold's numbers of members in the rotatable inclined lattice tower, defined in Table 9.1 for different wind speeds.

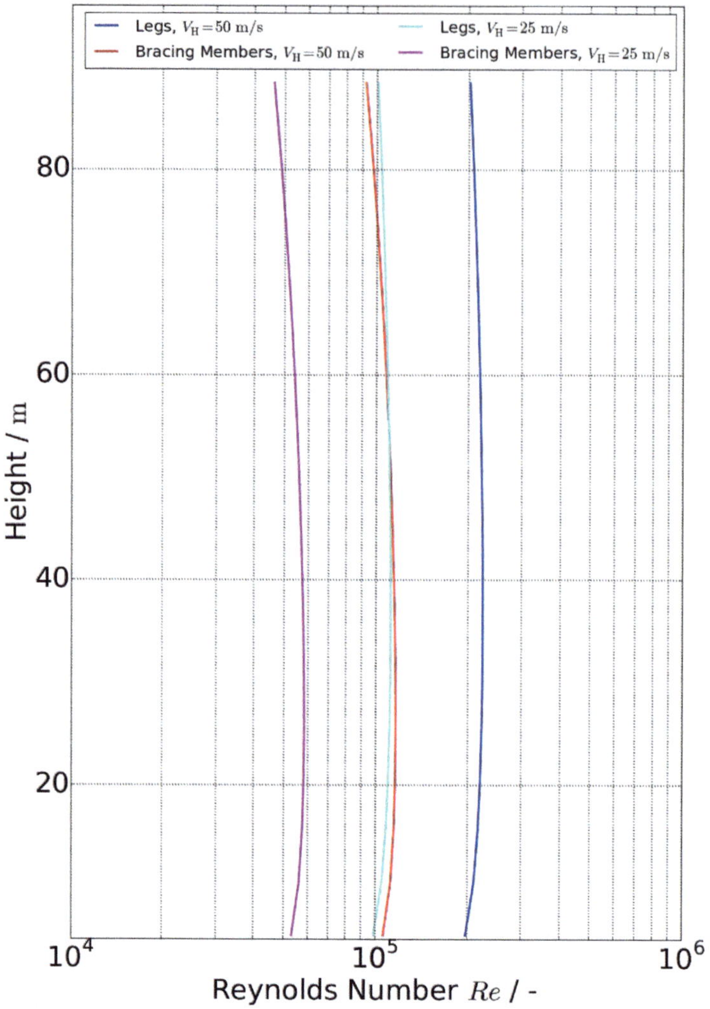

14.22 Rotatable Inclined Lattice Tower 87.6 m Member Parameters

The following figure shows all member parameters, of the rotatable inclined lattice tower with $H_T = 87.6$ m according to Table 9.1. All parameters are explained in the nomenclature of this work and its values are given in SI-units.

Front Leg Member Parameters

Segment	L_L	L_F	A	I_x	I_y	I_xy	I_min	i_min	L_cr	lambda_b	Phi	chi
1	0.3960e0	0.9000e-1	0.7014e-1	0.7802e-2	0.7802e-2	0.2037e-2	0.5765e-2	0.2867e0	5.3314e0	0.2434e0	0.5370e0	0.9846e0
2	0.2941e0	0.9000e-1	0.4153e-1	0.2669e-2	0.2669e-2	0.7147e-3	0.1954e-2	0.2169e0	5.2224e0	0.3151e0	0.5692e0	0.9585e0
3	0.2895e0	0.9000e-1	0.4098e-1	0.2558e-2	0.2558e-2	0.6861e-3	0.1872e-2	0.2137e0	5.1158e0	0.3132e0	0.5683e0	0.9592e0
4	0.2850e0	0.9000e-1	0.4045e-1	0.2453e-2	0.2453e-2	0.6588e-3	0.1795e-2	0.2106e0	5.0112e0	0.3113e0	0.5674e0	0.9599e0
5	0.2806e0	0.9000e-1	0.3999e-1	0.2361e-2	0.2361e-2	0.6358e-3	0.1726e-2	0.2077e0	4.9088e0	0.3093e0	0.5664e0	0.9607e0
6	0.2762e0	0.9000e-1	0.3701e-1	0.2125e-2	0.2125e-2	0.5729e-3	0.1552e-2	0.2048e0	4.8086e0	0.3073e0	0.5655e0	0.9614e0
7	0.2720e0	0.9000e-1	0.3654e-1	0.2040e-2	0.2040e-2	0.5507e-3	0.1489e-2	0.2019e0	4.7103e0	0.3054e0	0.5646e0	0.9621e0
8	0.2678e0	0.9000e-1	0.3608e-1	0.1958e-2	0.1958e-2	0.5295e-3	0.1429e-2	0.1990e0	4.6140e0	0.3034e0	0.5636e0	0.9628e0
9	0.2638e0	0.9000e-1	0.3562e-1	0.1881e-2	0.1881e-2	0.5093e-3	0.1371e-2	0.1962e0	4.5198e0	0.3015e0	0.5627e0	0.9636e0
10	0.2598e0	0.9000e-1	0.3518e-1	0.1807e-2	0.1807e-2	0.4900e-3	0.1317e-2	0.1935e0	4.4275e0	0.2995e0	0.5618e0	0.9643e0
11	0.2559e0	0.9000e-1	0.3243e-1	0.1621e-2	0.1621e-2	0.4403e-3	0.1181e-2	0.1908e0	4.3369e0	0.2975e0	0.5608e0	0.9650e0
12	0.2520e0	0.9000e-1	0.3203e-1	0.1558e-2	0.1558e-2	0.4239e-3	0.1135e-2	0.1882e0	4.2485e0	0.2954e0	0.5599e0	0.9658e0
13	0.2483e0	0.9000e-1	0.3164e-1	0.1499e-2	0.1499e-2	0.4082e-3	0.1090e-2	0.1856e0	4.1615e0	0.2934e0	0.5589e0	0.9665e0
14	0.2446e0	0.9000e-1	0.3126e-1	0.1441e-2	0.1441e-2	0.3933e-3	0.1048e-2	0.1831e0	4.0766e0	0.2913e0	0.5580e0	0.9673e0
15	0.2410e0	0.9000e-1	0.3088e-1	0.1387e-2	0.1387e-2	0.3789e-3	0.1008e-2	0.1807e0	3.9933e0	0.2893e0	0.5570e0	0.9680e0
16	0.2432e0	0.9000e-1	0.3111e-1	0.1420e-2	0.1420e-2	0.3877e-3	0.1032e-2	0.1822e0	3.9116e0	0.2810e0	0.5533e0	0.9710e0
17	0.2538e0	0.9000e-1	0.3221e-1	0.1587e-2	0.1587e-2	0.4313e-3	0.1155e-2	0.1894e0	3.8318e0	0.2648e0	0.5461e0	0.9769e0
18	0.2640e0	0.9000e-1	0.3565e-1	0.1885e-2	0.1885e-2	0.5105e-3	0.1375e-2	0.1964e0	3.7534e0	0.2501e0	0.5398e0	0.9822e0
19	0.2602e0	0.9000e-1	0.3523e-1	0.1815e-2	0.1815e-2	0.4921e-3	0.1323e-2	0.1938e0	3.6768e0	0.2483e0	0.5390e0	0.9828e0
20	0.2970e0	0.9000e-1	0.4188e-1	0.2740e-2	0.2740e-2	0.7331e-3	0.2007e-2	0.2189e0	1.2006e0	0.7178e-1	0.4808e0	1.0000e0

Back Leg Member Parameters

Segment	L_L	L_F	A	I_x	I_y	I_xy	I_min	i_min	L_cr	lambda_b	Phi	chi
1	0.3960e0	0.9000e-1	0.7014e-1	0.7802e-2	0.7802e-2	0.2037e-2	0.5765e-2	0.2867e0	5.3500e0	0.2442e0	0.5373e0	0.9843e0
2	0.3543e0	0.9000e-1	0.5765e-1	0.5213e-2	0.5213e-2	0.1373e-2	0.3840e-2	0.2581e0	5.2406e0	0.2657e0	0.5465e0	0.9766e0
3	0.3488e0	0.9000e-1	0.5695e-1	0.5007e-2	0.5007e-2	0.1321e-2	0.3686e-2	0.2544e0	5.1337e0	0.2641e0	0.5458e0	0.9771e0
4	0.3433e0	0.9000e-1	0.5618e-1	0.4797e-2	0.4797e-2	0.1267e-2	0.3530e-2	0.2506e0	5.0287e0	0.2626e0	0.5451e0	0.9777e0
5	0.3380e0	0.9000e-1	0.5252e-1	0.4356e-2	0.4356e-2	0.1152e-2	0.3203e-2	0.2470e0	4.9260e0	0.2610e0	0.5444e0	0.9783e0
6	0.3328e0	0.9000e-1	0.5182e-1	0.4175e-2	0.4175e-2	0.1106e-2	0.3070e-2	0.2434e0	4.8254e0	0.2595e0	0.5438e0	0.9788e0
7	0.3277e0	0.9000e-1	0.5113e-1	0.4004e-2	0.4004e-2	0.1061e-2	0.2942e-2	0.2399e0	4.7268e0	0.2579e0	0.5431e0	0.9794e0
8	0.3227e0	0.9000e-1	0.5046e-1	0.3840e-2	0.3840e-2	0.1019e-2	0.2821e-2	0.2364e0	4.6302e0	0.2563e0	0.5424e0	0.9800e0
9	0.3178e0	0.9000e-1	0.4704e-1	0.3480e-2	0.3480e-2	0.9249e-3	0.2555e-2	0.2331e0	4.5356e0	0.2547e0	0.5417e0	0.9805e0
10	0.3130e0	0.9000e-1	0.4651e-1	0.3350e-2	0.3350e-2	0.8925e-3	0.2458e-2	0.2299e0	4.4430e0	0.2530e0	0.5410e0	0.9811e0
11	0.3083e0	0.9000e-1	0.4591e-1	0.3216e-2	0.3216e-2	0.8579e-3	0.2358e-2	0.2266e0	4.3521e0	0.2513e0	0.5403e0	0.9817e0
12	0.3037e0	0.9000e-1	0.4533e-1	0.3089e-2	0.3089e-2	0.8248e-3	0.2264e-2	0.2235e0	4.2633e0	0.2497e0	0.5396e0	0.9823e0
13	0.2991e0	0.9000e-1	0.4476e-1	0.2967e-2	0.2967e-2	0.7933e-3	0.2174e-2	0.2204e0	4.1761e0	0.2480e0	0.5389e0	0.9829e0
14	0.2947e0	0.9000e-1	0.4160e-1	0.2684e-2	0.2684e-2	0.7186e-3	0.1965e-2	0.2173e0	4.0908e0	0.2463e0	0.5382e0	0.9835e0
15	0.3049e0	0.9000e-1	0.4549e-1	0.3123e-2	0.3123e-2	0.8337e-3	0.2289e-2	0.2243e0	4.0073e0	0.2338e0	0.5331e0	0.9880e0
16	0.3291e0	0.9000e-1	0.5132e-1	0.4049e-2	0.4049e-2	0.1073e-2	0.2976e-2	0.2408e0	3.9253e0	0.2133e0	0.5250e0	0.9953e0
17	0.3102e0	0.9000e-1	0.4616e-1	0.3270e-2	0.3270e-2	0.8718e-3	0.2398e-2	0.2279e0	3.8452e0	0.2208e0	0.5279e0	0.9926e0
18	0.3057e0	0.9000e-1	0.4559e-1	0.3145e-2	0.3145e-2	0.8393e-3	0.2305e-2	0.2249e0	3.7666e0	0.2192e0	0.5273e0	0.9932e0
19	0.3150e0	0.9000e-1	0.4677e-1	0.3409e-2	0.3409e-2	0.9077e-3	0.2501e-2	0.2313e0	3.6896e0	0.2088e0	0.5233e0	0.9969e0
20	0.2970e0	0.9000e-1	0.4188e-1	0.2740e-2	0.2740e-2	0.7331e-3	0.2007e-2	0.2189e0	1.2005e0	0.7177e-1	0.4808e0	1.0000e0

Bracing Member Parameters

Segment	H_L	H_F	A	I_x	I_y	I_xy	I_min	i_min	L_cr	lambda_b	Phi	chi
1	0.1950e0	0.7000e-1	0.9994e-2	0.9147e-4	0.1836e-3	0.00000e0	0.9147e-4	0.9567e-1	4.0464e0	0.5535e0	0.7133e0	0.8597e0
2	0.1212e0	0.7000e-1	0.6897e-2	0.3192e-4	0.4595e-4	0.00000e0	0.3192e-4	0.6803e-1	3.9636e0	0.7625e0	0.8863e0	0.7473e0
3	0.1133e0	0.7000e-1	0.6563e-2	0.2801e-4	0.3780e-4	0.00000e0	0.2801e-4	0.6533e-1	3.8828e0	0.7779e0	0.9008e0	0.7380e0
4	0.1145e0	0.7000e-1	0.6611e-2	0.2855e-4	0.3891e-4	0.00000e0	0.2855e-4	0.6571e-1	3.8034e0	0.7575e0	0.8817e0	0.7503e0
5	0.1160e0	0.7000e-1	0.6676e-2	0.2929e-4	0.4045e-4	0.00000e0	0.2929e-4	0.6624e-1	3.7257e0	0.7361e0	0.8621e0	0.7629e0
6	0.1220e0	0.7000e-1	0.6928e-2	0.3231e-4	0.4678e-4	0.00000e0	0.3231e-4	0.6829e-1	3.6496e0	0.6994e0	0.8295e0	0.7840e0
7	0.1280e0	0.7000e-1	0.7180e-2	0.3557e-4	0.5375e-4	0.00000e0	0.3557e-4	0.7039e-1	3.5750e0	0.6647e0	0.7999e0	0.8032e0
8	0.1340e0	0.7000e-1	0.7432e-2	0.3908e-4	0.6137e-4	0.00000e0	0.3908e-4	0.7252e-1	3.5019e0	0.6320e0	0.7732e0	0.8207e0
9	0.1390e0	0.7000e-1	0.7642e-2	0.4221e-4	0.6826e-4	0.00000e0	0.4221e-4	0.7432e-1	3.4305e0	0.6041e0	0.7512e0	0.8350e0
10	0.1450e0	0.7000e-1	0.7894e-2	0.4621e-4	0.7719e-4	0.00000e0	0.4621e-4	0.7651e-1	3.3603e0	0.5748e0	0.7289e0	0.8495e0
11	0.1500e0	0.7000e-1	0.8104e-2	0.4976e-4	0.8520e-4	0.00000e0	0.4976e-4	0.7836e-1	3.2917e0	0.5498e0	0.7106e0	0.8615e0
12	0.1560e0	0.7000e-1	0.8356e-2	0.5428e-4	0.9552e-4	0.00000e0	0.5428e-4	0.8060e-1	3.2245e0	0.5236e0	0.6921e0	0.8736e0
13	0.1610e0	0.7000e-1	0.8566e-2	0.5828e-4	0.1047e-3	0.00000e0	0.5828e-4	0.8248e-1	3.1585e0	0.5012e0	0.6768e0	0.8837e0
14	0.1660e0	0.7000e-1	0.8776e-2	0.6249e-4	0.1145e-3	0.00000e0	0.6249e-4	0.8438e-1	3.0940e0	0.4799e0	0.6627e0	0.8930e0
15	0.1710e0	0.7000e-1	0.8986e-2	0.6692e-4	0.1249e-3	0.00000e0	0.6692e-4	0.8630e-1	3.0309e0	0.4596e0	0.6498e0	0.9017e0
16	0.1760e0	0.7000e-1	0.9196e-2	0.7158e-4	0.1359e-3	0.00000e0	0.7158e-4	0.8823e-1	2.9689e0	0.4404e0	0.6378e0	0.9097e0
17	0.1810e0	0.7000e-1	0.9406e-2	0.7648e-4	0.1475e-3	0.00000e0	0.7648e-4	0.9017e-1	2.9083e0	0.4221e0	0.6269e0	0.9172e0
18	0.1860e0	0.7000e-1	0.9616e-2	0.8161e-4	0.1598e-3	0.00000e0	0.8161e-4	0.9212e-1	2.8488e0	0.4047e0	0.6167e0	0.9242e0
19	0.1900e0	0.7000e-1	0.9784e-2	0.8589e-4	0.1701e-3	0.00000e0	0.8589e-4	0.9370e-1	2.7906e0	0.3898e0	0.6082e0	0.9301e0
20	0.1950e0	0.7000e-1	0.9994e-2	0.9147e-4	0.1836e-3	0.00000e0	0.9147e-4	0.9567e-1	2.7336e0	0.3740e0	0.5995e0	0.9363e0

14.23 Rotatable Inclined Lattice Tower 87.6 m DLC Utilizations

Maximum utilizations of all DLCs over their different wind speeds, wind directions, and random seeds.

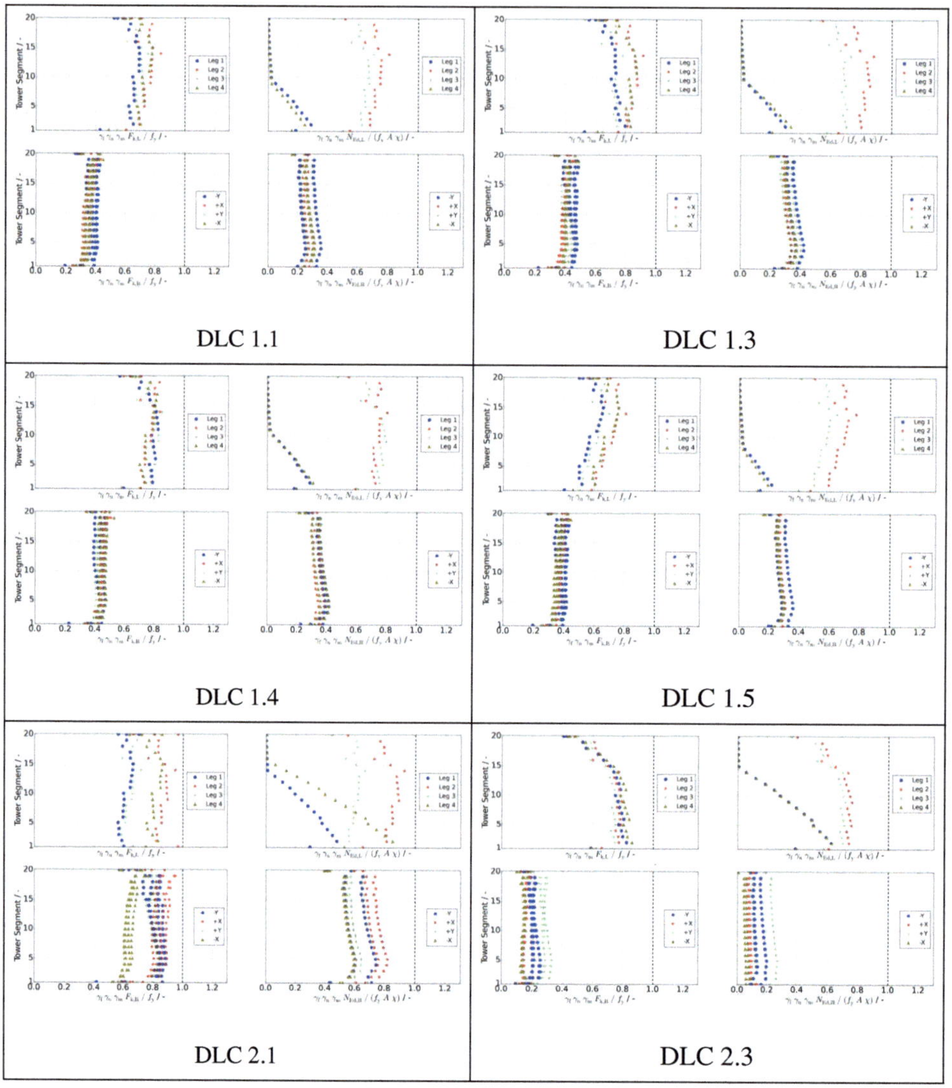

Table 14.2: Member yield strength and buckling utilizations of DLCs 1.1, 1.3, 1.4, 1.5, 2.1, and 2.3 of the 87.6 m high RILT

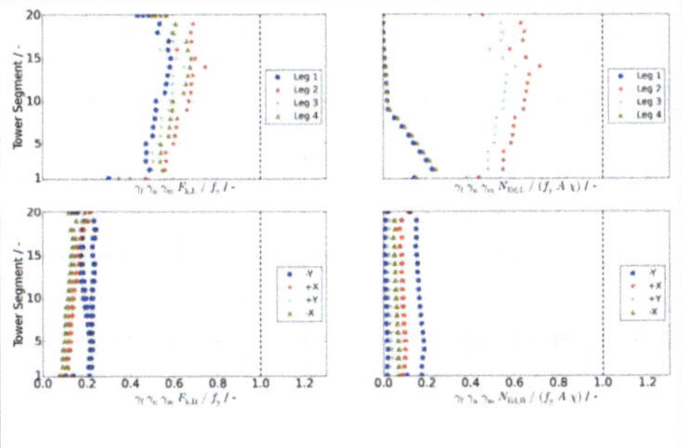

DLC 3.1

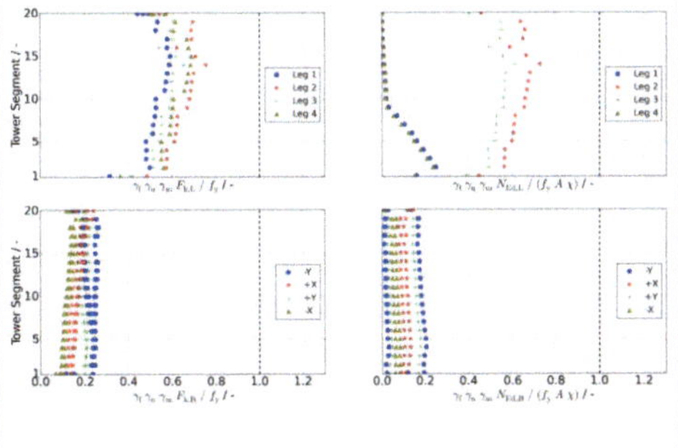

DLC 3.2

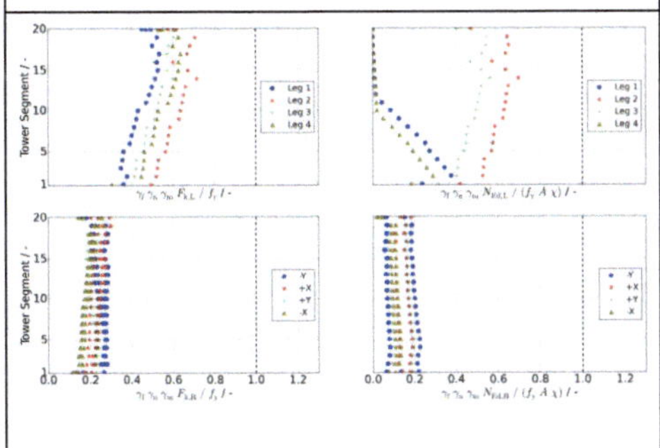

DLC 3.3

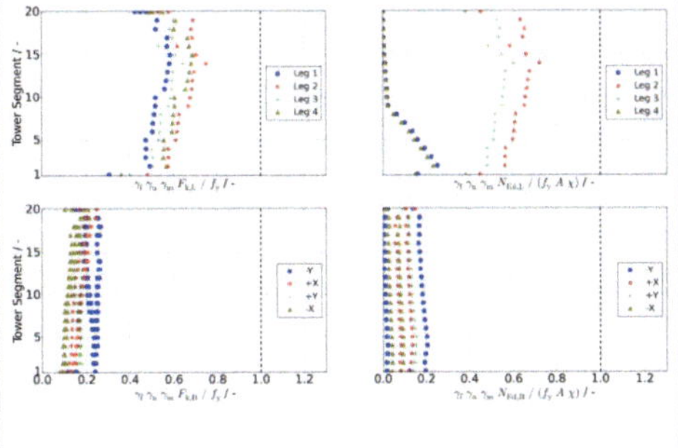

DLC 4.1

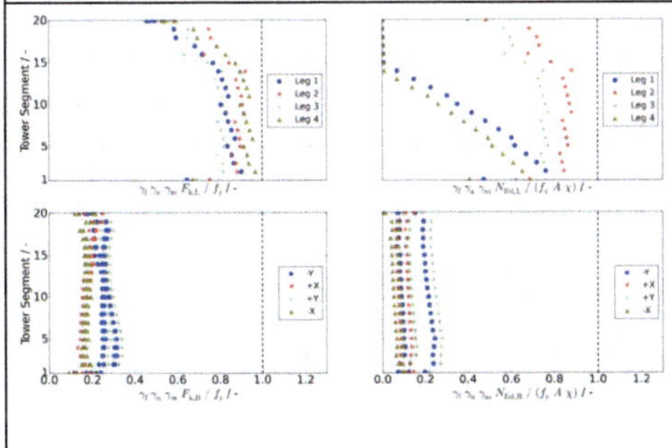

DLC 4.2

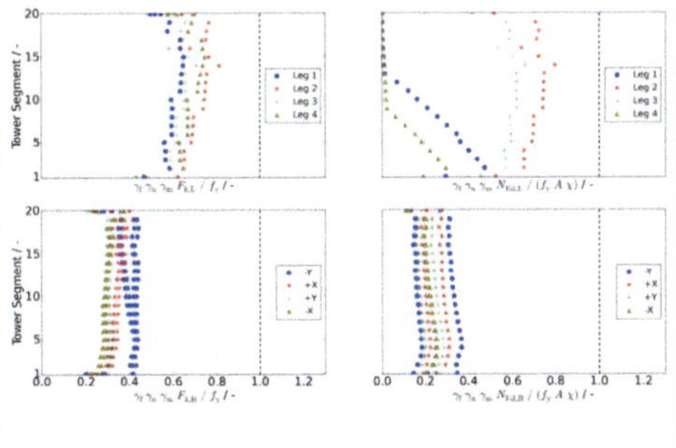

DLC 5.1

Table 14.3: Member yield strength and buckling utilizations of DLCs 3.1, 3.2, 3.3, 4.1, 4.2, and 5.1 of the 87.6 m high RILT

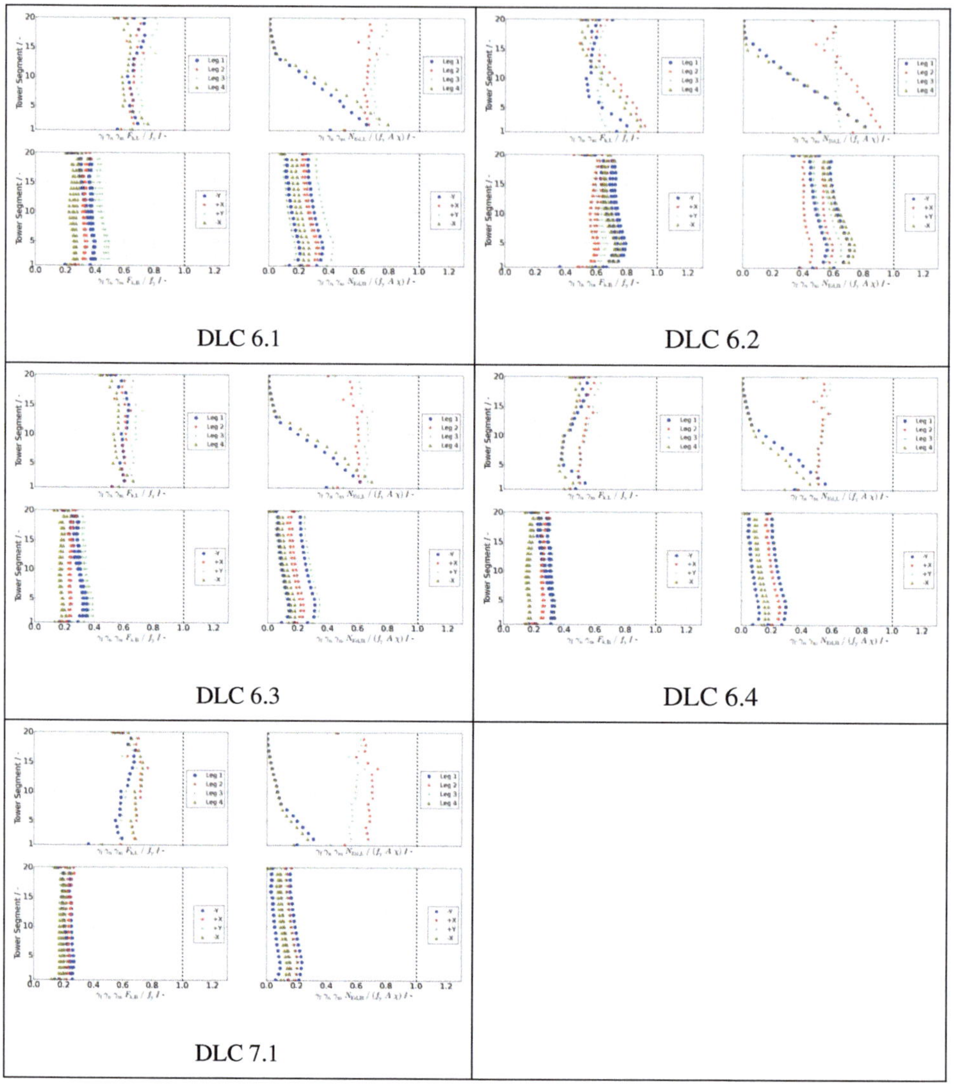

Table 14.4: Member yield strength and buckling utilizations of DLCs 6.1, 6.2, 6.3, 6.4, and 7.1 of the 87.6 m high RILT

14.24 Rotatable Inclined Lattice Tower 147.6 m Member Parameters

The following figure shows all member parameters, of the rotatable inclined lattice tower with H_{T} = 147.6 m according to Table 10.1. All parameters are explained in the nomenclature of this work and its values are given in SI-units.

Front Leg Member Parameters

Segment	L_L	L_F	A	I_x	I_y	I_xy	I_min	i_min	L_cr	lambda_b	Phi	chi
1	0.3410e0	0.9000e-1	0.5585e-1	0.4708e-2	0.4708e-2	0.1245e-2	0.3464e-2	0.2490e0	1.0225e1	0.5374e0	0.7017e0	0.8673e0
2	0.2556e0	0.9000e-1	0.3240e-1	0.1616e-2	0.1616e-2	0.4390e-3	0.1177e-2	0.1906e0	9.7379e0	0.6686e0	0.8032e0	0.8011e0
3	0.2539e0	0.9000e-1	0.3223e-1	0.1589e-2	0.1589e-2	0.4318e-3	0.1157e-2	0.1895e0	9.2749e0	0.6407e0	0.7801e0	0.8161e0
4	0.2523e0	0.9000e-1	0.3206e-1	0.1563e-2	0.1563e-2	0.4250e-3	0.1138e-2	0.1884e0	8.8348e0	0.6138e0	0.7587e0	0.8301e0
5	0.2508e0	0.9000e-1	0.3190e-1	0.1538e-2	0.1538e-2	0.4187e-3	0.1120e-2	0.1873e0	8.4164e0	0.5879e0	0.7388e0	0.8431e0
6	0.2493e0	0.9000e-1	0.3175e-1	0.1515e-2	0.1515e-2	0.4126e-3	0.1103e-2	0.1864e0	8.0185e0	0.5631e0	0.7203e0	0.8552e0
7	0.2480e0	0.9000e-1	0.3161e-1	0.1494e-2	0.1494e-2	0.4070e-3	0.1087e-2	0.1854e0	7.6402e0	0.5393e0	0.7031e0	0.8664e0
8	0.2467e0	0.9000e-1	0.3147e-1	0.1473e-2	0.1473e-2	0.4016e-3	0.1072e-2	0.1845e0	7.2804e0	0.5163e0	0.6871e0	0.8769e0
9	0.2454e0	0.9000e-1	0.3134e-1	0.1454e-2	0.1454e-2	0.3965e-3	0.1057e-2	0.1837e0	6.9383e0	0.4944e0	0.6722e0	0.8867e0
10	0.2442e0	0.9000e-1	0.3122e-1	0.1436e-2	0.1436e-2	0.3918e-3	0.1044e-2	0.1829e0	6.6128e0	0.4733e0	0.6584e0	0.8959e0
11	0.2431e0	0.9000e-1	0.3110e-1	0.1419e-2	0.1419e-2	0.3872e-3	0.1031e-2	0.1821e0	6.3032e0	0.4530e0	0.6456e0	0.9045e0
12	0.2420e0	0.9000e-1	0.3099e-1	0.1402e-2	0.1402e-2	0.3830e-3	0.1019e-2	0.1814e0	6.0087e0	0.4336e0	0.6337e0	0.9125e0
13	0.2410e0	0.9000e-1	0.3088e-1	0.1387e-2	0.1387e-2	0.3789e-3	0.1008e-2	0.1807e0	5.7285e0	0.4150e0	0.6227e0	0.9201e0
14	0.2400e0	0.9000e-1	0.3078e-1	0.1372e-2	0.1372e-2	0.3751e-3	0.9971e-3	0.1800e0	5.4618e0	0.3972e0	0.6124e0	0.9272e0
15	0.2391e0	0.9000e-1	0.3068e-1	0.1358e-2	0.1358e-2	0.3714e-3	0.9870e-3	0.1793e0	5.2083e0	0.3801e0	0.6028e0	0.9339e0
16	0.2382e0	0.9000e-1	0.3059e-1	0.1345e-2	0.1345e-2	0.3680e-3	0.9773e-3	0.1787e0	4.9667e0	0.3637e0	0.5940e0	0.9402e0
17	0.2373e0	0.9000e-1	0.3050e-1	0.1333e-2	0.1333e-2	0.3647e-3	0.9682e-3	0.1782e0	4.7369e0	0.3480e0	0.5857e0	0.9462e0
18	0.2365e0	0.9000e-1	0.2825e-1	0.1227e-2	0.1227e-2	0.3360e-3	0.8913e-3	0.1776e0	4.5182e0	0.3329e0	0.5780e0	0.9519e0
19	0.2358e0	0.9000e-1	0.2818e-1	0.1217e-2	0.1217e-2	0.3333e-3	0.8836e-3	0.1771e0	4.3100e0	0.3185e0	0.5709e0	0.9573e0
20	0.2407e0	0.9000e-1	0.3085e-1	0.1382e-2	0.1382e-2	0.3777e-3	0.1004e-2	0.1804e0	4.1117e0	0.2982e0	0.5612e0	0.9648e0
21	0.2428e0	0.9000e-1	0.3107e-1	0.1414e-2	0.1414e-2	0.3860e-3	0.1028e-2	0.1819e0	3.9229e0	0.2823e0	0.5538e0	0.9706e0
22	0.2477e0	0.9000e-1	0.3158e-1	0.1490e-2	0.1490e-2	0.4059e-3	0.1084e-2	0.1853e0	3.7433e0	0.2644e0	0.5459e0	0.9770e0
23	0.2527e0	0.9000e-1	0.3210e-1	0.1569e-2	0.1569e-2	0.4266e-3	0.1142e-2	0.1886e0	3.5721e0	0.2478e0	0.5388e0	0.9830e0
24	0.3220e0	0.9000e-1	0.5037e-1	0.3818e-2	0.3818e-2	0.1014e-2	0.2805e-2	0.2360e0	1.1363e0	0.6303e-1	0.4787e0	1.0000e0

Back Leg Member Parameters

Segment	L_L	L_F	A	I_x	I_y	I_xy	I_min	i_min	L_cr	lambda_b	Phi	chi
1	0.3410e0	0.9000e-1	0.5585e-1	0.4708e-2	0.4708e-2	0.1245e-2	0.3464e-2	0.2490e0	1.0361e1	0.5445e0	0.7068e0	0.8640e0
2	0.3079e0	0.9000e-1	0.4587e-1	0.3206e-2	0.3206e-2	0.8552e-3	0.2351e-2	0.2264e0	9.8676e0	0.5705e0	0.7257e0	0.8516e0
3	0.3059e0	0.9000e-1	0.4561e-1	0.3150e-2	0.3150e-2	0.8407e-3	0.2309e-2	0.2250e0	9.3984e0	0.5467e0	0.7084e0	0.8629e0
4	0.3040e0	0.9000e-1	0.4537e-1	0.3097e-2	0.3097e-2	0.8271e-3	0.2270e-2	0.2237e0	8.9524e0	0.5238e0	0.6922e0	0.8735e0
5	0.3022e0	0.9000e-1	0.4514e-1	0.3048e-2	0.3048e-2	0.8143e-3	0.2234e-2	0.2224e0	8.5284e0	0.5018e0	0.6772e0	0.8834e0
6	0.3004e0	0.9000e-1	0.4492e-1	0.3001e-2	0.3001e-2	0.8022e-3	0.2199e-2	0.2212e0	8.1252e0	0.4806e0	0.6632e0	0.8927e0
7	0.2988e0	0.9000e-1	0.4209e-1	0.2784e-2	0.2784e-2	0.7445e-3	0.2039e-2	0.2201e0	7.7419e0	0.4603e0	0.6502e0	0.9014e0
8	0.2972e0	0.9000e-1	0.4190e-1	0.2745e-2	0.2745e-2	0.7344e-3	0.2010e-2	0.2190e0	7.3773e0	0.4408e0	0.6381e0	0.9095e0
9	0.2957e0	0.9000e-1	0.4172e-1	0.2708e-2	0.2708e-2	0.7248e-3	0.1983e-2	0.2180e0	7.0307e0	0.4221e0	0.6268e0	0.9172e0
10	0.2943e0	0.9000e-1	0.4155e-1	0.2673e-2	0.2673e-2	0.7157e-3	0.1957e-2	0.2170e0	6.7008e0	0.4041e0	0.6163e0	0.9244e0
11	0.2929e0	0.9000e-1	0.4139e-1	0.2640e-2	0.2640e-2	0.7072e-3	0.1933e-2	0.2161e0	6.3871e0	0.3868e0	0.6066e0	0.9313e0
12	0.2916e0	0.9000e-1	0.4123e-1	0.2609e-2	0.2609e-2	0.6991e-3	0.1910e-2	0.2152e0	6.0887e0	0.3703e0	0.5975e0	0.9377e0
13	0.2904e0	0.9000e-1	0.4109e-1	0.2579e-2	0.2579e-2	0.6914e-3	0.1888e-2	0.2143e0	5.8048e0	0.3544e0	0.5891e0	0.9438e0
14	0.2892e0	0.9000e-1	0.4095e-1	0.2551e-2	0.2551e-2	0.6842e-3	0.1867e-2	0.2135e0	5.5345e0	0.3392e0	0.5812e0	0.9495e0
15	0.2881e0	0.9000e-1	0.4081e-1	0.2525e-2	0.2525e-2	0.6773e-3	0.1848e-2	0.2128e0	5.2776e0	0.3246e0	0.5739e0	0.9550e0
16	0.2870e0	0.9000e-1	0.4068e-1	0.2500e-2	0.2500e-2	0.6709e-3	0.1829e-2	0.2120e0	5.0328e0	0.3107e0	0.5671e0	0.9602e0
17	0.2860e0	0.9000e-1	0.4056e-1	0.2476e-2	0.2476e-2	0.6647e-3	0.1811e-2	0.2113e0	4.8000e0	0.2973e0	0.5607e0	0.9651e0
18	0.2850e0	0.9000e-1	0.4045e-1	0.2454e-2	0.2454e-2	0.6589e-3	0.1795e-2	0.2107e0	4.5783e0	0.2844e0	0.5548e0	0.9698e0
19	0.2841e0	0.9000e-1	0.4034e-1	0.2432e-2	0.2432e-2	0.6533e-3	0.1779e-2	0.2100e0	4.3674e0	0.2722e0	0.5493e0	0.9742e0
20	0.2832e0	0.9000e-1	0.4023e-1	0.2412e-2	0.2412e-2	0.6481e-3	0.1764e-2	0.2094e0	4.1664e0	0.2604e0	0.5442e0	0.9785e0
21	0.2823e0	0.9000e-1	0.4013e-1	0.2393e-2	0.2393e-2	0.6431e-3	0.1750e-2	0.2088e0	3.9752e0	0.2491e0	0.5394e0	0.9825e0
22	0.2815e0	0.9000e-1	0.4003e-1	0.2375e-2	0.2375e-2	0.6384e-3	0.1736e-2	0.2083e0	3.7931e0	0.2384e0	0.5349e0	0.9864e0
23	0.2807e0	0.9000e-1	0.4001e-1	0.2365e-2	0.2365e-2	0.6368e-3	0.1729e-2	0.2079e0	3.6196e0	0.2279e0	0.5307e0	0.9901e0
24	0.3220e0	0.9000e-1	0.5037e-1	0.3818e-2	0.3818e-2	0.1014e-2	0.2805e-2	0.2360e0	1.1364e0	0.6303e-1	0.4787e0	1.0000e0

Bracing Member Parameters

Segment	H_L	H_F	A	I_x	I_y	I_xy	I_min	i_min	L_cr	lambda_b	Phi	chi
1	0.1900e0	0.7000e-1	0.9784e-2	0.8589e-4	0.1701e-3	0.00000e0	0.8589e-4	0.9370e-1	1.0497e1	1.4663e0	1.7903e0	0.3549e0
2	0.1075e0	0.7000e-1	0.6319e-2	0.2540e-4	0.3250e-4	0.00000e0	0.2540e-4	0.6340e-1	1.0053e1	2.0752e0	2.9719e0	0.1961e0
3	0.1023e0	0.7000e-1	0.6101e-2	0.2324e-4	0.2820e-4	0.00000e0	0.2324e-4	0.6172e-1	9.6307e0	2.0421e0	2.8983e0	0.2018e0
4	0.1050e0	0.7000e-1	0.6214e-2	0.2434e-4	0.3039e-4	0.00000e0	0.2434e-4	0.6259e-1	9.2284e0	1.9296e0	2.6558e0	0.2232e0
5	0.1070e0	0.7000e-1	0.6298e-2	0.2519e-4	0.3207e-4	0.00000e0	0.2519e-4	0.6324e-1	8.8455e0	1.8305e0	2.4527e0	0.2448e0
6	0.1140e0	0.7000e-1	0.6592e-2	0.2834e-4	0.3847e-4	0.00000e0	0.2834e-4	0.6556e-1	8.4808e0	1.6929e0	2.1868e0	0.2800e0
7	0.1200e0	0.7000e-1	0.6844e-2	0.3128e-4	0.4460e-4	0.00000e0	0.3128e-4	0.6760e-1	8.1337e0	1.5746e0	1.9734e0	0.3162e0
8	0.1260e0	0.7000e-1	0.7096e-2	0.3446e-4	0.5135e-4	0.00000e0	0.3446e-4	0.6968e-1	7.8032e0	1.4655e0	1.7890e0	0.3552e0
9	0.1310e0	0.7000e-1	0.7306e-2	0.3730e-4	0.5748e-4	0.00000e0	0.3730e-4	0.7145e-1	7.4885e0	1.3717e0	1.6400e0	0.3939e0
10	0.1370e0	0.7000e-1	0.7558e-2	0.4094e-4	0.6545e-4	0.00000e0	0.4094e-4	0.7360e-1	7.1887e0	1.2784e0	1.5004e0	0.4374e0
11	0.1420e0	0.7000e-1	0.7768e-2	0.4417e-4	0.7263e-4	0.00000e0	0.4417e-4	0.7541e-1	6.9033e0	1.1981e0	1.3873e0	0.4792e0
12	0.1470e0	0.7000e-1	0.7978e-2	0.4760e-4	0.8033e-4	0.00000e0	0.4760e-4	0.7725e-1	6.6314e0	1.1235e0	1.2882e0	0.5213e0
13	0.1510e0	0.7000e-1	0.8146e-2	0.5049e-4	0.8686e-4	0.00000e0	0.5049e-4	0.7873e-1	6.3724e0	1.0593e0	1.2072e0	0.5599e0
14	0.1560e0	0.7000e-1	0.8356e-2	0.5428e-4	0.9552e-4	0.00000e0	0.5428e-4	0.8060e-1	6.1256e0	0.9947e0	1.1298e0	0.6004e0
15	0.1600e0	0.7000e-1	0.8524e-2	0.5746e-4	0.1028e-3	0.00000e0	0.5746e-4	0.8210e-1	5.8906e0	0.9390e0	1.0665e0	0.6361e0
16	0.1640e0	0.7000e-1	0.8692e-2	0.6078e-4	0.1105e-3	0.00000e0	0.6078e-4	0.8362e-1	5.6668e0	0.8869e0	1.0101e0	0.6696e0
17	0.1680e0	0.7000e-1	0.8860e-2	0.6424e-4	0.1186e-3	0.00000e0	0.6424e-4	0.8515e-1	5.4534e0	0.8382e0	0.9598e0	0.7006e0
18	0.1710e0	0.7000e-1	0.8986e-2	0.6692e-4	0.1249e-3	0.00000e0	0.6692e-4	0.8630e-1	5.2500e0	0.7962e0	0.9183e0	0.7268e0
19	0.1750e0	0.7000e-1	0.9154e-2	0.7063e-4	0.1337e-3	0.00000e0	0.7063e-4	0.8784e-1	5.0563e0	0.7533e0	0.8778e0	0.7528e0
20	0.1780e0	0.7000e-1	0.9280e-2	0.7351e-4	0.1405e-3	0.00000e0	0.7351e-4	0.8900e-1	4.8716e0	0.7163e0	0.8443e0	0.7744e0
21	0.1810e0	0.7000e-1	0.9406e-2	0.7648e-4	0.1475e-3	0.00000e0	0.7648e-4	0.9017e-1	4.6956e0	0.6815e0	0.8141e0	0.7940e0
22	0.1840e0	0.7000e-1	0.9532e-2	0.7953e-4	0.1548e-3	0.00000e0	0.7953e-4	0.9134e-1	4.5276e0	0.6487e0	0.7867e0	0.8118e0
23	0.2244e0	0.7000e-1	0.1123e-1	0.1297e-3	0.2775e-3	0.00000e0	0.1297e-3	0.1075e0	4.3679e0	0.5320e0	0.6979e0	0.8698e0
24	0.1900e0	0.7000e-1	0.9784e-2	0.8589e-4	0.1701e-3	0.00000e0	0.8589e-4	0.9370e-1	4.2154e0	0.5888e0	0.7394e0	0.8426e0

14.25 Rotatable Inclined Lattice Tower 147.6 m Utilizations

The following figure shows the extreme member utilization due to tower wind drag, of the rotatable inclined lattice tower with H_T = 147.6 m according to Table 10.1. Note that this tower has asymmetrical tower top dimensions, wherefore an additional case of sidewise wind attack was introduced.

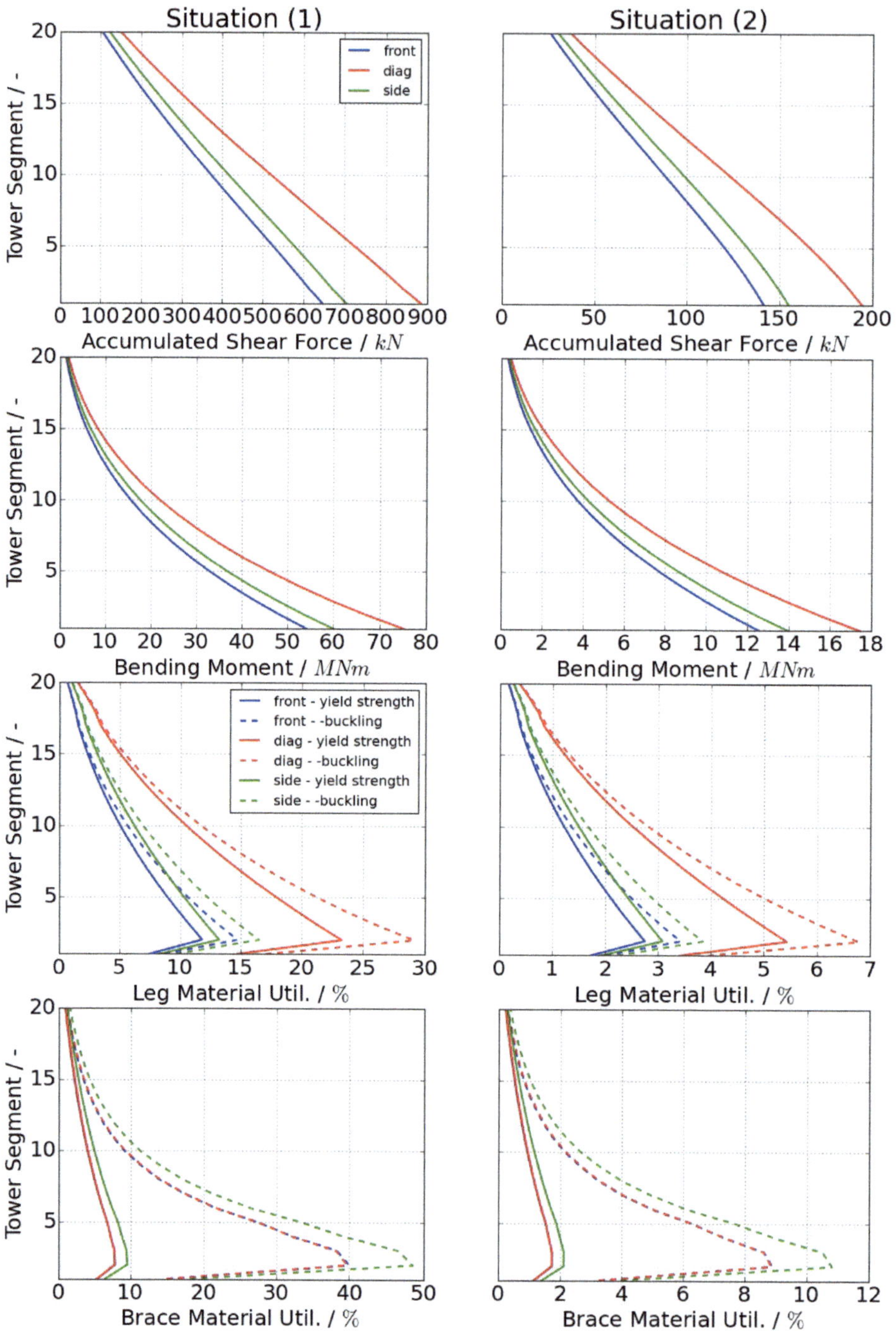

Situation (1)
Situation (2)
front
diag
side
Tower Segment / -
Accumulated Shear Force / kN
Bending Moment / MNm
front - yield strength
front - -buckling
diag - yield strength
diag - -buckling
side - yield strength
side - -buckling
Leg Material Util. / %
Brace Material Util. / %

Maximum utilizations of all DLCs over their different wind speeds, wind directions, and random seeds are presented in the following diagrams.

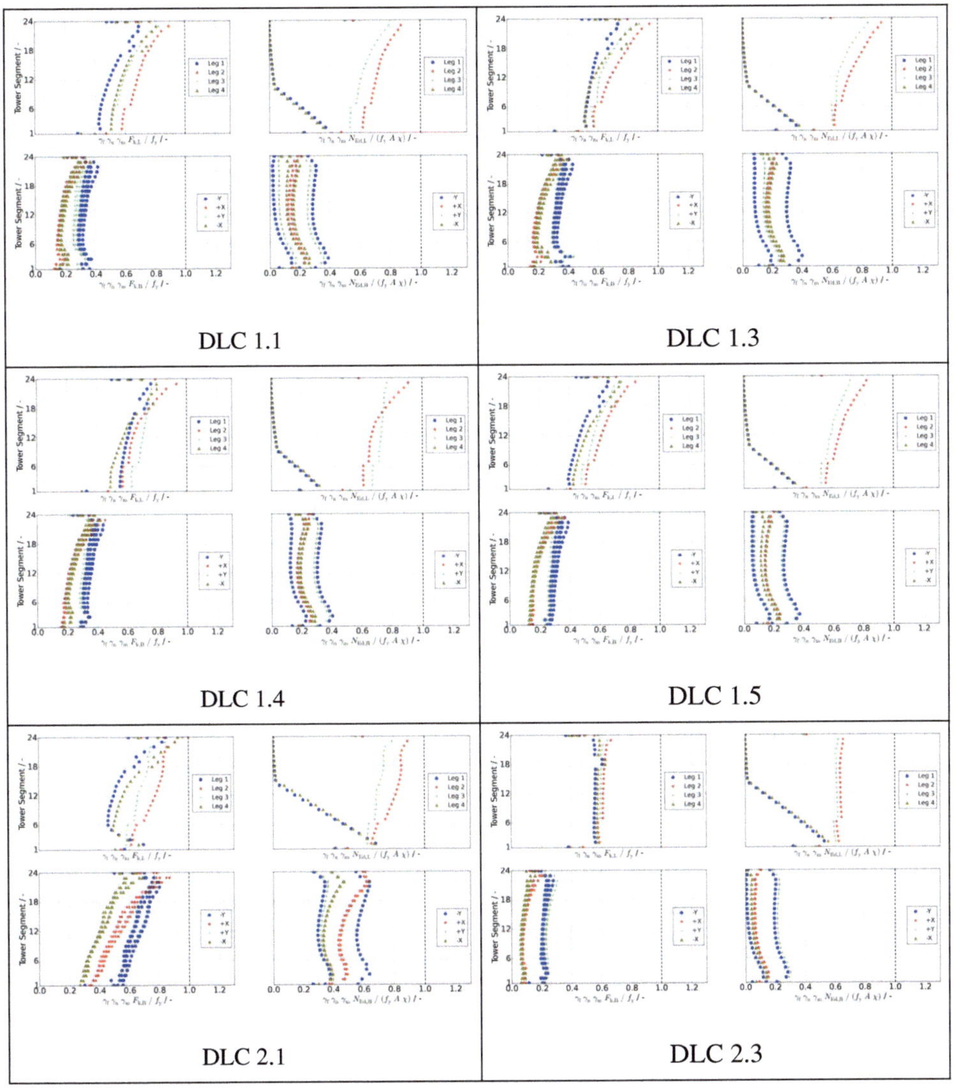

Table 14.5: Member yield strength and buckling utilizations of DLCs 1.1, 1.3, 1.4, 1.5, 2.1, and 2.3 of the 147.6 m high RILT

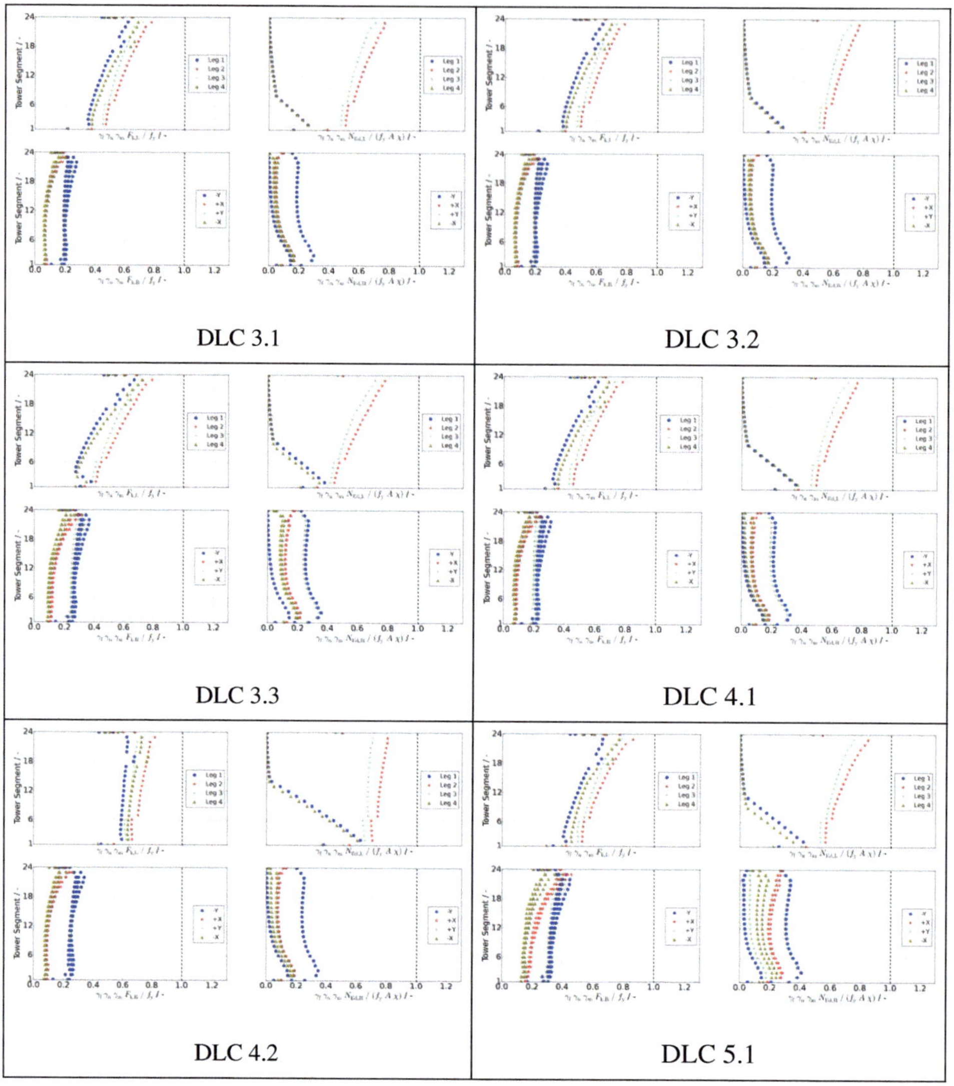

Table 14.6: Member yield strength and buckling utilizations of DLCs 3.1, 3.2, 3.3, 4.1, 4.2, and 5.1 of the 147.6 m high RILT

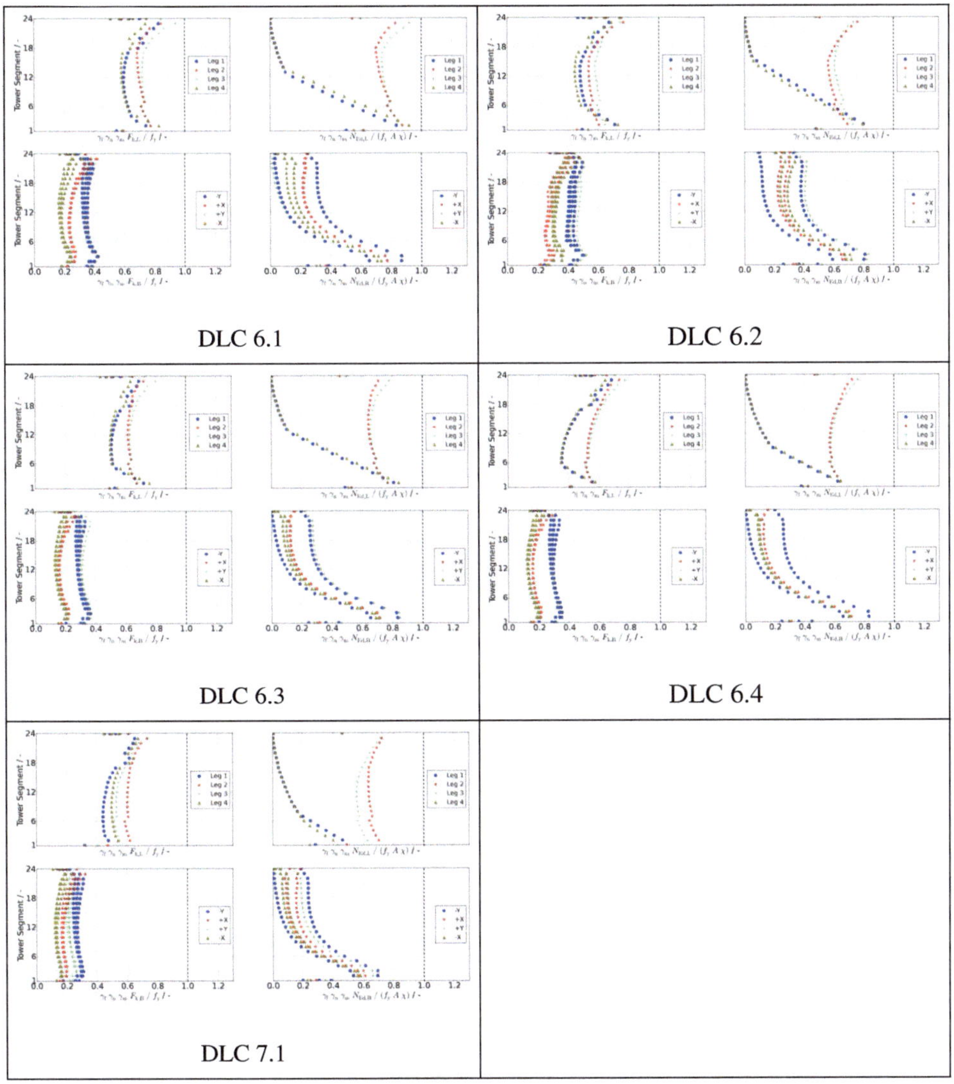

Table 14.7: Member yield strength and buckling utilizations of DLCs 6.1, 6.2, 6.3, 6.4, and 7.1 of the 147.6 m high RILT

The following figure shows the lifetime damage of the rotatable inclined lattice tower with H_T = 147.6 m according to Table 10.1. All assumptions were the same as for the RILT with H_T = 87.6 m height.

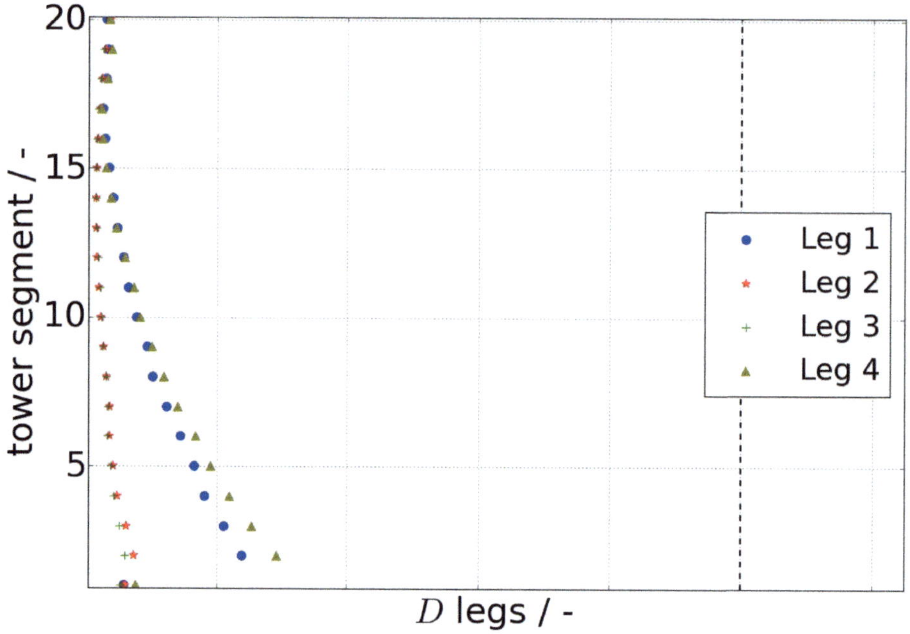

20
15
10
5
tower segment / -
D legs / -
Leg 1
Leg 2
Leg 3
Leg 4

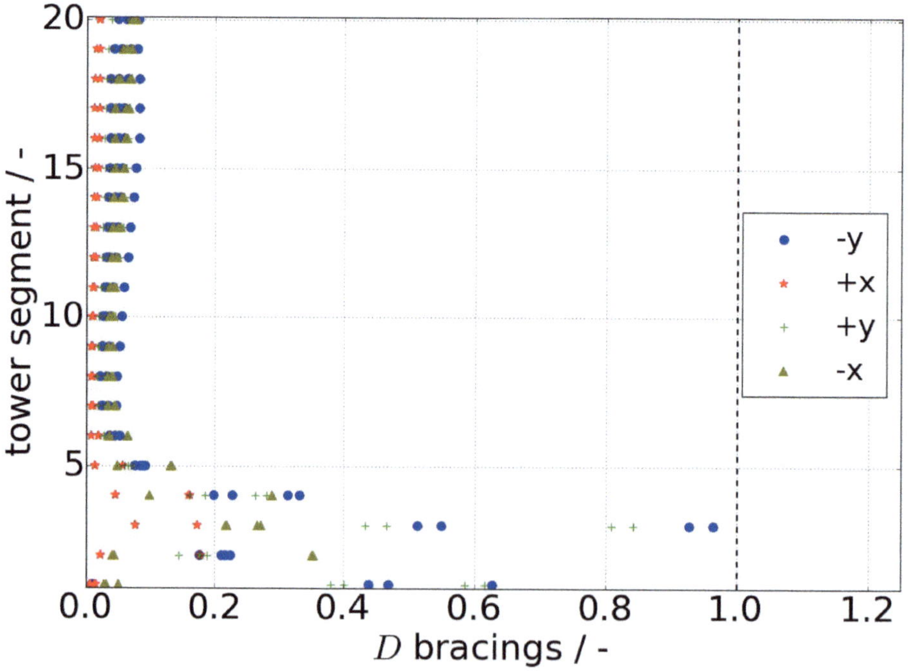

20
15
10
5
tower segment / -
0.0
0.2
0.4
0.6
0.8
1.0
1.2
D bracings / -
-y
+x
+y
-x

BERICHTE ZUM STAHL- UND LEICHTBAU

Versuchsanstalt für Stahl, Holz und Steine, Stahl- und Leichtbau
Karlsruher Institut für Technologie (KIT) | ISSN 2198-7912

Eine Übersicht der Berichte der Versuchsanstalt für Stahl, Holz und Steine ab dem Jahr 1963 finden Sie unter folgender URL: http://stahl.vaka.kit.edu/berichte.php

Band 1 **OLIVER FLEISCHER**
Axial beanspruchte K-Knoten aus dünnwandigen Rechteckhohlprofilen. 2014
ISBN 978-3-7315-0190-9

Band 2 **THOMAS REINKE**
Tragverhalten von biegebeanspruchten Stahlmasten mit polygonalen Querschnitten. 2015
ISBN 978-3-7315-0398-9

Band 3 **ROBIN MARC PLUM**
Fatigue crack detection on structural steel members by using ultrasound excited thermography. Erkennung von Ermüdungsrissen in Stahlbauteilen durch ultraschallangeregte Thermografie. 2015
ISBN 978-3-7315-0417-7

Band 4 **TIM ZINKE**
Nachhaltigkeit von Infrastrukturbauwerken – Ganzheitliche Bewertung von Autobahnbrücken unter besonderer Berücksichtigung externer Effekte. 2016
ISBN 978-3-7315-0509-9

Band 5 **MAX JONAS SPANNAUS**
Bemessung von Erzeugnissen aus Stahlguss unter vorwiegend ruhender Beanspruchung. 2016
ISBN 978-3-7315-0560-0

Band 6 **MATTHIAS FRIEDRICH ALBIEZ**
Zur statischen Tragfähigkeit geklebter Kreishohlprofilverbindungen im Stahlbau. 2016
ISBN 978-3-7315-0561-7

Band 7 **ANDREAS LIPP**
Kreishohlprofil-X-Knoten aus nichtrostenden Stählen unter Axialbeanspruchung. 2016
ISBN 978-3-7315-0569-3

Band 8 **PAUL DARIO TOASA CAIZA**
Consideration of runouts by the evaluation of fatigue experiments. 2019
ISBN 978-3-7315-0900-4

Band 9 **JENNIFER C. HRABOWSKI**
Ermüdungsverhalten von Schweißverbindungen aus höchstfestem Stahl im Kurzzeitfestigkeitsbereich. 2019
ISBN 978-3-7315-0931-8

Band 10 **STEFAN RACK**
Formadaptive Tensegrity-Strukturen – Ein Beitrag zur numerischen Simulation. 2019
ISBN 978-3-7315-0574-7

Band 11 **ACHIM STRUVE**
Analysis of a Rotatable Wind Turbine Tower by means of Aero-Servo-Elastic Load Simulations. 2021
ISBN 978-3-7315-1045-1